Economics and the Ecosystem

Edited by
Edward Fullbrook & Jamie Morgan

World Economics Association Books

World Economics Association BOOK SERIES

Economics and the Ecosystem

Edited by Edward Fullbrook and Jamie Morgan

Previously published as issue 87 of *Real-World Economics Review* in March 2019.

ISBN: 978-1-911156-46-8 (paperback)
978-1-911156-47-5 (eBook, Mobi)
978-1-911156-48-2 (eBook, PDF)

Cover image: NASA

Published by World Economics Association, Bristol, UK
www.worldeconomicsassociation.org

The **World Economics Association (WEA)** was launched on May 16, 2011. Already over 13,000 economists and related scholars have joined. This phenomenal success has come about because the WEA fills a huge gap in the international community of economists – the absence of a professional organization which is truly international and pluralist.

The World Economics Association seeks to increase the relevance, breadth and depth of economic thought. Its key qualities are worldwide membership and governance, and inclusiveness with respect to: (a) the variety of theoretical perspectives; (b) the range of human activities and issues which fall within the broad domain of economics; and (c) the study of the world's diverse economies.

The Association's activities centre on the development, promotion and diffusion of economic research and knowledge and on illuminating their social character.

The WEA publishes books, three open-access journals *(Economic Thought, World Social and Economic Review* and *Real-World Economics Review*), a bi-monthly newsletter, blogs, holds global online conferences, runs a textbook commentaries project and an eBook library.

Contents

Chapter 1

Introduction: Economics and civilization in ecological crisis

Jamie Morgan and Edward Fullbrook

What is the problem?

The essays collected here grapple with different aspects of what, if natural scientists are to be believed, is the most profound set of issues humanity has ever faced. The United Nations Framework Convention on Climate Change (UNFCCC) was created in 1992. Article 2 of the Convention, states its goal as the "stabilization of greenhouse gas concentrations in the atmosphere at a level that would prevent dangerous anthropogenic interference with the climate system" (UNFCCC, 1992: p. 4). The subsequent "Conference of the Parties" process is formally supposed to achieve this. However, during the period since then, despite the Kyoto Protocols signed at COP 3 in 1997, and the subsequent 2015 Paris Agreement, which will come into force in the new decade, annual global greenhouse gas (GHG) emissions have increased massively (UNFCCC, 2015; Morgan, 2016). The *rate of increase* of carbon emissions has markedly reduced in recent years, but the absolute annual figure remains huge. In 2017 total annual gigatonnes of CO_2 equivalent emissions reached a record high of 53.5 (UNEP, 2018). This is important because emissions remain in the atmosphere for long periods - CO_2 can remain in the atmosphere for well over a century. As the cumulative parts per million (ppm) in the atmosphere increase, then the warming effect created by the gases also increases which leads to positive feedbacks and the increased chance of a global ecosystem collapse.

The estimated pre-industrial revolution atmospheric carbon level ranged at considerably less than 300ppm over several hundred thousand years. In 2013 it exceeded 400ppm for the first time in human history and as of 2019 is averaging over 407ppm. Standard models collated by the Inter-Governmental Panel on Climate Change (IPCC) use 450ppm as the trigger level for a 2°C average warming. As a "carbon budget", this has translated into total cumulative emissions at the lower end of 3,000+ $GtCo_2$ to achieve the target of remaining below 2°C warming. Since the industrial revolution began we have already produced more than 2,000 $GtCo_2$. More recent work and observations by climate scientists indicate that prior standard models have likely underestimated the subsequent rate of warming and the thresholds at which positive feedback effects might begin and which could become irreversible. This resulted in the inclusion in the Paris Agreement of an "aspiration" to restrict warming to 1.5°C involving a further restriction of the remaining carbon budget. However, it is also the case that even this target is no guarantee that "Hothouse Earth" irreversible effects can be avoided (see Steffen et al, 2018; Hansen et al, 2017). And yet the current emissions trends and country commitments stated as "Nationally Determined Contributions" (NDCs) to reductions under the Paris Agreement look set to exceed the 3000 target in a matter of a few short years and this leads, even under previous standard models, to estimated warming between upper 2 and over 4°C over the second half of this century and into the next. This *entire range* is extremely serious in terms of its consequences for our species.

As the NASA global temperature database makes clear, almost all the hottest years on record have occurred since the beginning of the twenty-first century, and erratic weather effects are observably increasing around the world.[1]

Earth temperature differential 2018

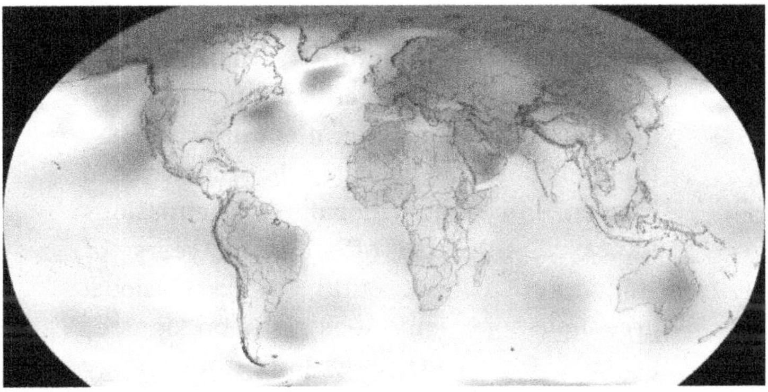

Data source: NASA/GISS Credit: NASA Scientific Visualization Studio

[1] https://climate.nasa.gov/vital-signs/global-temperature/

The identified consequences include: heatwaves, droughts, flooding, loss of landmass, inability of species to adapt exacerbating extinction rates, falling yields and rising crop failures, food and water insecurity, famine, loss of life from "natural" disaster, increasing poverty and escalating problems of induced safety-seeking mass migration. So, adverse climate change is already here and we seem to be sleepwalking towards catastrophe. The needed changes were stated quite starkly by the IPCC in October 2018 and in the UNEP 9th *Emissions Gap Report* in the following month:

> "Current commitments expressed in the NDCs are inadequate to bridge the emissions gap in 2030. Technically, it is still possible to bridge the gap to ensure global warming stays well below 2°C and 1.5°C, but if NDC ambitions are not increased before 2030, exceeding the 1.5°C goal can no longer be avoided. **Now more than ever, unprecedented and urgent action is required by all nations. The assessment of actions by the G20 countries indicates that this is yet to happen**; in fact, global CO_2 emissions increased in 2017 after three years of stagnation. (UNEP, 2018, p. xiv) Global greenhouse gas emissions show no signs of peaking. Global CO_2 emissions from energy and industry increased in 2017, following a three-year period of stabilization. Total annual greenhouse gases emissions, including from land-use change, reached a record high of 53.5 $GtCO_2e$ in 2017, an

> increase of 0.7 $GtCO_2e$ compared with 2016. In contrast, **global GHG emissions in 2030 need to be approximately 25 percent and 55 percent lower than in 2017** to put the world on a least-cost pathway to limiting global warming to 2°C and 1.5°C respectively.... **Global peaking of emissions by 2020 is crucial** for achieving the temperature targets of the Paris Agreement" (UNEP, 2018, p. xv; bold added).

To be clear, our best estimates based on the scientific consensus is that emissions *must* fall to 55% of their current level by 2030 and then (if one reads both reports) reduce further to *net zero* by mid-century.

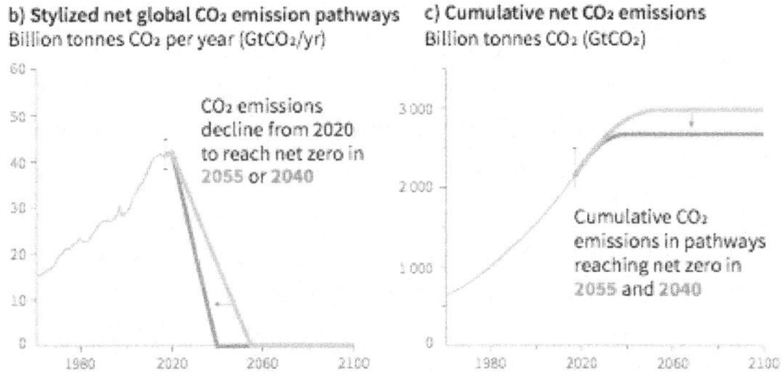

IPCC, 2018: p. 8

As has started to become obvious, this is a deep problem of political economy. Mayer Hillman and others have previously stated and the IPCC now confirms that the problem seemingly requires a fundamental reorientation of how we organize and live based on mass mobilization to an extent never previously seen outside of a war setting.

However, as the essays set out here address, the very form and function of our political economies resists recognizing the seriousness of the situation and resists translating any recognition into concrete and immediate action. Events and published progress at COP 24 in Katowice, Poland, confirm this (see UNFCCC, 2018). We have rhetoric and some limited changes... We have had decades of global agreements such as Kyoto, discourse concerning sustainable development, education for environmental awareness, recycling and policies focused on carbon trading and efficiency enhancing innovations. At the same time, our economies as currently operative have been and are dependent on material expansion and growth, and remain configured to foster continual expansion and diffusion of industrialisation and consumption. The planet is deaf to good intention, it responds only to what we actually do, rather than merely what we say. Moreover, we have been socialised to conflate larger economies with necessarily better economies and to consider expansionary economies as a predicate of technological solutions to induced problems of economic activity. At the same time, we have been discouraged from thinking about the basic incompatibly of an ever-expanding material economy *within* a finite world.

The essays collected here serve to confirm various insights. There has never been a zero-carbon industrial-consumption economy, and we are now in a situation where we must *really recognise* that (as Meadows et al stated years ago) there are limits to growth and that one cannot blithely hope expansion + technologies is a simple source of salvation that requires no sacrifice. We are also in a situation where we must realise that our economies have powerful actors of one kind or another, so markets are currently not "free", and yet there is no single centre of decision making that has demonstrated it is currently capable of dictating what needs to be done to solve a global problem: states and corporations are resisting recognition or action based on the true scale of the problem, and many of the citizenry remain unaware of the true nature of the problem. This is extraordinary. Moreover, climate change based on carbon emissions is merely one consequence of a broader set of changes induced by the way we are changing the planet.

Emissions are a by-product of our socio-economic practices. It is these broader practices that account for cumulative consequences that have been catalogued in (so far) five *Global Environmental Outlook Reports* from the UNEP. Those practices and consequences include: deforestation, extension of intensive agriculture, industrial scale fishing, extraction of minerals, gas and oil, the proliferation of energy production, transportation, and manufacturing, and consumption patterns that treat our environment as a bottomless disposal site for plastics, pesticides, cosmetics, fertilizers, food waste, heavy metals, medicines and more, cumulatively resulting in, *in addition to global warming*, water table depletion, desertification, eutrophication and rising toxicities in soil and air, sea level rises, rapid species extinction and general loss of biodiversity on land and sea, ultimately creating pressure on food chains and culminating in progressive ecosystem collapse (UNEP, 2012). The global population has increased from about 1.6 billion in 1900 to over 7.5 billion in 2018. The global economy has increased from about US$1.1 trillion in 1900 to over US$80 trillion in 2018. Along with these changes has come amazing transformations in the way we live, but the issue is not whether there are aspects of the way (some of us) live that we like, the *real* issue is whether our design for life is survivable.

The situation, then, is ultra-serious. With this in mind, *Real-World Economics Review* has invited a range of experts to consider how we arrived in our current predicament and to what degree there is scope to address the challenges now confronted.

The essays

The essays collected here are predominantly informed by an ecological economics perspective. As such, they are implicitly or explicitly critical of the theory and role of both mainstream economics, in its general neglect of environmental issues, and environmental economics in so far as it has operated as a sub-discipline in ways highly constrained by mainstream economic concepts that have been antithetical to a more realistic approach to recognizing environmental limits and remaining within them. A primary concern is that economics has contributed to complacency and has helped to limit the collective imagination or proper context for solutions that might (starting *far earlier* than today) have steered humanity along a different path. This, for example, is basic in different ways to the essays by Herman Daly, Max Koch, Peter Söderbaum, Ted Trainer and Samuel Alexander, Clive Spash and Tone Smith, and Richard Smith. Each builds on well-established prior work (see, for example, Smith, 2016; Daly, 2014; Söderbaum, 2018; Spash, 2017; Koch, 2012). This basic commitment is, in many ways, a sub-set of a general theme that contributors to the *Real-World Economics Review* have repeatedly expressed. That is, though mainstream economics frequently celebrates its use of data and statistical analysis, the trajectory of mainstream economics has typically been pseudo-scientific and antithetical to progress in both normative-ethical terms and realistic theory terms (and these are in fact aspects of the same issue in so far as social reality is normatively informed). By contrast the goals of humanity might better be fostered by a pluralistic and realistically founded economics (see Fullbrook, 2016, 2009, 2008).

In general, ecological economics recognizes that an economy is a materially significant activity (Georgescu-Roegen, 1971). It is embedded in an environment of physical and biological processes that create limits. Fundamentally, one must recognize that economic activity is a processing of "throughput" that involves energy use, resource transformation and depletion, waste creation and an underlying entropy. It *may* be possible to produce more with less, it *may* be possible to recycle, but there *are* limits to the degree of circularity and of energy (re)use, resource use and scale. An economy is *not* an isolated system that can be treated as a mere circular flow of ever-expanding output and values. It is a component in a social system that is dependent on and causally related to the environment it is operative within. Ultimately the whole is an open system with path dependent characteristics. No reasonable vision of the future can neglect this (for example, Morgan, 2017). The adverse evidence from the IPCC and UNEP is making this ever clearer. Positive feedback, thresholds of transition and irreversibility of consequences all apply in the real world and there is no reason to think that an economic system is better suited than science to identifying points of no return and rowing back from them, not least because all of the individual incentive structures in an economic context of decision making are against this.

As all the contributors are aware, from an open system point of view, it is deeply concerning that conventional economics *does not* begin from material processes and real limits, but rather notional concepts in models of subjective values of goods and services in idealised market equilibrating situations. Issues of the environment are delegated to sub-discipline specialists to grapple with as and when markets "fail". This foundational position has by default committed economic theory to reckless expansion, and, as a consequence (unintended or otherwise), limited the scope of solutions, since solutions have tacitly been required to take expansion as given. Within this framing, solutions only become relevant where markets fail in terms of pricing structures. Solutions mainly orient on correcting market failures, applying state intervention as a limited last resort. By elimination, prohibition is thus a thought that is mainly prevented. This framing takes it as given that corrected markets mainly solve the problems they create, as the system as a whole inexorably expands... The very idea of limits is, therefore, peripheralized.

Of course, at the same time, common sense tells economists of all hues, as it does all other citizens, that *everything* has its limit. And yet, economics as an ideational resource has worked against common sense. In mainstream economics, *some* applications have limits but there is always scope to substitute and transfer to other market situations of exploitation and development. In working against common sense, mainstream economics has followed a direction of travel of convenient compatibility that has been convergent with the concerns of corporations to expand and profit, and with the overriding competitive interest of countries to grow and materially develop. As citizens of wealthy countries, we have collectively been mainly ok with this because it has been easy to accept the logic of least cost future adaptations based mainly on solutions (efficiencies of innovation and technology) simply emerging from market processes (perhaps with some behavioural nudging, a few minor tax changes and seed investment for new transformative markets). For most of us this has become a rather dull somnambulant-inducing and eminently ignorable concern with externalities, discount rates, backstop resources and matters of marginal abatement and mitigation.

And so here we are, now requiring solutions that *begin* from a 55% cut in emissions by 2030 and a fall to net zero by the middle of the century. The challenge is great. However, the essays collected here provide insight and points of departure for the problem. Many, for example, will find inspiration in Richard Smith's work that takes the idea of the Green New Deal forward.[2] Others will see value in Ted Trainer and Samuel Alexander's anarcho-syndicalism, with its practical localism. It is to be hoped that all readers will respond with a sense that they need to take personal responsibility for their (our) own actions, and for the actions of those who claim to represent us. If there is one outcome we cannot currently afford it is a sense that the problem is overwhelming. Doing something is liable to initiate a sense that it is not enough and that every (in)action involves some degree of hypocrisy. However, perhaps the more we are sensitized to this, the more we are progressively prone to practical reorientation.

Conclusion

History is replete with harbingers of apocalyptic civilizational crisis. The very fact there is *history* to attest to this indicates that vocal dread has typically been ill-founded. However, we have now entered a phase where we have truly placed ourselves in jeopardy. The overwhelming weight of evidence regarding our collective environmental consequences as a species suggests that the common sense of the last forty years is now forced to confront its own complacency. "Extinction Rebellion" and other new civil society and expert activist movements have begun to appear.[3] Encouragingly the young seem particularly attuned to the scale of the

---

[2] https://en.wikipedia.org/wiki/Green_New_Deal, https://www.greennewdealgroup.org.
[3] https://rebellion.earth, https://www.sunrisemovement.org/%20

problem they have been bequeathed. A call to sanity has been initiated. Are we sufficiently rational as a species to respond?

References

Daly, H. (2014) *Beyond Growth: The Economics of Sustainable Development.* Boston: Beacon Press.

Fullbrook, E. (2016) *Narrative Fixation in Economics.* Bristol: World Economics Association Books.

Fullbrook, E. (Ed.) (2009) *Ontology and Economics.* London: Routledge.

Fullbrook, E. (Ed.) (2008) *Pluralist Economics.* London: Zed Books.

Georgescu-Roegen, N. (1971) *The Entropy Law and the Economic Process.* Cambridge and London: Harvard University Press.

Hansen, J. et al. (2017) "Young people's burden: requirement of negative CO_2 emissions." *Earth System Dynamics*, 8, pp. 577-616.

IPCC (2018) *Global Warming of 1.5°C: Summary for policymakers.* Geneva: Author.

Koch, M. (2012) *Capitalism and Climate Change. Theoretical Discussion, Historical Development and Policy Responses.* London: Palgrave Macmillan.

Morgan, J. (2017) "Piketty and the growth dilemma revisited in the context of ecological economics." *Ecological Economics*, 136, pp. 169-177.

Morgan, J. (2016) "Paris COP21: Power that speaks the truth?" *Globalizations* 13(6), pp. 943-951.

Smith, R. (2016) *Green Capitalism: The God that failed.* Bristol: World Economics Association Books.

Söderbaum, P. (2018) *Economics, ideological orientation and democracy for sustainable development,* Bristol: World Economics Association Books, second edition.

Spash, C. (2017) "Social Ecological Economics." In Spash, C. (Ed.) *Routledge Handbook of Ecological Economics: Nature and Society.* New York: Routledge, pp. 3-16.

Steffen, W. et al. (2018) "Trajectories of the Earth System in the Anthropocene." *PNAS*, pp. 1-8.

UNEP (2018) *Emissions Gap Report 2018* (9th edition), November. New York: Author.

UNEP (2012) *Global Environmental Outlook Report 5: Environment for the Future We Want.* New York: Author.

UNFCCC (1992) *United Nations Framework Convention on Climate Change.* May 9. New York: Author.

UNFCCC (2015) *Adoption of the Paris agreement and annex: Paris agreement.* December 12. Paris: Author.

UNFCCC (2018) *Katowice texts.* December 14. Katowice: Author.

Chapter 2

Growthism: its ecological, economic and ethical limits

Herman Daly

We have many problems – poverty, unemployment, environmental destruction, climate change, financial instability, etc. – but only one solution for everything, namely economic growth. We believe that growth is the costless, win-win solution to all problems, or at least the necessary precondition for any solution. This is growthism. It now creates more problems than it solves.

A journey of no return, not a circular economy

The economic process is not a mechanical analog that can be run forward and backward, nor a circular process that can return to any previous state. Rather it is an irreversible and irrevocable process moving in the direction of time's arrow of increasing entropy.[1] Finitude

[1] Nicholas Georgescu-Roegen (1972)*The Entropy Law and the Economic Process*, Harvard University Press.

and entropy guarantee that the economic life of our species will be a journey of no return. Therefore even a stationary economy, in the classical sense of constant population and constant capital stock, is ultimately a journey of no return, because the metabolic throughput of matter and energy required to maintain constant stocks of people and physical capital, in the face of depreciation and death, is an entropic flow from ever less concentrated sources to ever filling sinks – and both sources and sinks are finite. Consequently, technology must change qualitatively to adapt to entropy increase, to depletion and pollution of the environment, even in the stationary, or "steady-state economy" as it has been more recently called. Relative to the growth economy the steady-state economy is a *slower* journey of no return, one that values longevity with sufficiency, and seeks qualitative improvement rather than quantitative increase. The many advantages of a slower journey were emphasized by John Stuart Mill[2], the champion of the classical stationary state:

> "I know not why it should be a matter of congratulation that persons who are already richer than anyone needs to be, should have doubled their means of consuming things which give little or no pleasure except as representative of wealth...."
>
> "The density of population necessary to enable mankind to obtain in the greatest degree, all the advantages both of

[2] John Stuart Mill (1857) *Principles of Political Economy*, vol. 2 (London: John W. Parker), pp. 320–326, with omissions.

> cooperation and of social intercourse, has, in all the most populous countries been attained...."
>
> "It is scarcely necessary to remark that a stationary condition of capital and population implies no stationary state of human improvement. There would be as much scope as ever for all kinds of mental culture, and moral and social progress; as much room for improving the Art of Living and much more likelihood of its being improved, when minds cease to be engrossed by the art of getting on."

In contrast to Mill's vision of the steady state, the reality of today's growthist economy is one of harried drivenness, of frantic adaptation to the unforeseen, unwilled, and out of control consequences of maximized, subsidized growth, pushed by ever larger scale and more dangerous technologies. Such growth is now threatening the capacity of earth to support life.

Many are not content with a slower more careful journey of no return. They want a so-called "circular economy" that can presumably live, and continue to grow, by ingesting only its own waste products. They assume that what they consider desirable must therefore be possible.

For anyone who has taken the first course in economics the recently revived term "circular economy" calls to mind the famous diagram of the circular flow of exchange value between firms and households found in the first pages of the standard textbooks. That diagram shows goods and factors of production flowing in a closed circle between firms and households with money flowing in the opposite direction. The economy is represented as an *isolated system* –nothing enters from the outside, nothing exits to the outside. There are no natural resources entering from the ecosphere, no wastes exiting back to the ecosphere. Indeed there is no ecosphere, no containing and constraining environment of any kind. This abstract vision is useful for studying exchange (supply, demand, prices, and national income), but worthless for studying environmental costs of economic growth because there is no finite environment to constrain growth.

This picture however is *not* what most advocates today mean by "circular economy", but it has a similar name of long standing, and is a source of confusion. By "circular economy" they mean an economy that recycles material natural resources to a high degree, and increases product lifetimes, and uses mainly renewable resources – all good policies, but destined to fall short of their goal of "sustainable growth". It might better have been called a "recycling economy" or an economy that maximizes natural resource productivity rather than labor or capital productivity. Increased resource efficiency is also referred to as "decoupling" as in disconnecting the output of goods and services from the throughput of resources. In the limit a totally "decoupled economy" would take us back to the neo classical circular flow representation of the economy as an isolated system. For this reason I prefer to avoid this reborn notion of "circular economy," and the related term "decoupling" because" they greatly overstate the degree of separability of production from resource throughput, further encouraging the unrealistic quest for "sustainable growth" in physical scale of the economic subsystem relative to the biosphere.

The heavy emphasis on circularity casts a deep shadow over the more basic fact that the metabolic throughput is fundamentally a linear one-way entropic flow. Yes, the overall linear flow can contain important countercurrents and reverse eddies of recycling, and it is important to take advantage of that. But the river itself flows from the mountains to the sea, and never backwards True, the hydrologic cycle powered by the sun, can evaporate the water to rain again in the mountains, but that happens in the ecosphere, *outside the economy.* If the "circular economy" relies on *natural* biophysical cycles powered by the sun, and does not grow in scale beyond the regenerative and absorptive capacities of the containing biosphere, then it approximates a steady-state economy – not a sustainable growth economy. In addition to a circulatory subsystem (recognized since the Physiocrats' analogy with blood circulation) the economy also has a digestive tract that ties it to its environment at both ends. That second more basic metabolic analogy has been neglected in economic theory.

Recycling is limited, first because it costs energy to carry out the recycle of materials; and second because energy itself is not subject to recycling (entropy means that it always takes more energy to effect the recycle than the amount of energy recycled – regardless of the price of energy!). The extra energy for the recycling also requires material instruments, trucks etc. So materials can be reduced, but at the cost of an increase in energy (and material) throughput, which after some number of cycles (how many?) becomes prohibitive, as remaining materials are ever more dispersed. Even expensive metals like gold, silver, and copper are currently only about one-third recycled and two-thirds newly depleted. Writers who expound the circular economy seem to be aware of this fact, but do not give it sufficient emphasis. Also it is important to distinguish prompt materials recycling that is internal to the economic subsystem, from long run external recycling through the containing ecosphere. While increased reliance on renewable resources is a good feature of the "circular economy", one must remember that, when exploited beyond sustainable yield, renewable resources effectively become nonrenewable. There is always a scale limit to a sustainable economic subsystem, beyond which growth, even in a "circular" economy, breaks down and sustainability requires a steady-state economy.

The basic issue of limits to growth that the Club of Rome did so much to emphasize in the early 1970s needs to remain front and center, with recycling considered as a useful accommodation to that limit, but not a path by which the growth economy can continue. Well before becoming physically impossible the growth of the economic subsystem becomes *uneconomic* in the sense that it costs more in terms of sacrificed ecosystem services than it is worth in terms of extra production. That richer is better than poorer is a truism. No dispute there. But is growth in GDP in wealthy countries really making us richer by any inclusive measure of wealth? That is the question. I think it is likely making us poorer by increasing unmeasured "illth" faster than measured wealth. Even a steady-state economy can be too big relative to the ecosphere. The neoclassical circular flow picture can never be too big by virtue of its being an isolated system. However, neoclassical economists do recognize that the economy can grow *too fast* (over-allocation of resources to investment relative to consumption), even though its scale can never be *too big.*

Inevitably national growth economies reach a point where many citizens begin to suspect that growth is no longer worth the cost of excessively rapid adaptation to an accelerating economy of no return – that so-called *economic* growth has in reality become *uneconomic* growth. John Stuart Mill recognized that long ago. Why have not more recognized it? Why is growth still the *summum bonum* of economists and politicians? Probably because growth is our substitute for sharing as a cure for poverty. And because our national accounts (GDP) are incapable of even registering uneconomic growth because they count only value added by labor and capital, and omit entirely the cost of using up that to which value is added, namely the entropic flow of natural resources, the very sap of life and wealth.

Globalization as an extension of growthism

Those of us old enough to remember the Cold War know that it was basically a contest between Socialism and Capitalism to see who could grow faster, and thereby accumulate more wealth and military power. The audience was the uncommitted countries of the world who would supposedly adopt the economic system of the winner of the growth race. What happened? Basically, Socialism collapsed, and Capitalism won by default. The losers (Russia, China, Eastern Europe) got back in the growth race by adopting State Capitalism, and China has become the growth champion. The present system of world growthism, in the broadly capitalist mode, is triumphant. But growthism itself has turned out to be a false god because growth in our finite and entropic world now increases ecological and social costs faster than production benefits, making us poorer, not richer (except for the top few percent). Recognition of this reversal is obscured by the fact that our national accounts (GDP), do not subtract the costs of growth, but effectively add them by counting the expenditures incurred to defend ourselves from the un-subtracted costs of growth. Even more egregiously, GDP counts the consumption of natural capital as income. Growthism is consuming the life support capacity of the biosphere for the benefit of a small minority of the present generation, while shifting the real but uncounted costs on to the poor, future generations, and other species.[3]

As national economies confront limits to their growth aspirations imposed by the carrying capacity of their territory and the extent of their national markets, they strive, by globalization, to grow into the ecological and economic space of all other countries, as well as into the remaining global commons. While this certainly provides extra degrees of freedom for individual nations to continue growing for a while, it does not remove global

[3] For more see www.greattransition.org/publication/economics-for-a-full-world.

limits. It simply ensures that those limits will be met more simultaneously and less sequentially. Consequently there will be less opportunity for one country to learn from the experience of others in adapting to limits. Furthermore, the ability of nations to enact independent policies for coming to terms with limits is undercut, because the net result of globalization is to convert many difficult, but tractable, national problems into one simultaneous intractable global problem, by speeding up and generalizing the economic journey of no return. At the same time, however, increasing energy costs will raise the cost of transport which acts as a general tariff on international trade and will promote national and local production, thereby weakening somewhat long distance trade and globalization.

The key to understanding globalization, I believe, is to clearly distinguish it from internationalization:

Internationalization refers to the increasing importance of relations between nations: international trade, international treaties, alliances, protocols, etc. The basic unit of community and policy remains the nation, even as relations among nations, and among individuals in different nations, become increasingly necessary and important.

Globalization refers to global economic integration of many formerly national economies into one global economy, by free trade, especially by free capital mobility, and also more recently by easy or uncontrolled migration. *Globalization is the effective erasure of national boundaries for economic purposes.* National boundaries become totally porous with respect to goods and capital, and increasingly porous with respect to people, viewed in this context as cheap labor, or in some cases cheap human capital.

In sum, globalization is the economic integration of the globe. But exactly what is "integration"? The word derives from "integer", meaning one, complete, or whole. Integration means much more than "interdependence" – it is the act of combining separate albeit related units into a single whole. Interdependence is to integration as friendship is to marriage. Since there can be only one whole, only one unity with reference to which parts are integrated, it follows that global economic integration logically implies national economic *dis*integration –parts are torn out of their national context (dis-integrated), presumably to be re-integrated into the new whole, the globalized economy. As the saying goes, to make an omelette you have to break some eggs. The disintegration of the national egg is necessary to integrate the global omelette. The benefits of global integration are extolled while the costs of national disintegration are neglected.

Of course globalization is far from complete, but the tendency is well advanced. What we have now is a collection of disintegrating national economies whose policies regarding international trade, capital mobility, and migration are taken over by monopoly global corporations, giant international banks, and a free-for-all of illegal migration of both cheap labor and human capital.

All that I have just said was expressed with admirable clarity, honesty, and brevity by Renato Ruggiero[4], former director-general of WTO: "We are no longer writing the rules of interaction among separate national economies. We are writing the constitution of a single global economy." This is a clear affirmation of globalization and rejection of internationalization as just defined. It is also a radical subversion of the Bretton Woods Charter. Internationalization is what the Bretton Woods Institutions were designed for, not globalization.

Everyone recognizes the desirability of community for the world as a whole-- but we have two very different models of world community: (1) a federated community of real national communities (internationalization), versus (2) a cosmopolitan direct membership of individuals in a single global abstract community (globalization).

If the IMF-WB-WTO are no longer serving the interests of their member nations as per their charter, then whose interests are they serving? The interests of the integrated "global economy" we are told. But what concrete reality lies behind that grand abstraction? Not real individual workers, peasants, or small businessmen, but rather giant pseudo-individuals, the transnational corporations.

---

[4] From a speech to the United Nations Conference on Trade and Development (UNCTAD) October, 1996.

Consequences of growth-driven globalization

Consider a few pattern-changing consequences of globalization, of the erasure of national boundaries for economic purposes. Briefly, they include: (1) standards-lowering competition to externalize social and environmental costs to achieve a competitive price advantage--a race to the bottom in terms of both efficiency in cost accounting and equity in income distribution; (2) increased tolerance of mergers and monopoly power in domestic markets in order to be big enough to compete internationally; (3) more intense national (regional) specialization according to the dictates of competitive advantage, with the consequence of reducing the range of choice of ways to earn a livelihood, and increasing dependence on other countries. Free trade and intense specialization negate the freedom not to trade; (4) world-wide enforcement of a muddled and self-serving doctrine of "trade-related intellectual property rights" in direct contradiction to Thomas Jefferson's dictum that "knowledge is the common property of mankind". Let us look at each of these in a bit more detail.

1. Standards lowering competition

The country that does the poorest job of internalizing all social and environmental costs of production into its prices gets a competitive advantage in international trade. More of world production shifts to countries that do the poorest job of counting costs-- a sure recipe for reducing the efficiency of global production. As uncounted, externalized costs increase, the positive correlation between GDP growth and welfare disappears, or even becomes negative.

Another dimension of the race to the bottom is the increasing inequality in the distribution of income in high-wage countries, such as the US, fostered by globalization. In the US there has been an implicit social contract established to ameliorate industrial strife between labor and capital. Specifically, a just distribution of income between labor and capital has been taken to be one that is more equal within the US than it is for the world as a whole. Global integration of markets necessarily abrogates that social contract. US wages will fall drastically because labor is relatively much more abundant globally than nationally. It also means that returns to capital in the US will increase because capital is relatively scarcer globally than nationally. US distribution of income then tends to the more unequal global distribution, thus breaking the implicit social contract.

Free trade, and by extension globalization, is often defended by appeal to Ricardian comparative advantage. The logic of comparative advantage assumes that factors of production, especially capital, are immobile between nations. Only products are traded.[5] With capital mobility now the major defining feature of globalization we have left the world of *comparative* advantage and entered a regime of *absolute* advantage, which guarantees gains from trade *to the world as a whole*, but does not guarantee that *each nation* will share in those gains, as was the case

[5] For a discussion see Chapter 18 in H. Daly and J. Farley (2011) *Ecological Economics*, Second Edition, Island Press, Washington D. C.

under comparative advantage. Global gains under absolute advantage are theoretically greater than under comparative advantage, but there is no reason to expect these gains to be shared by all trading partners. Mutual gain could be restored under absolute advantage by redistributing some of the global gains from trade. But I have never heard that idea discussed by globalization advocates. Often they appeal, quite illogically, to the doctrine of comparative advantage as a guarantee of mutual benefit, conveniently forgetting that the logic of comparative advantage requires immobile capital, and that capital is not immobile.[6] Indeed, some even argue for free capital mobility by extension of the comparative advantage argument – if free trade in goods is mutually beneficial then why not also have free trade in capital? However, one cannot use the conclusion of an argument to abolish one of the premises upon which the argument is based! Similar illogical arguments are made in defense of free labor mobility between nations.

2. *Tolerance of corporate power*

Fostering global competitive advantage is used as an excuse for tolerance of corporate mergers and monopoly in national markets so that domestic firms are big enough to compete globally (we now depend on international trade as a substitute for domestic trust busting to maintain competition). It is ironic that this is done in name of deregulation and the free market.

[6] To be clear, this refers primarily to the financial sense of capital; capital in the sense of already existing produced means of production can be highly immobile and is often destroyed by mobile "capital" (as the problems of the American rustbelt illustrate).

Chicago School economist and Nobel laureate Ronald Coase[7] said " – *Firms are islands of central planning in a sea of market relationships".* The islands of central planning become larger and larger relative to the remaining sea of market relationships as a result of merger. More and more resources are allocated by within-firm central planning, and less by between-firm market relationships. And this is hailed as a victory for markets! It is no such thing. It is a victory for corporations relative to national governments, which are no longer strong enough to regulate corporate capital and maintain competitive markets in the public interest. Of the 100 largest economic organizations roughly 52 are corporations and 48 are nations. Approximately one-third of the commerce that crosses national boundaries does not cross a corporate boundary, i.e. is an intra-firm, non-market transfer. The distribution of income within these centrally planned corporations has become much more concentrated. The ratio of salary of the Chief Executive Officer to low-level employees has passed 500 on its way to infinity--what else can we expect when central planners set their own salaries!

3. Intensified specialization

Free trade and free capital mobility increase pressures for specialization according to both comparative and absolute advantage. Therefore the range of choice of ways to earn a livelihood becomes greatly narrowed. In Uruguay, for example, everyone would have to be either a shepherd or a cowboy in conformity with the

---

[7] Ronald Coase (1937) "The Nature of the Firm", *Economica*, 4(16), pp. 386-405.

specialization dictated by competitive advantage in the global market. Everything else should be imported in exchange for beef, mutton, wool, and leather. Any Uruguayan who wants to play in a symphony orchestra or be an airline pilot should emigrate. Uruguayans have sensibly resisted such excessive specialization.

Most people derive as much satisfaction from how they earn their income as from how they spend it. Narrowing that range of choice is a welfare loss uncounted by trade theorists. Globalization assumes either that emigration and immigration are costless, or that narrowing the range of occupational choice within a nation is costless. Both assumptions are false.

While trade theorists ignore the range of choice in *earning* one's income, the range of choice in *spending* one's income receives exaggerated emphasis. For example, the US imports Danish butter cookies and Denmark imports US butter cookies. The cookies cross each other somewhere over the North Atlantic. Although the gains from trading such similar commodities cannot be great, trade theorists insist that expanding the range of consumer choice to the limit increases the welfare of cookie connoisseurs. Perhaps, but could not those gains be had more cheaply by simply trading recipes? One might think so, but *recipes* (trade related intellectual property rights) are the thing that free traders most want to protect.

4. The inconsistencies of intellectual property

Of all things knowledge is that which should be most freely shared, because in sharing it is multiplied rather than divided. Knowledge is a non-rival good and should be also non excludable. Yet, as already noted, our trade theorists have rejected Thomas Jefferson's dictum that "*Knowledge is the common property of mankind*" in exchange for a muddled doctrine of "trade related intellectual property rights" by which they are willing to grant private corporations monopoly ownership of the very basis of life itself--patents to seeds (including the patent-protecting, life-denying terminator gene) and to knowledge of basic genetic structures.

The argument offered to support this enclosure of the knowledge commons is that, unless we provide the economic incentive of monopoly ownership for a significant period of time, little new knowledge and innovation will be forthcoming. Yet, as far as I know, James Watson and Francis Crick, who discovered the structure of DNA, do not share in the patent royalties reaped by the second rate gene-jockeys who are profiting from their monumental discovery. Nor of course did Gregor Mendel get any royalties – but then he was a monk motivated by mere curiosity about how Creation works! Nor did Jonas Salk try to patent the polio vaccine. He thought it would be like trying to patent the sun.

Once knowledge exists, its proper allocative price is the marginal opportunity cost of sharing it, which is close to zero, since nothing is lost by sharing it. Yes, of course you do lose the *monopoly* on the knowledge, but then economists have traditionally argued that monopoly is inefficient as well as unjust because it creates an artificial scarcity of the monopolized item. Furthermore, the main input to the production of new knowledge is existing knowledge, and keeping the latter artificially expensive is bound to slow down the production of the former.

Of course the cost of production of new knowledge is not zero, even though the cost of sharing it is. This allows biotech corporations to claim that they deserve a fifteen or twenty year monopoly for the expenses they incur in research and development, even though they spend more on advertising than research. Of course they deserve a profit on their efforts, but not on Watson and Crick's contribution without which they could do nothing, nor on the contributions of Gregor Mendel, and all the great scientists of the past who made the fundamental discoveries.[8] As economist Joseph Schumpeter emphasized, being the first with an innovation already gives one a temporary monopoly. In his view these recurring temporary monopolies were the source of profit in a competitive economy whose theoretical tendency is to compete excess profits down to zero.

As the great Swiss economist, Sismondi, argued long ago, not all new knowledge is a benefit to mankind. We need a social and ethical filter to select out the beneficial knowledge. Motivating the search for knowledge by the purpose of benefiting mankind rather than by securing

[8] Similarly, it radically under-estimates the role of the state; its many contributions become invisible in much of mainstream economic theory; see the *Real-World Economics Review* special issue number 84: "The public economy and a new public economics"
http://www.paecon.net/PAEReview/issue84/whole84.pdf

monopoly profit provides a better filter. Perhaps the greatest virtue of the steady-state economy is that because it is a slow rather than a fast journey of no return, we would have time to evaluate and experiment with new technologies, rather than blindly accepting anything in order to keep growth from slowing.

This is not to say that we should abolish all intellectual property rights – that would create more problems than it would solve. But we should certainly begin restricting the domain and length of patent monopolies rather than increasing them so rapidly and recklessly. And we should become much more willing to share knowledge. Shared knowledge increases the productivity of all labor, capital, and resources. International development aid should consist far more of freely shared knowledge, and far less of foreign investment and interest-bearing loans.

John Maynard Keynes,[9] one of the founders of the recently subverted Bretton Woods Institutions, recommended the following pattern for our international economy:

> "I sympathize therefore, with those who would minimize, rather than those who would maximize, economic entanglement between nations. Ideas, knowledge, art, hospitality, travel – these are the things which should of their nature be international. But let goods be homespun

[9] J. M. Keynes (1933) "National Self-Sufficiency", in D. Muggeridge , ed., *The Collected Writings of John Maynard Keynes*, vol. 21, London: Macmillan and Cambridge University Press.

> whenever it is reasonably and conveniently possible; and, above all, let finance be primarily national."

Growth-driven globalization will maximize economic entanglement between nations in pursuit of trading advantage, of monopoly power, of privatizing the remaining commons, especially that of knowledge, and of concentrating income to an extreme degree. These are the patterns that growthism solves for by way of globalization. Globalism is not the realization of world community. Rather it is individualism writ large – corporate feudalism in a global open-access commons.

On the importance of boundaries in life and logic

John Lennon asked us to imagine a world without boundaries, singing wistfully "imagine there's no countries", and we all know what he meant – a world of human solidarity, peace, and cooperation. Conflicts and war usually involve disputes over borders. So why not just get rid of these troublesome boundaries? Let's have globalization – deregulated trade, capital mobility, and migration – only let's bless them each with the adjective "free" rather than "deregulated".

Neoclassical economists assure us that this will lead to peace and prosperity among rational utility-maximizing individuals, minimally governed by a benevolent World Democracy, dedicated to the post-modern values of scientific materialism, eloquently communicated in Esperanto. This vision has its serious appeal to many, but not so much to me, as the reader will by now have guessed.

Economic and political boundaries are necessary to achieve both national community, and a global federation of national communities living in peace and ecological sustainability. Boundaries are both biologically and logically necessary. Skin and membranes are organic boundaries. Within-skin versus outside-skin is a basic boundary condition for life. The skin boundary must be permeable, but not too permeable. If nothing enters or exits the organism it will soon die. If everything enters and exits, then the organism is already dead and decaying. Life requires boundaries that are neither completely closed nor completely open. A nation's borders are in many ways very different from the skin of an organism, yet neither permits complete closure or complete openness. Both must be qualitatively and quantitatively selective in what they admit and expel, if their separate existence is to continue rather than be dissolved into entropic equilibrium with its environment.

Logically boundaries imply both inclusion and exclusion. A world without boundaries includes everything and is often therefore thought to be warm and friendly. But "everything" must include the cold and the unfriendly as well, or it is not everything. Also, without boundaries, B can be both A and non-A, which makes definition, contradiction, and analytical reasoning impossible. So both life and logical thinking require boundaries. While "a world without boundaries" may be a poetic expression of a desired unity, and while it is possible to reason dialectically with overlapping boundaries, it is a major delusion to think that boundaries are not necessary.

It is understandable, yet ironic, that the most fundamental and dramatic boundary of all – that separating the earth from outer space – made clear in the iconic photo of the earth from the moon – seems to have led to a reaction against the very concept of boundaries on our spherical planet, since it is so obviously one whole and unified thing. Yet that beautiful and powerful vision of overall unity hides a world of diversity and difference. And we live on the earth, within that complex living diversity, not on the dead moon with no need for life-defining boundaries.

The illth of nations and the weakness of policy

Our traditional economic problems (poverty, over-population, unemployment, unjust distribution) have all been thought to have a common solution – namely an increase in wealth. All problems are easier if we are richer. The way to get richer has been thought to be by economic growth, usually as measured by GDP. I do not here question the first proposition that richer is better than poorer, other things equal. But I do question whether what we persuasively label "economic growth" is any longer making us richer. I suggest that physical throughput growth is, at the present margin and in the aggregate, increasing illth faster than wealth, thus making us poorer rather than richer. Consequently our traditional economic problems become more difficult with further growth. The correlation between throughput growth and GDP growth is sufficiently strong historically so that in the absence of countervailing policies even GDP growth increases illth faster than wealth.

What we conventionally call "economic growth" in the sense of "growth of the economy" has ironically become "uneconomic growth" in the literal sense of growth that increases costs by more than it increases benefits. I am thinking here of the North rather than the South, because in many poor countries where the majority lives close to subsistence the benefits of production growth, even if badly distributed, justify incurring large costs. But since the South is striving, with encouragement from the IMF and World Bank, to become like the North, I am not really neglecting the South by focusing on the North, but rather raising a caution for the South.

One will surely ask how do I know that growth has become uneconomic for many Northern countries? Some empirical evidence is referenced below.[10] But more

---

[10] For critical discussion and the latest revision of the ISEW, see, Clifford W. Cobb and John B. Cobb, Jr., et al., *The Green National Product*, University Press of America, New York, 1994. For a presentation of the ISEW see Appendix of *For the Common Good*, H. Daly and J. Cobb, Boston: Beacon Press, 1989; second edition 1994. See also Clifford W. Cobb, et al., "If the GDP is Up, Why is America Down?,_*Atlantic Monthly*, October, 1995. See also Manfred Max-Neef, Economic Growth and Quality of Life: A Threshold Hypothesis, *Ecological Economics*, 15, (1995), pp. 115-118. More recently the Lancet medical journal (NYT, Oct. 19, 2017) finds that the financial costs from pollution are some $4.6 trillion annually, about 6.2% of the global economy. If annual growth in Gross World Product is around 2.2%, and cost due to pollution is 6.2%, then with reasonable accounting we would have a net financial decline of some 4% annually. If that financial decline represents welfare loss, and it surely does since we are talking about reduced health and life expectancy, then the benefits of production growth are being more than cancelled out by the costs of the pollution generated by that growth. In other words, so-called "economic" growth has become uneconomic. That seems to have escaped the notice of economists.

convincing to me is the simple argument that as the scale of the human subsystem (the economy) expands relative to the fixed dimensions of the containing and sustaining ecosphere, we necessarily encroach upon that system and must pay the opportunity cost of lost ecosystem services as we enjoy the extra benefit of increased human scale. As rational beings we presumably satisfy our most pressing wants first, so that each increase in scale yields a diminishing marginal benefit. Likewise, we presumably would sequence our takeovers of the ecosystem so as to sacrifice first the least important natural services. Obviously we have not yet begun to do this because we are just now recognizing that natural services are scarce. But let me credit us with capacity to learn. Even so, that means that increasing marginal costs and decreasing marginal benefits will accompany increasing human scale. The optimum scale, from the human perspective, occurs when marginal cost equals marginal benefit. Beyond that point growth becomes uneconomic in the literal sense of costing more than it is worth.

It is interesting to know empirically if we have reached that point (I think we have, both globally and in many countries), but even if we have not, it is obvious that continued growth of a dependent subsystem relative to a finite sustaining total system will inevitably reach such an optimal scale. If we add to the limit of finitude of the total system the additional limits of entropy and complexity of ecological interdependence, then it is clear that the optimal scale will be encountered sooner rather than later. Additionally, if we expand our anthropocentric view of the optimum scale to a more biocentric view, by which I mean one that attributes not only instrumental but also intrinsic value to other species, then it is clear that the scale of the human presence should be further limited by the duty to reserve a place in the sun for other species, even beyond what they "pay for" in terms of their instrumental value to us. And of course the whole idea of "sustainability" is that the optimal scale should exist for a very long time, not just a few generations. Clearly a sustainable scale will be smaller than an unsustainable scale. For all these reasons I think that for policy purposes we do not need exact empirical measures of the optimal scale. If one jumps from an airplane it may be nice to have an altimeter, but what one really needs is a parachute.

So what policies constitute a parachute? Briefly, they are policies that limit aggregate throughput, while allowing the market to allocate that limited throughput – assuming the market is competitive and confined to some limited degree of inequality in the distribution of wealth and income. Such policy instruments are evolving now – e.g., cap-auction-trade systems for extraction rights, pollution emission rights, fishing rights, etc. Also ecological tax reform limits throughput by making it more expensive. It shifts the tax base from value added (something we want more of) on to "that to which value is added", namely the resource throughput (something we want to use less of). In differing ways each of the above "parachutes" would limit throughput and expansion of the scale of the economy into the ecosystem, and also provide public revenue. I will not discuss their relative merits, having to do with price versus quantity interventions in the market, but rather emphasize the advantage that both have over the currently favored strategy. The currently favored strategy might be called “efficiency first” in distinction to the “frugality first” principle embodied in both of the throughput-limiting mechanisms mentioned above, but more stringently in the second.

“Efficiency first” sounds good, especially when referred to as “win-win” strategies, or more picturesquely as “picking the low-hanging fruit”. But the problem of “efficiency first” is with what comes second. An improvement in efficiency by itself is equivalent to having a larger supply of the factor whose efficiency increased. The price of that factor will decline. More uses for the now cheaper factor will be found. We may end up consuming more of the resource than before, albeit more efficiently. Scale continues to grow. This is sometimes called the “Jevons effect”. A policy of “frugality first”, however, induces efficiency as a secondary consequence; “efficiency first” does not induce frugality – it makes frugality less necessary, nor does it give rise to a scarcity rent that can be captured and redistributed by tax or auction.

So far I have briefly outlined what I take to be the problem of the “illth of nations” (apologies to both Adam Smith and John Ruskin), and indicated some policy guidelines for avoiding the uneconomic growth that increases illth faster than wealth. These views do not find favor with mainstream economists. The concepts of throughput, of entropy, and even of optimal scale of the macroeconomy are foreign to them. The last is especially odd since in microeconomics the concept of the optimal scale of each micro activity is central. Yet the sum of all micro activities, the macro economy, is not thought to have an optimal scale relative to its sustaining ecosystem. Probably this is because macroeconomists think of the macroeconomy as the Whole, not as a Part of some larger Whole. For them nature is not a containing envelope, but just a sector of the macroeconomy – mines, wells, croplands, pastures, and fisheries. When the Whole grows it expands into the Void encroaching on nothing and incurring no opportunity cost. But of course the real economy is a Part and it grows not into the Void, but into the rest of the biosphere, and really does incur opportunity costs. I have long considered this Whole versus Part difference to reflect different pre-analytic visions (Schumpeter) or different paradigms (Kuhn). Different pre-analytic visions cannot, of course, be reconciled by further analysis, and they have different policy implications.

Presuppositions of policy

Even if we could agree on the right pre-analytic vision of the basic way the world is, would we then be able to enact and follow effective policies? So far, our capacity to enact policies of "frugality first" seems very weak. Indeed, even "efficiency first" policies are still resisted. So let us turn our attention to the question of policy in general, and policy fecklessness in particular.

What are the presuppositions we must make before we can reasonably and seriously discuss policy – policy of any kind? There are two that I can see.

First we must believe that there are real alternatives

among which to choose. If there are no alternatives, if everything is determined, then it hardly makes sense to discuss policy--what will be will be. No options, no responsibility, no need to think.

Second, even if there were real alternatives, policy dialogue would still make no sense unless there was a real criterion of value by which to choose from among the alternatives. Unless we can distinguish better from worse states of the world then it makes no sense to try to achieve one state of the world rather than another. No value criterion, no responsibility, no need to think.

In sum, serious policy must presuppose: (1) *nondeterminism* – that the world is not totally determined, that there is an element of freedom which offers us real alternatives; and (2) *nonnihilism* – that there is a real criterion of value to guide our choices, however vaguely we may perceive it.

To be sure, not every conceivable alternative is a real alternative. Many things really arc impossible. But the number of viable possibilities permitted by physical law and past history is seldom reduced to only one. Through our choices, value and purpose lure the physical world in one direction rather than the other. *Purpose is independently causative in the world.*

This seems pretty obvious to common sense – so what is the point of stating the obvious? The point is that many members of the intelligentsia deny one or both presuppositions, and yet want to engage in policy dialogue. I don't mean that we disagree on exactly what our alternatives are in a particular instance, or about just what our value criterion implies for a concrete case. That is part of the reasonable policy dialogue. I mean that determinists who deny the effective existence of alternatives, and nihilists or relativists who deny the existence of value beyond the level of subjective personal tastes, have no right to engage in policy dialogue – and yet they do! This is my cordial invitation to them to shut up – at least about policy.

Who are these people? In the sciences I am thinking about the materialist neo-Darwinists and socio-biologists; in the humanities, the post-modern deconstructionists; and in the social sciences, those economists who reduce value to subjective individual tastes any one of which is as good as another.

No one can in practice live by the creed of determinism or nihilism. In this sense no one takes them seriously, so we tend to discount any effect on policy of these doctrines. We tend to dismiss them as academic posturings. However, we may halfway suspect that the many learned people who publicly proclaim these frequently unopposed views might be right--and that is sometimes enough to enfeeble policy. For example, many people tell me that globalization is inevitable; any attempt to counter global economic integration is futile, or "on the wrong side of history", etc. If I manage to convince them that globalization is the result of past policy choices, and therefore might not be inevitable, the next line of defense is, how do we know that globalization will be any worse than the alternative? We cannot tell, we don't really know that globalization won't be good for us (because we don't know what is good in the first place), so there is no point in opposing it. Either it is inevitable, or if not then we can have no reason to believe that any alternative would be better. Forget policy, go back to sleep.

Perhaps I can clarify this controversial point by distinguishing four categories based on acceptance or non-acceptance of each of the two presuppositions identified.

(1) **The traditional Judeo-Christian view** – there exist real alternatives from which to choose by reference to objective criteria of value.

(2) **Criterionless choice** – alternatives are real options, but there is no objective criterion for choosing among them. (Existentialist angst)

(3) **Providential determinism** – there are no real options, but there is an objective criterion of value by which to choose, if only we had a choice. Fortunately providence has chosen for us according to the objective criterion, which we would not be wise or good enough to have followed on our own. (Theological predestination; technological providentialism)

(4) **Criterionless determinism** – there are no real alternatives to choose from, and even if there were, there is no objective criterion of value by which to choose. All is mechanism – random variation and natural selection, as claimed by the neo-Darwinist materialists.

People engaged in policy, yet holding to positions (2), (3), or (4) are in the grip of a severe and debilitating inconsistency. Their participation in policy dialogue should be subject to the injunction of "estoppel" – a legal restraint to prevent witnesses from contradicting their own testimony. It should be applied in academia as well as in the courtroom!

Some conclusions

Avoiding the uneconomic growth that is increasing the illth of nations will require clear and forceful policy to limit growth. All policy, especially such a radical one, requires a belief in both objective value and real alternatives. The fact that many people engaged in discussing and making policy reject one or both of these presuppositions is, in A. N. Whitehead's term,[11] "the lurking inconsistency", a contradiction at the basis of the modern worldview which enfeebles thought and renders action feckless. If we even halfway believe that purpose is an illusion foisted on us by our genes to somehow make us more efficient at procreation, or that one state of the world is, for all we can tell, as good as another, then it is hard to get serious about policy. Whitehead noted, "Scientists animated by the purpose of proving that they are purposeless constitute an interesting subject for study". He went on to say that, "It is not popular to dwell on the absolute contradiction here involved".

I think, 85 years later, that it is high time we dwelt on

[11] A.N. Whitehead, *The Function of Reason*, Princeton, N.J.: Princeton University Press, 1929, p.12.

this absolute contradiction. We pay a price for ignoring contradictions – in this case the price is feebleness of purpose and half-heartedness in policy. Citizens really must affirm that the world offers more than one possibility to choose from, and that some choices really are better than others. Determinists and nihilists have a right to exist, but an obligation to remain silent on policy!

This willful neglect has allowed the lurking inconsistency to metastasize into the marrow of modernity. The Enlightenment, with its rejection of teleology, certainly illuminated some hidden recesses of superstition in the so-called Dark Ages. But the angle of its cold light has also cast a deep shadow forward into the modern world, obscuring the reality of purpose. To conserve Creation we will first have to reclaim purpose from that darkness. I say Creation with a capital "C" advisedly, and certainly not in denial of the established facts of evolution. If our world and our lives are not in some sense a Creation, but just a purposeless happenstance – a random statistical fluke of multiplying infinitesimal probabilities by an infinite number of trials – then it is hard to see from where we will get the will and inspiration to care for it.

Indeed, our decision-making elites may already tacitly understand that growth has become uneconomic. But apparently they have also figured out how to keep the dwindling extra benefits for themselves, while "sharing" the exploding extra costs with the poor, the future, and other species. Why not, if it is all just a purposeless happenstance? The elite-owned media, the corporate-funded think tanks, the kept economists of high academia, and the World Bank – not to mention Goldman-Sachs and Wall Street – all sing hymns to growth in harmony with class interest and greed. The public is bamboozled by technical obfuscation, and by the false promise of growthism that one day we will all be rich. Intellectual confusion is real, but moral nihilism, abetted by naturalistic scientism, is the more basic problem. Such nihilism is hard to counter without strong appeal to the idea of purpose, of telos, and without raising its cosmic and religious implications. Many policies are being offered. But until the presuppositions of policy have been met they will remain just academic exercises.

Chapter 3

Producing ecological economy

Katharine N. Farrell

Introduction: Georgescu-Roegen unheeded

Economic analyses and conclusions are intimately bound up with judgements regarding the human condition. They are concerned with the study of what Marshall (1947, p. 1) referred to as "[hu]mankind in the ordinary business of life," and dedicated to examine "that part of individual and social action which is most closely connected with the attainment and with the use of the material requisites of wellbeing," (Ibid.). In that respect, economics is, from first principles, a normative enterprise.

While the idea that wellbeing and monetary wealth are so tightly correlated that the latter may be used as a proxy for the former was not an original premise of the early versions of economic analysis in European academic circles, and is today, increasingly brought into question not only from without but also from within mainstream economics, the presumption continues to influence the analytical apparatus used at all times by most, and at least sometimes by almost all scholars who understand themselves to be aligned with this field of enquiry. This is, as Georgescu-Roegen (1971) has noted, closely related to methodological choices made by some of the most important founding thinkers of this modern, euro-descendent, discipline: not least among them Pareto, Walras and Marshall himself. He argues (Georgescu-Roegen, 1971, Introduction) that their aspiration to secure economics a place at the table of the "hard sciences", led them to adopt an analytical approach of arithmetic fetishism (my words, not his) that leaves unattended the qualitative aspects of purposiveness and biodynamic transformation that lie at the heart of economic process: ignoring, thereby, aspects central to defining what constitutes the material requisites of wellbeing and to identifying viable means on the basis of which these may be attained and effectively used.

Notwithstanding the notable contributions of Herman Daly, an early student of his, and consistent engagement within the trans-discipline of ecological economics, Georgescu-Roegen's life work, like that of other heterodox economists, while taken up in part, within a variety of discourses, has generally been marginal to mainstream economics in the 20th and 21st Centuries. While the general disposition toward heterodox economics arguments has warmed considerably since the 2008 international financial crisis, adoption of radically distinct modelling approaches, such as those proposed by Georgescu-Roegen (1971) has not been forthcoming. The response has been, instead, mainly one of tweaks, focused either on correcting failures in the construction of GDP measures, through satellite accounts, the addition of compensatory sector variables or, in perhaps its most extreme form, the reactionary discourse on degrowth or, as in the case of post-Keynesianisms and much of behavioural and evolutionary economics, on the introduction of recalibrations, additional variables, reconfigurations and the incorporation of non-linearities into models that remain, nonetheless, at their core, closely aligned with the conventional structures of Walrasean analyses.

Georgescu-Roegen's response to this, which he calls "wholesale arithmetization" (Georgescu-Roegen, 1971, p. 15), constructed through the elaboration of a wide range of arguments, over a period of decades (Georgescu-Roegen, 1960; 1965a; 1965b; 1966; 1968; 1969; 1975; 1976; 1977; 1981; 1986; 1988; 1999[1971]), includes detailed attention to two interrelated points:

1. That economic processes are essentially biological in character.
2. That institutions constitute a core and critical aspect of human biology.

These provide, in my view, an excellent reference structure for considering all three questions that have been posed for this brief intervention:

1. How and to what degree is the economy changing the ecosystem?
2. How must economics change if it is to become a force for leading us away from catastrophe rather than toward it?
3. How can the global economy be changed so as avoid ecological collapse?

How and to what degree is the economy changing the ecosystem?

I would argue that our ability, from within an economics based approach, merely to grasp the information required to address this question is severely constrained, precisely by the two limitations observed by Georgescu-Roegen. That is to say, this question cannot be answered without structured reference to the biophysical characteristics of economic process and due attention to the role played, within those processes, by the human characteristic of using institutions to organise economic activity.

Again, with reference to the complexity of the problem, here, I think it is also important to distinguish, before proceeding, between different economies and different ecosystems. Some ecosystems, such as sustainably harvested temperate forests, are in quite good condition, in spite of having been changed dramatically by human economic processes; others, such as the tropical belt of mangrove forests, are in grave condition, in part due to changes caused by humans, but also in part due to their unfortunate positioning, at the mouths of rivers, where the ecological stress of upstream changes is concentrated and amplified. Similarly, not all economies are changing ecosystems in the same ways, and finally, not all changes are reified in the immediate surroundings of the economies that are causing them to occur. This give rise to a plethora of related social justice questions which fall not within the remit of the economist but of the social theorist and the body politic proper and cannot be addressed in an appropriate way here. Nonetheless, they should not be overlooked. Happily, there are a growing number of examples across the world of economic activity leading to ecological recuperation, not only to destruction. That said, destruction is clearly still the norm.

Taking then, rather a broad view, and working from within a social, historical frame of reference, I would suggest, following on from arguments presented in the late 1970s by the German Democratic Republic dissident Rudolf Bahro (1977; 1987), that the most far reaching and deeply seated way in which the contemporary global, late-industrial economy is changing the ecosystems within which it is embedded is by systematically and collectively ignoring its biological and social relationships to them. Following Faber *et al.* (1995; 1996), we might refer to this as an extreme deficiency in what they refer to as the third *tele* of living organisms: service. The label "third *tele*" is based on a teleological taxonomy, borrowed from biology, which they employ to help make sense of the blatant disregard that industrialised humans seem to have for the negative ecological impacts they cause, while going about the "ordinary business of life." Drawing on the Aristotelean concept of *entelecheia*, which means, literally, to have one's *telos* (or final cause) as a characteristic of one's self (e.g. it is in the *entelecheia* of a bird to fly), they propose "a teleological terminology to characterise living beings (i.e., organisms)... [which enables them] - to emphasise the uniqueness of a living being; - to consider the relationship of a living being to its species; - to represent its integration into the oneness of Nature" (Faber et al., 1996, p. 45). They propose that the fulfilment of purpose of a living organism, and by association, with a few logical degrees of differentiation, of a biological species, can be described through reference to the internalizing of three basic *tele* (plural of *telos*), which pertain to "What aims (tele) can we ascribe to a living being?" (Ibid.):

1. Self-maintenance, development and self-realisation;
2. Replication and renewal;
3. Service to other species and or the whole of nature.

They then go on to argue, much in keeping with Bahro (1977), that deficiency in the third *telos* – service – is a basic a feature of industrial societies, which have become disassociated from the biological systems that surround them, leading to ecological imbalance, as the ecological impacts of industrialised humans fail to contribute toward the flourishing of the ecological systems of which they form a part. One clear example of this is the excessive entropy production of the industrial economy. A necessary correlate to the massive production rate exhibited during the 20th and now 21st centuries, this implies a problem of system overload, where the entropy production associated with human activity has exceeded the entropy processing capacity of the ecosystems upon which we are dependant. Resolving this will require more than improved efficiency, which would carry with it yet more entropy production. It will require that we are able to understand and improve our relationships with the entropy processing systems of the planet (Mayumi, 1995; Tsuchida and Murota, 1985) and perhaps that we discover new ways of processing entropy and/or rediscover ones that industrialised humans have ceased to practice.

The lack of attention to the contribution that human actions make toward maintaining or diminishing the wellbeing of our non-human neighbours is bound up with the logic and history of industrialisation. The aim to liberate man, and I do mean man, from the caprices of nature, implies that the whims of nature can thus be ignored. While such disregard could be maintained for some time, during the early stages of industrialisation, as both ecosystems and human populations adjusted to the changes in their relationship, the now accelerating cascade of global impacts (Steffens et al., 2018) illustrates the temporary character of that charmed position. On an optimistic note, if one of the main problems is our lack of attention to impacts, this would seem to imply that increased awareness, combined with moral motivation to act appropriately, might help to address the problem. Unfortunately, awareness, in humans, is a rather complicated affair, which implies engaging with everything from public education, to the business models of Google and Facebook.

On the question of degree, I am inclined to demure, referring the reader instead to the myriad of documentation, which, sadly, is readily available, concerning the extent to which human economic activity is compromising the viability of many forms of life across the planet earth, including, all too often, human life. That said, the simple answer would seem to me to be: to an unacceptable degree. However, bearing in mind that humans, like our biological companion species, the rat, the pigeon, the dog and the cockroach, are remarkably versatile, we should take into account that "unacceptable to humans" might well be a degree of change far beyond the level of contamination and habitat destruction that other species can support. So, I would settle here then on the following: wildly beyond that which the ecosystems of the planet can reasonably support while continuing to generate habitat suitable for humans.

How must economics change if it is to become a force for leading us away from catastrophe rather than toward it?

Georgescu-Roegen's call, echoed by many of his contemporaries, and today paid lip service to by most, if not all economist, was to give serious analytical attention to representing the role of biological dynamics in economic process. It was expressed in large part through his detailed and repeated reference to the second law of thermodynamics, which served as the basis for his proposal to radically reconfigure the mathematical foundations of economic analysis: because economic process is *intended* to bring about qualitative change, which is frequently irreversible and which "eludes arithmomorphic schematization" (Georgescu-Roegen, 1971, p. 63). This means that accurate representation of the *dynamics* of economic process must include theory that addresses the structure of the *relationship between* qualitative and quantitative elements. While there is not sufficient space to unpack the point here in detail, that position, which includes postulates regarding the relationship between time, space and human intentionality, is closely linked to a second position that underpins his elaboration of an alternative analytical economics methodology – the flow-fund theory.

Using flow-fund theory, which replaces the stock, flow, fund distinction used in conventional economic analysis, with a flow-fund distinction that depends on the spatial and temporal boundaries of the economic process in question (Farrell and Mayumi, 2009; Silva-Macher and Farrell, 2014; Farrell and Silva Macher, 2017), makes it possible to construct complex, functional analyses that continue to represent the basic features of economic process, while making explicit the role of intentionality in their delimitation and also providing a means to include ecological elements and dynamics, which cannot be accurately represented in monetary units. The two propositions at the heart of Georgescu-Roegen's flow-fund theory, to make analytical space: 1) for the representation of biodynamics and 2) for the role of purpose in delimiting the boundaries of an economic process, rest at the core of what he referred to as "bioeconomics", (Georgescu-Roegen, 1986; Mayumi and Gowdy, 1999). Mayumi (2009, p. 1237) describes this as "a new style of scientific thought... that combines elements of evolutionary biology, institutional economics and biophysical analysis associated with energy and mineral resources." At a most basic level, I would say, the work of Georgescu-Roegen needs to be taken far more seriously by mainstream and conventional heterodox economists than it has been to date. Precisely because it implies the need for a radical break with convention, it has been left to the side or cherry picked. It is well past time for that to change.

More generally, following on from these observations, a further suggestion, regarding how economics must change, if it is to become a force for leading us away from catastrophe, rather than toward it, is that arithmetic fetishism must be jettisoned, and way made for the development of completely new types of inter-and transdisciplinary models, in which economic analysis is subordinated to a larger goal: representing the social-ecological and biophysical complexity of the human driven biological processes currently wreaking havoc across the planet earth.

While it has become fashionable to blame economic growth for the current ecological woes of the planet, and there is, of course, much evidence to support that position, I believe the problem is not so simple. Growth is a natural biological process, employed by all living organisms on the planet earth in order to resist the inevitable and constant deterioration that is implied by their inherently entropic nature (Schroedinger, 1944). It comes in many forms, from maturation, to regeneration to cancer, which are distinguishable through reference to their qualitative differences. It seems illogical to me to propose that growth, in itself, is inherently a problem and irresponsible to attempt to analyse economic processes without having a plausible theory regarding the role and function of growth within them. Rather it is the pursuit of growth for growth's sake, and the associated construction of models that presume the realization of growth to be a suitable measure of utility, that seem to me to be the problem. This implies a need to redesign economic models in a way that situates growth as *one among multiple economic phenomena* involved in regulating the viability of an economy: others being, for example, ecological impact and social acceptability. Taken as an end in itself, as opposed to being treated as a means to a

more humane end, growth serves growth, not society (Raine et al., 2006). Considered in the absence of attention to the associated phenomena of waste production and death, analyses focused only on the quantity, as opposed to also including attention also to the *quality* of growth *and* deterioration, are incomplete.
While Georgescu-Roegen's work is discussed at present, more often than not, in the context of the contemporary discourse on degrowth, which claims him as a founding thinker, his position on the question was arguably more conservative than is often assumed and is perhaps better described as advocacy of "agrowth" (Missemer, 2017) or balanced development (Georgescu-Roegen, 1965b). His position can be understood in terms of the simple matter of resource allocation trade-offs, where economic actors, in choosing what economic process(es) they undertake, are situated somewhere along a kinky, multi-dimensional, production possibility frontier, where the allocation and distribution of available resources may be configured to produce final goods and services, productive capacity or some combination of the two (Georgescu-Roegen, 1965b; 1999[1971], pp. 239-240 and 274-275; Scheidel and Farrell, 2015, p. 231).

By retaining reference to both the purpose of the economic actors in question and the limiting factor of resource availability, the preceding conceptualisation of growth can include, for example, recovery and transformation, both of which would imply a shift deep into the domain of producing productive capacity but not necessarily an increase in the quantity of deterioration or in the quality of an economy's ecological impact. Linking that position directly to his flow-fund theory,

Georgescu-Roegen (1968) would appear to have been most concerned with identifying the conditions required to ensure *balanced growth* of living economic processes, i.e. *to develop theory that would make it possible to explicitly link the rate of growth and productivity of an economy to the rate of growth and productivity of the biological systems upon which an economy's own productivity is inevitably, if only ultimately, dependent* (O'Hara, 1999; 2016). This, I would posit, is another aspect of how economics must change: a more nuanced and contextualised approach to growth is required on both sides of the growth / degrowth divide.

Taking up the idea of pursing balanced, ecologically viable, embedded growth, which implies also taking into account deterioration and death, we can speak of a two strand research agenda which I would propose to call producing ecological economy:

1. Identifying local, regional and international modes of production, consumption, sharing and exchange that are both economically and ecologically viable;
2. Changing local, regional and international regulations and practices to facilitate the development and maintenance of these types of economic activity.

Both strands imply a need for economic research to open up to what Max-Neef (2005) calls strong inter-disciplinarity, where multiple disciplines are involved not only in the execution but also in the configuration of analyses and in the specification of analytical problems to be addressed. This, I believe, may be the most pressing and most challenging change that needs to be brought about in economics. In contrast to subsuming knowledge from other disciplines to serve the ends and means of conventional modern economic analysis, as is done, for example, in the fields of neuro- and behavioural economics, this implies situating economics as a contributor toward the collaborative project of developing multi-dimensional, complex representations of the social-ecological relationships and processes that both underlie and are impacted by late-industrial economic activity.

Producing ecological economy has both a descriptive and a normative aspect, regarding, in the first instance, the identification of social and material criteria suitable for establishing ecologically beneficial economic activities across the entire planet and in the second, the specification of means for realizing their operationalisation under humane and ethical terms, across cultures and social-ecological contexts. Much of that work is of a political, rather than a scientific nature. And although I will address here only the latter, it should be noted that the former is also of vital importance for achieving lasting social change of any sort, not least such as might serve to halt the steady march through calamity in which humanity would appear to be engaged at present. The multiple statuses of politics in this process - within and across inter-disciplinary teams and between research teams and their clients, in some instances the public - must be taken into account when developing comprehensive models and analyses. This too implies a radical reconfiguration of the analytical basis upon which economic models are constructed.

Farrell and Silva Macher (2017, p. 167) have described attention to this contextualised and relational character of economic process as work focused on the ecological economic *Gestalt*: i.e. on the relationship *between* ecological and economic systems. Such work requires effective integration of insights deriving from a myriad of disciplines and applied to contextualised research questions related to the ordinary lives of many different types of economic and ecological communities. Here there is some ground for optimism, as there are a growing number of examples of such work (Bischi, 2018; Farrell and Silva Macher, 2017; Farley and Malghan, 2016; Moreau et al., 2017; Rincón Ruiz, et al., 2018; Wilson and Kirman, 2016). Nonetheless, this is still a project in the early stages of development and much of the attention of environmental and ecological economists continues to be dedicated to identifying ways to estimate the "real" costs and benefits of environmental externalities and to develop strategies to internalize them into price based decision processes. The persistence of such work illustrates the momentum of arithmetic fetishism, in which processes that do not easily lend themselves to quantification are arithmetized for the purpose of forcing them into the existing, quantitative analytical rubric. It is, I would posit, largely a waste of time and resources, as the resulting data are not only meaningless but also distracting (Farrell, 2007).

Work reaching beyond that fetishism, into the conceptual domain of the ecological economic *Gestalt*, has tended, up to now, to be in the area of institutional economics, where there is more openness to structural critiques of conventional modelling approaches. In the case of

Mayumi (1995; 2001; 2009; 2017), Georgescu-Roegen's last student, the focus has been on questions of epistemology and mathematical formalisation. Both Gowdy (1994; Gowdy and Mesner, 1998; Mayumi and Gowdy, 1999), picking up on Georgescu-Roegen's attention to the exosomatic evolutionary dynamics of technological and institutional change, and O'Hara (1997; 1999; 2016), picking up on his attention to the relationship between economic process and both social and ecological context, have developed interpretations of his work that can be linked with contemporary institutional economics and I would suggest that this is an important way forward for changing economics. Here the early work of Nobel Laureate Elinor Ostrom (1990) and more recent works by Vatn (2005) and Hodgson (2015) provide quite a comprehensive, environment-oriented, complement to the existing body of Classical Institutional Economics contributions, suggesting a promising route for developing the situated economic theory that is needed.

How can the global economy be changed so as avoid ecological collapse?

This question, I think is basically impossible to answer I find it decidedly uncomfortable to even attempt to answer such a general and far reaching question directly and so will proceed through reference to a metaphor. Many years ago, in conversation with a colleague, at a conference, we imagined the following image to represent this challenge: what would it imply, to transform a jet airliner, full of passengers, into a flock of birds, in mid-flight? That is to say, to transform, while running, a mechanical system, dependant on inputs of fossil fuel and an individualist based organising principle, into a biological one, employing biodynamic energy sources and structured around an organising principle of cooperation and attention to one's relations to others. The level of coordination required to avoid a catastrophic collapse of the system, in-full-flight, is, on its own, daunting: not to mention the massive amount of diverse technical expertise that would be required to realize such a transformation. Then there is the magical element, of realising some form of biomechanical metamorphosis, transforming human beings using machine, into birds.

Taken lightly, for illustrative purposes, our metaphor suggests a few concrete criteria that might be applied to address this final question. First, handle with care. The chances of a misstep leading to a total system collapse are high. Looking into the specifics, we could say that there is a clear need to effectively manage the transition from a mechanically based to a biologically based operating system. This implies holding on to the knowledge that is presently available regarding how the mechanical system (i.e. the industrial, accumulation driven economy) functions and working with that knowledge, to identify ways of coupling that systems with a biologically based one, in order to maintain momentum and avoid system failure. It also implies, as has been mentioned above, a need for the coordinated effort of diverse inter-disciplinary teams, comprised of experts in everything from human behaviour to fluid dynamics, so once again, strong-interdisciplinarity. And finally, it implies a need to adopt a posture of humility in front of the life-giving capacity of the natural world, which modern industrial science has yet, for all its achievements, to replicate.

While I do not agree with all the propositions contained therein, I believe one would be hard pressed to find a more succinct and coherent articulation concerning how the global economy not only could but indeed, must, be changed if humanity is to even hope to be able achieve the transformation to an ecological economy, than the following statement, issued at Rio +20, by a coalition of leaders of indigenous communities from across the Americas and the world:

> Mother Earth is the source of life which needs to be protected, not a resource to be exploited and commodified as a "natural capital". We have our place and our responsibilities within Creation's sacred order. We feel the sustaining joy as things occur in harmony with the Earth and with all life that it creates and sustains. We feel the pain of disharmony when we witness the dishonor of the natural order of Creation and the continued economic colonization and degradation of Mother Earth and all life upon her. Until Indigenous Peoples rights are observed and respected, sustainable development and the eradication of poverty will not be achieved (Kari Oca II Declaration, 2012).

References

Bahro, R., (1977) *Die Alternative: zur Kritik der real existierenden Sozialismus*. Frankfurt am Main: Europäische Verlaganstalt.

Bahro, R., (1987) *Logik der Rettung: Wer kann die Apokalypse aufhalten? Ein Versuch über die Grundlagen ökologischer Politik*. Stuttgart: Weitbrecht Verlag.

Bischi, G.I., (2018) Wilson, David S. and Alan Kirman (Eds) "Complexity and evolution: toward a new synthesis for economics." *Journal of Economics* 123, pp. 89-91.

Faber, M., Manstetten, R., Proops, J. (1995) "On the conceptual foundations of ecological economics: a teleological approach." *Ecological Economics* 12, pp. 41–54.

Faber, M., Manstetten, R., Proops, J. (1996) *Ecological Economics: concepts and methods*. Cheltenham, UK: Edward Elgar.

Farley, J., Malghan, D. (2016) *Beyond Uneconomic Growth, Vol. 2: A Festschrift in Honor of Herman Daly*. Burlington, VT: University of Vermont.

Farrell, K.N. (2007) "Living with living systems: The co-evolution of values and valuation." *International Journal of Sustainable Development and World Ecology*, 14, pp. 14-26.

Farrell, K.N., Mayumi, K. (2009) "Time horizons and electricity futures: An application of Nicholas Georgescu-Roegen's general theory of economic production." *Energy*, 34, pp. 301-307.

Farrell, K.N., Silva-Macher, J.C. (2017) "Exploring Futures for Amazonia's Sierra del Divisor: An Environmental Valuation Triadics Approach to Analyzing Ecological Economic Decision Choices in the Context of Major Shifts in Boundary Conditions." *Ecological Economics*, 141, pp. 166-179.

Georgescu-Roegen, N. (1960) "Economic Theory and Agrarian Economics" (with Postscripts from 1966 and 1975). In: Georgescu-Roegen, N. (Ed.), *Energy and Economic Myths: institutional and analytical economic essays*. Oxford: Pergamon Press, pp. 103-145.

Georgescu-Roegen, N. (1965) "The Institutional Aspects of Peasant Communities." In: Georgescu-Roegen, N. (Ed.), *Energy and Economic Myths: institutional and analytical economic essays.* Oxford: Pergamon Press, pp. 199-231.

Georgescu-Roegen, N. (1965) "Process in Farming versus Process in Manufacturing: a problem of balanced development." In: Georgescu-Roegen, N. (Ed.), *Energy and Economic Myths: institutional and analytical economic essays.* Oxford: Pergamon Press, pp. 71-102.

Georgescu-Roegen, N. (1966) *Analytical Economics: Issues and Problems.* Cambridge, MA: Harvard University Press.

Georgescu-Roegen, N. (1968) "Structural Inflation-Lock and Balanced Growth." In: Georgescu-Roegen, N. (Ed.), *Energy and Economic Myths: institutional and analytical economic essays.* Oxford: Pergamon Press, pp. 149-197.

Georgescu-Roegen, N. (1969) "The Economics of Production." In: Georgescu-Roegen, N. (Ed.), *Energy and Economic Myths: institutional and analytical economic essays.* Oxford: Pergamon Press, pp. 61-69.

Georgescu-Roegen, N. (1999[1971]) *The Entropy Law and the Economic Process.* Harvard University Press, Distributed by Oxford University Press, London.

Georgescu-Roegen, N. (1975) "Energy and Economic Myths." *Southern Economic Journal*, 41, pp. 347-381.

Georgescu-Roegen, N. (1976) *Energy and Economic Myths: institutional and analytical economic essays.* Oxford: Pergamon Press.

Georgescu-Roegen, N. (1977) "Inequality, Limits and Growth from a Bioeconomic Viewpoint." *Review of Social Economy*, 35, pp. 361-375.

Georgescu-Roegen, N. (1981) "Energy, matter, and economic valuation: where do we stand?" In: Daly, H.E., Umaiia, A. (Eds.), *Energy, Economics and the Environment.* Boulder, CO: Westview Press.

Georgescu-Roegen, N. (1986) "The Entropy Law and the Economic Process in Retrospect." *Eastern Economic Journal*, XII.

Georgescu-Roegen, N. (1988) "The Interplay Between Institutional and Material Factors: The Problem and its Status." In: Kregel, J.A., Matzner, E., Roncaglia, A. (Eds.), *Barriers to full employment*. London: Macmillan Press, pp. 297-326.

Gowdy, J. (1994) *Co-evolutionary Economics*. The Netherlands: Kluwer Academic Publishers.

Gowdy, J., Mesner, S. (1998) "The Evolution of Georgescu-Roegen's Bioeconomics." *Review of Social Economy*, LVI, 136-156.

Hodgson, G.M. (2015) *Conceptualizing Capitalism: Institutions, Evolution, Future*. Chicago: Chicago University Press.

Marshall, A. (1947[1920]) *Principles of Economics* 8th Edition. London: Macmillan & Co.

Max-Neef, M.A. (2005) "Foundations of transdisciplinarity." *Ecological Economics*, 53, pp. 5-16.

Mayumi, K. (1995) "Nicholas Georgescu-Roegen (1906-1994): An Admirable Epistemologist." *Structural Change and Economic Dynamics*, 6, pp. 261-265.

Mayumi, K. (2001) *The Origins of Ecological Economics: The Bioeconomics of Georgescu-Roegen*. London: Routledge.

Mayumi, K., (2009) "Nicholas Georgescu-Roegen: His Bioeconomics Approach to Development and Change." *Development and Change*, 40, pp. 1235-1254.

Mayumi, K., Gowdy, J. (1999) *Bioeconomics and Sustainability: Essays in Honor of Nicholas Georgescu-Roegen*. Cheltenham: Edward Elgar.

Missemer, A. (2017) "Nicholas Georgescu-Roegen and degrowth." *The European Journal of the History of Economic Thought*, 24, pp. 493-506.

Moreau, V., Sahakian, M., Griethuysen, P., Vuille, F. (2017) "Coming Full Circle: Why Social and Institutional Dimensions Matter for the Circular Economy." *Journal of Industrial Ecology*, 21, pp. 497-506.

O'Hara, S. (1997) "Toward a sustaining production theory." *Ecological Economics*, 20, pp. 141-154.

O'Hara, S. (1999) "Economics, Ecology and Quality of Life: Who Evaluates?" *Feminist Economics*, 5, 83-89.

O'Hara, S. (2016) "Production in Context: The Concept of Sustaining Production." In: Farley, J., Malghan, D. (Eds.), *Beyond Uneconomic Growth, Vol. 2: A Festschrift in Honor of Herman Daly*. Burlington, VT: University of Vermont, pp. 75-106.

OECD, (2009) *Defining the Bioeconomy, The Bioeconomy to 2030: Designing a Policy Agenda*. Paris: OECD Publishing, pp. 19-30.

Ostrom, E. (1990) *Governing the commons: The evolution of institutions for collective action*. Cambridge: Cambridge University Press.

Raine, A., Foster, J., Potts, J. (2006) "The new entropy law and the economic process." *Ecological Complexity*, 3, 354-360.

Rincón-Ruiz, A., Arias-Arévalo, P., Núñez Hernández, J.M., Cotler, H., Aguado Caso, M., Meli, P., Tauro, A., Ávila Akerberg, V.D., Avila-Foucat, V.S., Cardenas, J.P., Castillo Hernández, L.A., Castro, L.G., Cerón Hernández, V.A., Contreras Araque, A., Deschamps-Lomeli, J., Galeana-Pizaña, J.M., Guillén Oñate, K., Hernández Aguilar, J.A., Jimenez, A.D., López Mathamba, L.Á., Márquez Pérez, L., Moreno Díaz, M.L., Marín Marín, W., Ochoa, V., Sarmiento, M.Á., Tauro, A., Díaz Timote, J., Tique Cardozo, L.L., Trujillo Acosta, A., Waldron, T. (2019) "Applying integrated valuation of ecosystem services in Latin America: Insights from 21 case studies." *Ecosystem Services*, 36, p. 100901.

Scheidel, A., Farrell, K.N. (2015) "Small-scale cooperative banking and the production of capital: Reflecting on the role of institutional agreements in supporting rural livelihood in Kampot, Cambodia." *Ecological Economics*, 119, pp. 230-240.

Schrödinger, Erwin (1948[1944]) *What is Life?* Cambridge: Cambridge University Press.

Silva-Macher, J.C., Farrell, K.N. (2014) "The flow/fund model of Conga: exploring the anatomy of environmental conflicts at the Andes–Amazon commodity frontier." *Environment, Development and Sustainability*, 16, pp. 747-768.

Tsuchida, Atsushi; Murota, Takeshi (1985) "Fundamentals in the Entropy Theory of Watercycle, Ecocycle, and Human Economy. Entropy Studies on Ecology and Economy." Unpublished booklet of two papers presented to the Institut de la Vie symposium Man's Coevolution with the Bioshpere in the Age of Advanced Technologies. 21-25 January, 1985, Strong College, Toronto Canada.

Vatn, A. (2005) *Institutions and the Environment.* Cheltenham, UK: Edwar Elgar.

Wilson, D.S., Kirman, A. (2016) *Complexity and evolution: toward a new synthesis for economics.* Cambridge, MA: MIT Press.

Zawojska, A., Siudek, T. (2016) *Bioeconomics as an Interdisciplinary Science.* Proceedings of the 2016 International Conference "Economic Science for Rural Development" 21-22 April 2016 (Volume 41), Jelgava, LV, pp. 273-280.

Chapter 4

Economics 101: Dog barking, overgrazing and ecological collapse

Edward Fullbrook

> "the collapse of our civilisations and the extinction of much of the natural world is on the horizon" (David Attenborough).

Today's economics, especially Economics 101, is a major source of humankind's denial of the possibility of the calamity of all calamities which ***our economy*** is engineering. Annually millions of students around the world are forced to study textbooks that indoctrinate them in to thinking that there is no significant causal connection running from our economy to ***the ecosphere***. Once upon a time there wasn't. Although from the first forest-clearing onwards, the economy has caused environmental damage and at an increasing rate, it was only when in the 19th century the economy began the big switch away from muscle energy that it began to acquire the means to cause lethal damage to the ecosphere.

It has now been over half a century since the natural sciences began to discover that ***the economy was causing*** fundamental and irreversible changes to the ecosphere ***by which we and the economy exist.*** Given that economics is the study of the economy, a more radical change in a science's empirical realm is unimaginable.

In 50 years, what has economics done about it? Virtually nothing. There have been brilliant and intellectually brave economists, some of whom are contributors to this *Real-World Economics Review* special issue, who have created "ecological economics". But that work remains ignored by over 90 per cent of the profession and in nearly 100 per cent of its classrooms. In today's teaching of economics, 19th century theory continues to hold sway. Students are given a picture of the economy that blocks from view the fundamental facts ***about the economy*** that natural science has discovered. Let's take a look at how this censorship is achieved.

Gregory Mankiw's *Principles of Economics* is said to be the world's most used economics textbook and is the prototype of nearly all the others. It is a huge book. Its fourth edition index is 18 pages long with over 2,500 entries. This index illustrates how comprehensive the censorship is. Here are 11 key and now common terms pertaining to *the economy's effect and dependency on our life-support system*:

> biosphere, climate change, climate science, climatology, ecosphere, ecosystem, emissions, global warming, greenhouse gas, threshold, tipping-point.

How many of these terms appear in Mankiw's 2,500 entries? None. Nor do any of them appear in the book's 13 section titles, 36 chapter titles and over 700 sub-chapter titles. Why?

Because *the basic theoretical structure* of the economics that is taught to millions of university students every year will not accommodate the bidirectional causal link between the economy and the ecosphere. In the 19th century, when today's mainstream economics was invented, the global economy was too small to have observable effects on the ecosphere and none were anticipated. Of course even then economies had negative effects on their immediate environment, but they were small enough to make it seem reasonable to ignore them when considering how an economy works. So economists conceptually dumped an economy's negative effects into a broad category they called "***externalities***", and today in Economics 101 that is where they remain under the name "negative externalities"

Chapter 10 of Mankiw's textbook is titled "Externalities". It defines "negative externalities" as all those not so nice things that happen when market "equilibrium fails to maximize the total benefit to society as a whole"; (204) and Mankiw gives two examples:

- "The exhaust from automobiles... because it creates smog that other people have to breath",
- "Barking dogs... because neighbors are disturbed by the noise" (p. 204).

Further on, Mankiw explains to students that today's "environmental degradation" is analogous to the problem of overgrazing in the Middle Ages (pp. 231-234).

Climatologists see the problem of "externalities" as more serious than barking dogs and overgrazing. Here, for example, are quotes from a 2015 paper in the journal *Science*

> "There is an urgent need for a new paradigm that integrates the continued development of human societies and the maintenance of the Earth system (ES) in a resilient and accommodating state."

and

> "The relatively stable, 11,700-year-long Holocene epoch *is the only state of the ES that we know for certain can support contemporary human societies.* There is increasing evidence that human activities are affecting ES functioning to a degree that threatens the resilience of the ES – its ability to persist in a Holocene-like state in the face of increasing human pressures and shocks" [emphasis added].

In an interview the eminent climatologist Will Steffen sums up the economy versus the ecosphere problem informally:

> "It's clear the economic system is driving us towards an unsustainable future and people

> of my daughter's generation will find it increasingly hard to survive,"

and

> "History has shown that civilisations have risen, stuck to their core values and then collapsed because they didn't change. That's where we are today."

It is because humanity has engaged with today's Economics 101 fantasy – that the connection between the ecosphere and the economy is unidirectional – that we are now in this dire threshold situation. As John Maynard Keynes noted, "The ideas of economists..., both when they are right and when they are wrong, are more powerful than is commonly understood." And the ideas of economists have never been more wrong nor nearly so powerful at doing wrong as those force-fed to the students of today's Economics 101. We now know thanks to natural scientists, that the longer this mass indoctrination into this fantasy world continues, the more likely that the ultimate disaster will happen. It is not only with bombs and gas that crimes against humanity can be committed. Everyone connected with economics, perhaps most of all its students, need to ask themselves what they can do. It is hoped that this special issue of the *Real-World Economics Review* will help this questioning to take place.

Chapter 5

Addressing meta-externalities: investments in restoring the Earth

Neva Goodwin

1. Necessary changes in economic theory

Ecology teaches that everything is connected to everything else. Economics teaches that the market is a – some say *the* – great connector. Its specialty is to connect demand (what people want) to supply (what people produce), via prices.

There are, of course, known problems in the use of prices as a society's key connector. For one thing, those with more money have more of what is sometimes called "effective demand"; they can send louder, more effective signals to suppliers to produce the goods and services they want. Those with very little money can hardly get their needs and wants noticed. Aside from this translation of unequal purchasing power into unequal impact, the other most notable problem with markets as connectors is the presence of externalities, when something that matters simply is not picked up in market signals.

Ecologists sometimes complain that economists dismiss such important issues as "just" externalities – implying that these issues are regarded as unworthy of consideration. Good economists do not do this: they recognize full well that where there are externalities (where an economic actor produces effects that do not translate as signals to other economic actors) there is a market failure. Unfortunately, in situations of significant market failures markets do not produce the optimal outcomes that are expected in standard economic theory. This does make mainstream economists squeamish about admitting to externalities, since the optimality of market outcomes is one of their main boasts, and they don't have an alternative theory to pull out of the hat. Some economists who have positions of influence in academic or policy circles have begun to grapple with Stern's famous remark that "Climate change is a result of the greatest market failure the world has seen".[1] However, in the absence of a widely accepted alternative theory, the growing acceptance of climate change reality by economists simply creates cognitive dissonance without resolution.

The existing dominant system of economic theory is used to justify the current conformation of the economy of the United States, and of much of the world. It is becoming increasingly clear that it is producing very sub-optimal results for most of society, though benefitting the short-

---

[1] Quoted by *Alison Benjamin* in the *Guardian*, London, *29 November 2007*. *"Stern: Climate change a 'market failure'"*.

term gains of the rich and powerful. It has permitted and sometimes encouraged economic actors – especially powerful corporations and governments – to ignore the harms they impose on people and other parts of nature having little political/economic power. These harms are not trivial; they have included the murder of indigenous people for the value of their lands or of the minerals under their lands; toxic wastes dumped in oceans and in the neighbourhoods of poorer people; schemes to cover-up the harms of profitable products like tobacco and fossil fuels; and, over many decades, effective prevention of public education about the dangers of climate change, and of ways to avert it – until it is too late to prevent a future of ever more catastrophe.

In order for economic theory, teaching, and policy application to provide useful guidance in this time of great danger, it must change in many ways. The following is a brief summary of the most critical changes that are needed.

Recognition of context and history; there is not much that can true and useful in economic theory that is relevant to all times and places. This is because the economy is not a stand-alone system, but is embedded in, and dependent on, the contexts of the natural world and the built environment as well as human social systems – the latter including history, culture and ethics. Realities of power, politics, institutions and history are essential parts of the human fabric in which the economy is woven. Beyond direct externalities, and even less visible to current economic theory, are *meta-externalities – unwanted side-effects of the whole economic system upon* *its physical and social contexts*. To give just one example: the idealized concept of "free markets" along with an ethic of individualism have degraded political culture by belittling the roles that need to be played by collective action, aside from the particular kind of collective expressed through individual purchases of industrial products. This is not a small thing; the ability of governments to produce public goods and services has been seriously weakened in many nations due to effective campaigns to denigrate their effectiveness and efficiency (see Sekera, 2016).

An economic theory appropriate to our time must also address the **normative aspects of economic thinking**, starting with a recognition that any economy is a human creation, designed to support the goals of some people. *Whose* goals get addressed most effectively is of course a critical question to which we have just seen the answer: they are the ones with the money to have a lot of effective demand.

Sometimes it is not just money that determines how the goals of some economic actors are most emphatically expressed in the design of an economy; there is also the matter of **political power** – an issue that is so closely associated with Marxian economics that it has been a third rail for mainstream economists of the West. However, if we look at the economy in which we live, expecting to see that it works best for those with the most power and money, we will find that, for the most part, this is the reality. An economic theory that ignores this reality cannot expect to be useful for the real world.

Another critical issue is the prevalence of **idealization and idealized assumptions** in 20th century textbooks: human beings are idealized as perfectly rational and endowed with extraordinary capacities for maximizing calculation, while markets are idealized as the way to achieve the best of all possible worlds. If an economy is designed to support the goals of some participants, economics, as a discipline, is also the result of design, with a purpose. Idealization is a way of ignoring critical realities that are inconvenient for those in power. A more useful economics must include a thorough rethinking of the standard 20th century economic idealizations and associated assumptions.

One purpose for the idealizations just mentioned was the desire to make economics a respectable "hard" science. This was in part a result of consternation over the ejection of economics from the Royal Academy of Sciences in England at the beginning of the 20th century. As has been noted by a number of observers (e.g, Mirowski, 1989) the science that then appeared to have the most respectability, the most purity, was physics. Accordingly, economics adopted assumptions that allowed heavy dependence on **mathematical modelling** in order to emulate 19th century physics; for example, the use of various mathematical, scientific-looking techniques is made possible by the assumption of highly simplified, maximizing behaviour on the part of economic actors.[2]

---

[2] As physics evolved over the decades, to look less and like the early object of economists' physics envy, the discipline of economics also evolved, but retained many of its commitments to 19th century physics methodology, including a fundamental reliance on idealization. In the mid twentieth century Paul Samuelson, a man with a very tidy mind, took on the project

The other reason for the idealizations and increasing mathematization adopted in economic theory was that all of this could be used to support the idea that "**free markets**" would bring about the best possible outcomes (this can be more or less 'proven' – *given acceptance of all the assumptions and idealizations*). From this emerged the truly suspect idea that market actors – especially large, powerful or rich economic actors – should be free to do whatever they choose; any meddling from non-market forces (such as governments) would divert the economy away from the best possible outcome.

A related issue is the kind of quantification employed to assess the performance of an economy through the single metric, **Gross Domestic Product** (GDP). Among many problems with current uses of GDP, they are used to support policies that emphasize growth in throughput over increase in well-being. They ignore the contributions of unpaid workers (especially women, especially but not only in household work) as well as the cost of environmental damage – unless that damage requires compensatory activity, in which case it is listed as an *addition* to GDP.

Methodological change is thus required if economic theory is to contribute to addressing the challenges of

---

of cleaning up the fuzziness that was left over from the work of such as the great early theorizer, Alfred Marshall (who openly expressed opinions on the purpose of economics – namely, to alleviate the misery of the poor and elevate the moral lives of all), and from John Maynard Keynes. Keynes was a believer in mathematization, but when he had to face a trade-off between the simplifications required for mathematical modelling vs. addressing real world issues, he generally went for the latter. Samuelson's so-called "Keynesian synthesis" in fact got rid of all the parts of Keynes' thinking that could not be fitted into the Procrustian bed of ever-more complex modelling techniques.

climate change and other forms of environmental loss and degradation. A new balance must be found between simplifications imposed for the purpose of making sense of the economy, and attempts to recognize the actual complexity of the world. A more relevant and useful theory than that now prevailing would pull this balance somewhat away from methodologies that require extreme simplification, towards a richer understanding of the nature of economic actors and economic activity. This requires a broader conception of "the economy," to include economic activity that occurs not only in the business sector, but also within households and communities, and in governments and other public purpose organizations.

An economic theory that can help to mitigate the ecological and civilizational catastrophes bearing down on us must differ significantly from that which currently dominates policy-making. The points made above are only a few of the changes that are needed – more are illustrated in the textbooks I and my colleagues have been writing over the last few decades[3].

To summarize very briefly:

- Economics for the 21st century must describe and understand the workings of any economy within its physical and social contexts.

[3] These include *Microeconomics in Context*, *Macroeconomics in Context*, *Principles of Economics in Context*, *Environmental and Natural Resource Economics*, and, forthcoming, *Essentials of Economics in Context* – all published by Routledge.

- It must openly recognize how imbalances of power and resources affect the economic decisions that are taken at every political level.
- It must openly embrace a normative stance that recognizes that there are better and worse states of the world. Such a stance opens up the discipline to untidy discussions about what is better and worse: for example, how much inequality is tolerable, what trade-offs are necessary in order to reduce inequality, and how important it is to prevent harm to future generations.
- Regarding methodology, it must address the question of how much our understanding of economic matters is harmed by accepting the simplifications and idealizations that have been found necessary in order to use sophisticated, mathematized methods of analysis.

Such changes will require drastic alterations in economics curricula as well as in the honesty that should be encouraged in politicians by economic analysts who can see how political-economic decisions affect the world.

The foregoing begs many questions, such as: How can we change the mindset of economists? And: Who will ensure that those in power listen to people who care about the future, rather than only about their own enrichment? The first is the question I have spent my working life trying to address, through the production of better teaching materials – but these have not yet swept the field. As students and teachers see the relationship between economic behaviour and ecological outcomes, and as students increasingly give voice to their dissatisfaction with the irrelevance of most existing economics curricula, it is to be hoped that teachers of economics will be swayed by the force of demand.

The second question has to do with public education, and also with the hope that earlier cultural beliefs in morality will return. It is about time for such a shift; one can perceive, looking back over centuries and cultures, times in which a more caring culture has prevailed, as well as times when that was swamped by a culture of greed and selfishness. We may be at about the lowest point on the latter cycle; and enough people are witnessing and feeling the harms of climate change, ecological destruction and growing economic inequality that many small voices may coalesce into an effective force against the currently centralized and empowered voices united to enrich the few at the expense of the many.

2. Investment in ecological restoration

This paper has addressed, very briefly, the question of how the discipline of economics must change if it is to play a constructive role in addressing the huge challenges of the 21st century. A practical application of this question is a second one: how can the global economy be altereded so as avoid ecological collapse? This second question is strongly motivated by a growing awareness that we have reached or surpassed several indicators of collapse. The most enormous is the set of disasters attendant upon the amount of climate change already occurring and on track to increase. No less critical is the degradation of virtually the whole of the natural world through human action; by some estimates only 5% of ice-free land areas on the globe is still in pristine condition.[4]

These two issues are closely connected in their causes: human activities. They also share an important remedy. Ecological restoration has the potential to turn many ecological systems and sites away from net export of carbon and other greenhouse gasses into the atmosphere. Other sites that are still net importers of carbon etc. can import and store even more. It is increasingly believed that even the most optimistic projections regarding other kinds of change in human activity cannot be enough to keep the earth's atmosphere from warming well above what has been deemed the lowest safe global temperature increase. Ecological restoration, by capturing large amounts of greenhouse gasses, has the potential to close this gap. At the same time, massive efforts at ecological restoration will be essential if the earth is to provide the goods and services needed for humanity to survive and thrive in the future.

This paper began with the ecologists' insight that everything is connected to everything else, and the economic observation that markets have the capacity to make some important kinds of connections. One of the most important connections they can make are between

---

[4] "Assessments of the percentage of ice-free land affected by human action vary from 20% to 100%.... Ellis and Ramankutty (2008) concluded that more than 75% of Earth's ice-free land area could no longer be considered wild. Of Earth's ice-free land area, 83% is likely directly influenced by human beings (Sanderson et al., 2002). Our pollutants affect plant and animal physiology worldwide (McKibben, 1989, e.g., p. 38, 58)". Hooke, 2012

the present and the future, via investments. I will focus the rest of this paper on the question of where to find the resources – especially financial resources – for investments in ecological restoration.

Ecological restoration is defined as the process of assisting the recovery of ecosystems that have been degraded, damaged or destroyed (Society for Ecological Restoration, 2004). Although it is vital to protect existing intact ecosystems, conservation alone is insufficient, given the extent to which ecological degradation has proceeded and continues to expand. Inherent adaptive characteristics of ecosystems make it possible for humans to help change their trajectory away from degradation and toward recovery, repair, and self-sustaining regeneration and renewal. In some cases this requires large, expensive efforts, while other cases require little more than a cessation of harmful activities.

In general, investments offering clear, short-term benefits that can be readily captured by the investor are easy to fund through private market action. More difficult are the investment opportunities that include positive externalities, such that the investor cannot capture all of the benefits; or where the return will occur over a relatively long time. In the latter cases investors, using a present value calculation, will demand higher returns for the outer years. Investments in ecological restoration are very often deterred by one or the other of these characteristics. Restorative activities can provide enormous benefits, including carbon sequestration, biodiversity, support for livelihoods, and support for critical ecosystem services such as the production of nutritious food and of healthy trees and plants, as well as water filtration and absorption. However these activities are often characterized by relatively long time scales to produce the desired benefits, while the benefits generally include values that may be realized by groups aside from the investors. To give a few examples:

- Where farmers preserve strips of uncultivated land to attract pollinators and beneficial insects, neighboring farmers who do not do this may also benefit. Thus there is a positive externality, inviting free-riders and unlikely, through normal markets, to attract investment commensurate with the social benefit.
- City-dwellers benefit from the added nutritional value of food grown organically. Some are able and willing to pay extra for this food, making it easier for farmers to get through the years when the switch from conventional to organic farming has not yet fully paid off in productivity. However the demand for higher priced, more nutritious food is limited, and many farmers lack the resources to tide themselves over to a future in which more nutritious food will be cost-effective to produce. When that transition is complete, the result will be a huge externalized health benefit to large populations, but some kind of assistance is normally required to help farmers make the early investments.

Such cases argue for non-market interventions, to support a social good. This paper will first emphasize solutions that fall more within the purview of economics, per se, and will then list some "market-plus" and non-market approaches that have the potential to fill the gaps left by market action alone.

Market-led restoration activities

In fact it is hard to find cases in which private investors or commercial enterprises invest in restoration programs simply based on an expectation of competitive risk-adjusted returns. Endowments with long time horizons, such as that of Harvard University, may hold ownership positions in forests with the expectation of receiving returns with a lower present value than some more risky investments, but that are considered competitive because of their low risk. These holdings have not generally been regarded as restoration projects to start with, but, as investors and other stakeholders have increasingly pushed for "triple bottom line" accounting, there has been pressure to demonstrate social and environmental as well as financial returns. Over the past few decades there has been a cultural shift that makes businesses increasingly aware of the need to maintain a good public image. Where an institution can, without too much effort or expense, add some evidence of environmental and social responsibility, this is increasingly considered a sensible thing to do. Thus large investors with a stake in the Northern forests of New England have felt and responded to local pressures to keep the logging roads open for local people to use for camping, hunting, or snow-mobiling; they have also been aware of more dispersed sources of pressures to pay increasing attention to how and where trees are cut, and what is left on the ground as habitat for fauna and to allow regrowth of native flora, including trees.

The preceding paragraph is not a description of a brilliant ecological restoration project. If we want to find examples – and they exist – of corporations that put an ecological restoration motive high enough in their priorities so that they are willing to search out best practices, we must include in our understanding of market motivations psychological issues such as the wish of executives to be admired. That wish has been operative all along; it is becoming increasingly relevant as the cultural definition of "admirable" moves to take into account markers of success other than Milton Friedman's singular emphasis on high profits.

Seeking restoration projects carried out by corporations that understand the full meaning of meta-externalities (more on that below), I turn to the World Business Council for Sustainable Development (WBCSD), which describes itself as "a global, CEO-led organization of over 200 leading businesses and partners working together to accelerate the transition to a sustainable world." In a 2018 report, "The Business Case for Investing in Soil Health", it is noted that

> "The health of our soil is a critically important component in the production of food and the promotion of water and food security. Organic matter in soil increases its ability to store water and is directly related to the biodiversity in the soil. When soil biodiversity decreases so does its capacity to infiltrate and thus capture and store water... Sustainably managing land can... reduce surface erosion of soil into reservoirs, energy

> generating systems such as hydropower dams, public water supply, and industrial uses. This helps to reduce operation and maintenance costs and improves the longevity of investments.
>
> ...global soils contain two to three times more carbon than the atmosphere. If this carbon level was increased by 0.4%, or 4‰ per year, in the top 30-40 cm of soils, the annual increase in CO_2 in the atmosphere would be stopped" (WBCD, 2018).

This commentary may be read as a list of some of the positive meta-externalities that may be created by, and that can benefit, businesses that have any kind of interaction with the natural world. In the broadest sense, this includes all businesses, because all production derives from the natural world, which supplies humans (including workers) with the essentials for life: nourishing food, drinkable water, and breathable air. From the natural world, also, we derive all the materials – animal, mineral, or vegetable — that go into the production of every other conceivable good that goes through a market.

When a modern corporation builds its corporate headquarters, or any other structure in which its business will be carried out, it tries to avoid barriers to the flow of communication or movement among its functions. It seeks to maximize the efficient use of resources, minimizing the amount that is turned into waste. It pays attention to how lighting, air quality and other environmental issues affect human productivity. Corporations that are serious about accounting for the triple bottom line of their business treat a broader part of their environment as the structure within which they will need to continue doing business for as long as they exist. To some extent, thus, they internalize the meta-externality of their interaction with the social and physical environment. Here are two examples provided by WBCSD:

> "With the launch of the Good Growth Plan in 2013, Syngenta committed to improving the fertility of 10 million ha of degraded farmland by 2020. The company has subsequently worked with partners to develop and promote local solutions farmers could easily adopt. One early success was in Andalusia in Spain, where the company supported olive farmers in adopting vegetative cover in their orchards. Syngenta partnered with Asaja Sevilla (a farmer association), the Andalusian environmental authorities and a group of producers to introduce vegetative ground covers. The practice helped producers reduce erosion by up to 70% by replacing bare ground with vegetation and improving soil water-holding capacity. Local environmental subsidies linked to the region's rural development plans were also leveraged."

This account is especially noteworthy because Syngenta acknowledged the necessity of involving local people in the process of a restoration program that emphasized the needs of the people at least as much as the reversal of soil and crop destruction. Another account from WBCSD demonstrates another aspect of what is critically needed in order to keep a multinational corporation's intentions credible, and increase the efficiency and effectiveness of ecological restoration work; namely collaboration with independent scientists, in this case at a world-class university:

> "In South Africa, [the paper pulp, paper, and plastic wrapping company] Mondi manages large plantations interwoven with conservation corridors and nodes of natural habitat. The management of this mosaic – what Mondi calls "ecological networks" – is central to protecting biodiversity and the ecosystem services that sustain long-term productivity. Mondi has cooperated with Stellenbosch University for more than 10 years to understand how biodiversity can be conserved in ecological networks. More recently, the collaboration has started to investigate the role of soil health in plantation landscapes. The aim is to understand how different harvesting and silvicultural practices impact soil biodiversity, on the premise that soil biota are essential for soil health and long-term production."

Early readers of this paper have protested that WBCSD may be giving too much credit to corporations. Without inside knowledge of the particular cases cited here, thus unable to assess whether they are as good as suggested, the points to be made are that companies do sometimes see it as in their interest, however defined, to engage in ecological restoration; and that much of the world's financial and other resources are controlled by corporations – therefore, their engagement is essential. As the world economies are now structured it is unlikely that, in fact, corporations will step up to this task to the required degree. However popular awareness is growing regarding corporate corruption and their ability to assign an ever greater share of global output and wealth to an ever smaller fraction of the population. If, or as, this popular sentiment translates into opportunities for restructuring and redirecting corporate power and resources toward the common good, one of the more promising ideas is the 're-chartering' movement, whereby transnational corporations, in particular, would "have an obligation to define and pursue a self-declared social mission." (Great Transition Initiative, 2010).

Internalizing meta-externalities

The world's major insurance and re-insurance companies have more reason than most companies to be aware of meta-externalities, and have shown this awareness in a number of actions. ForestRe is an insurance company with a focus on forests and ecosystems. It is active in the region of the Panama Canal, where deforestation has increased the amount of sediments and nutrients reaching the canal, requiring expensive dredging. ForestRe "has been working with other insurance companies to underwrite a bond to finance watershed reforestation. It has been proposed that companies heavily dependent on the canal buy the bond and receive a reduction in their insurance (against closure of the canal) in exchange" (de Groot et al., 2007).

Like insurance companies, some of the very large pension funds have also shown a serious awareness of meta-externalities: they are responsible for supporting the lives of their pensioners into the future, and if their investments degrade the possibility for a good future quality of life, high investment returns are not a full excuse. This seems to have had a place in the thinking of the huge California pension funds, CalPers and CalSters, as well as some British funds that have been leaders in early attention to social and environmental investments impacts. This has largely shown up in the form of negative screens, for example against tobacco and fossil fuels – but it is a starting point.

A 2005 article in *The Economist* discussed the relative benefits of conservation and restoration of natural capital as compared to the industrial approach to land that assumes a gain in efficiency when natural diversity is replaced with single crops, whether of trees or other plants. The conclusion: "every dollar invested [in nature conservation and restoration] saves anywhere between US$7.5 and US$200 in avoided damage and repair costs" (quoted in de Groot et al.). However, many business entities are unaware of their dependence on a healthy natural world; and the meta-externality argument cannot, on its own, be expected to make all corporations into environmentalists, in part because of the issues of free-riders and of short-termism. A variety of other motivators, such as regulatory requirements, costs reduction or avoidance, and reputation, can be called upon to reinforce such constructive behaviour.

Some observers see a hopeful trend in which private sector investment models increasingly take sustainability issues into account. These trends can be further promoted through economic as well as regulatory and other incentives to producers and consumers. Market-based economic instruments to this end include taxes, subsidies, permits, and payments for ecosystem services. Taxes and subsidies are fairly obvious. For example, taxes on products sold with excessive packaging are a way of using the connective/communicative aspect of markets to affect consumer choice. Subsidies can be given to companies or individuals that employ sustainable fishing methods, including compensation for increasing the time that boats are idle, or limiting catch. Permitting is also a tool commonly used by governments to prevent undesirable behaviors or encourage desirable ones.

Payment for Ecosystem Services (PES) programs (also sometimes known as "markets for environmental services") are generally less familiar. The most widely known example is the US$1.5 billion investment made by New York City, when it paid upstate landowners and other stakeholders to engage in watershed conservation. In 1997 a memorandum of agreement was signed by the City and State of New York, the US EPA, 73 local municipalities and eight counties in the watersheds, as well as five environmental organizations. This allowed the city to receive filtration avoidance determinations from the EPA, avoiding the necessity to invest in a

Catskill/Delaware filtration facility, whose cost had been estimated at approximately $US 6 billion for design and construction and $US 300 million in annual operating expenses (Pires, 2004). Among other examples is the Grain to Green Project in China which is paying farmers to convert steeply sloping cropland to forest and pasture.

PES programs may be used to solicit private investments in restoration. For example, they may offer financial credits to parties for actions that maintain or increase the provision of ecosystem functions. These credits can then be purchased, sold, or traded. This is the idea behind many proposals for carbon markets, in which economic actors that increase the storage of carbon (or prevent its release), as compared to a benchmark, receive carbon credits that have a monetary value. Often such schemes allow the producers of carbon credits to sell their credits to entities that are causing carbon emissions. An example that depends on individual concerns about climate change involves people who feel guilty about the emissions associated with their jet travel; they may make voluntary contributions to foresters who are allowing their forests to regenerate, sequestering carbon in the wood and in the soil. Programs with more teeth include carbon trading systems in which governments impose a cap on businesses, for each one mandating how much greenhouse gasses they are allowed to emit during a year. If a business exceeds this cap, in order to avoid financial penalties it must purchase carbon credits from other businesses whose emissions were below their cap.

"Carbon farming" is an idea that is gaining popularity in the United States, where some farmers anticipate that the losses borne in the early years of transitioning from industrial to organic farming models can be covered by payments from states or other entities that have established a scheme for measuring and compensating the gain in carbon storage in the soil's increasing organic content[5]. The primary intent of such schemes is to counteract climate change, but their promoters see enormous additional values in arresting and reversing soil degradation, improving nutritional value of crops, and increasing biodiversity. Carbon farming depends on market functioning, but it must first create a market where one has not previously existed. There are, indeed, a vast array of requirements for a carbon trading system of any kind to work, not least the technical and administrative issues of credible measuring and monitoring of carbon storage.

Australia boasts the first soil credits worldwide to be eligible under the Paris Agreement; the first Australian Carbon Credit Units were just issued to a soil carbon project under the government's Emissions Reduction Fund (ERF). This is the result of a collaboration involving the government, which has created a soil carbon market with assured credits, and several business including Corporate Carbon Advisory, a multi-sector carbon contractor with over 40 million tonnes of abatement (100 projects) contracted under the ERF; and Bootstrap Environmental Services, offering field based soil sampling services and consultancy to the carbon farming

[5] The increase of carbon in soils with higher organic material is not accounted for only by the storage of carbon in the organic materials themselves. These materials also encourage the multiplication of biota in the soil biome, which also play as yet not fully understood roles in capturing and storing carbon.

sector. The Bootstrap media release of 14th March 2019 notes that

> "Soil organic carbon is in effect a scorecard for regenerative farming practices. The broader remit beyond carbon is to commercialise regenerative practices to enable producers to access a premium supply chain while the public gains access to more nutritious food and a regenerating environment. Australia is uniquely placed for its land sector to commercialise large scale carbon abatement. It has the measurement methods in place, the mechanisms to build soil carbon and a market for soil carbon credits including the $2.55B ERF, the Safeguard Mechanism and a growing voluntary market."
> https://www.bootstrap.net.au/carbonfarming

Where governments take the lead

This discussion has edged farther and farther into the area of government regulation and other "market-plus" interventions. Some impressive commitments have been made at all different levels of governments. On the international level, there are in existence international commitments for ecosystem restoration that add up, remarkably, to one-quarter of the world's arable land (Strassburg et al., 2018). Several international and national laws hold parties liable when they damage ecosystems. For example, in the US the Endangered Species Act and Clean Water Act hold responsible parties liable for restoring certain types of ecosystems. Under the US Conservation Reserve Program (CRP) "landowners are paid to retire agricultural land of high conservation value (primarily next to streams) and restore it to a grassland or forested ecosystem. The amount paid to the farmer for undertaking this restoration is based on the profit that could have been made had the land remained in production" (de Groot et al.). Since 1996, CRP has created nearly 2.7 million acres of restored wetlands. Other governments programs include a system of environmental assurance bonding, whereby extractive industries or other environmentally damaging activities post bonds that can be used to pay restoration costs if they default on their commitment to return an ecosystem to some specified condition. This approach is widely used, for example, by the Australian government to promote restoration of lands that have suffered from mining activities.

While the Australian government's interactions with the mining sector has endured for a long time, with results that are generally considered at least acceptable, and the USDA's CRP is protecting more than 170,000 stream miles with riparian forest and grass buffers, "enough to go around the world 7 times" (USDA, 2005), the examples given here also remind us of the weakness of governmental commitments, whether local or international. The laws may be made at a time that a particular government is strongly pro-environment, and then ignored or weakened by a subsequent administration, such as that of the Trump presidency in the US, which has other priorities. Too often, as well, governments lack the capacity to enforce treaties signed onto at an international level.

These are among the reasons that many have turned to public-private partnerships in an effort to combine private resources and capacity with the goals of the public purpose sector (which includes governments and NGOs). Turning again to examples provided by the World Business Council for Sustainable Development, "a public-private partnership of Rabobank and UN Environment has created a USD $1 billion facility to finance forest protection and sustainable agriculture projects beyond what is commercially viable. It identifies and minimizes barriers to sustainable agriculture on existing degraded land, improving productivity and thus avoiding deforestation." Another example is the Land Degradation Neutrality Fund (LDN Fund), which was launched in 2017 with the support of UNCCD.

> "The LDN Fund is an impact investment fund blending resources from the public, private and philanthropic sectors. It seeks to channel resources towards land-based private sector projects contributing to addressing land degradation through sustainable land management and land restoration projects" (WBCSD, 2018).

Recap: what are the current sources for investment in ecological restoration?

After surveying the role of government, the *Business Case for Soil Health* adds that "Other capital sources that could be further incentivized to invest in soil include pension, insurance, and sovereign wealth funds, public expenditures and foundations." Some philanthropic foundations have been moving into 'impact investing,' where the effort is not merely to avoid doing harm, but to actively promote their mission values through their investments. Leaders have been relatively small foundations, such as the F.B. Heron and Nathan Cummings Foundations; the latter intends to "align 100 percent of our nearly half-billion dollar endowment with our mission" (Calhoun, 2014). The Ford Foundation has put aside $1 billion of its endowment for impact investing. The John D. and Catherine T. MacArthur Foundation, with the Rockefeller Foundation and the Omidyar Network, have jointly allocated $150 million to help address financing gaps in impact investing.

To put this in perspective, "foundations hold only around $800 billion in their endowments. But there are trillions and trillions in private capital markets, while the 400 richest Americans alone have a net worth of $3 trillion" (Ibid.). Some modern billionaires, such as Bill Gates, Michael Dell and Pierre Omidyar, have invested heavily in new, sustainable energy technologies. Michael Bloomberg is working with a consortium of individuals and family foundations to show the irrationality of continuing to invest in coal as an energy source. This is the low-hanging fruit; energy is big business, and, while many energy investments are risky, there is also a reasonable prospect for high returns.

So far, it has been much more difficult to make a case for market-rate returns to be expected from investments in ecological restoration. Thus, as indicated above, such investments have been made through a variety of market-related or non-market approaches. To recapitulate: some companies, moved by a variety of motives, have made significant investments in more sustainable farming methods and related aspects of soil protection; it is hoped that this trend will grow. Governments sometimes put their funds into activities such as the extensive efforts, in South Africa, toward removal of invasive plants or, in Rwanda[6] (with assistance from the UNDP), towards rebuilding the organic content of soils. More often governments seek to work with the private sector, either through public-private partnerships, or through instruments such as regulations, subsidies, permits, and payments for environmental services. And foundations and philanthropically-minded individuals have been moving cautiously towards ecological restoration, asking themselves whether they can accept 2-3% returns in the portions of their portfolios that are invested in this way.[7]

---

[6] Overall project cost is 1,562,000US$, with 600,000US$ from GEF, and 962,000$ from co-finance: 300,000$ from UNDP, 397,000$ parallel co-finance from the ICRAF (World Agroforestry Centre), and 265,000$ in - kind from the Government of Rwanda. UNDP/GEF-MSP PROJECT ON LAND DEGRADATION IN RWANDA, proposal 2007: http://www.rw.undp.org/content/dam/rwanda/docs/operations/Projects/Environment%20and%20energy/RW_EE_Sustainable_land_use_and_management.pdf

[7] The focus of this paper on investment should not exclude the banking sector. However, very few banks are willing to lend funds for the rates of repayment that can normally be expected to come from restoration projects. It is relevant here to mention the Community Development Financial Institutions (CDFIs) in the United States, which bundle together federal funds with private sector capital (some of it at market rates, some in the form of very low-interest loans or gifts from the philanthropic community) to generate economic growth and opportunity in economically disadvantaged communities. In the past these have focused heavily on housing, but are occasionally supportive of food coops and

This list leaves out a segment of society that is critically important for ecological restoration. Before offering some comments on the role of local communities, I will take a brief detour to look at the issue of prioritization.

Prioritizing restoration projects

In November, 2017, The Brazilian government announced the National Plan for the Recovery of Native Vegetation (Planaveg). The initiative, announced during the 23rd Conference of the Parties to the UNFCCC (COP23) in Bonn, Germany, represents an important step by Brazil towards the recovery of its native vegetation and the fulfilment of its commitments under the Paris Agreement. The goal is to recover the native vegetation of at least 12 million hectares of area by 2030.[8] While this piece of news may, in the 2019 context, be a depressing reminder of a point made earlier – that commitments by one government may be dismissed by a new, anti-environmental party – it is still instructive to look behind the scenes at some details in how Brazilian Ministry of Environment devised the particulars of its plan. (These details are elaborated on in Strasburg et al., 2018)

The preexisting Brazilian Native Vegetation Protection Law "requires Atlantic Forest farmers to keep at least 20% of their farms under native vegetation. Farmers currently below this threshold must comply either by implementing restoration in their own farms or by

---

other activities that are relevant to regenerative farming. See the quotation from Jacob Israelow.

[8] http://www.brazil.gov.br/about-brazil/news/2017/11/government-launches-national-plan-for-the-recovery-of-native-vegetation

financing conservation or restoration offsets elsewhere within the biome" (Strasburg et al., 2018). Strasburg's team used a linear programing model with multicriteria spatial planning to reveal and manage the trade-offs and synergies involved in selecting areas for conservation/ restoration offsets. The restoration benefits to be maximized included reduction in projected extinctions of native flora and fauna, and potential aboveground carbon sequestration. (They did not include below-ground carbon storage, as the relevant soil science was not yet sufficiently advanced.) The constraints included restoration implementation costs (expected to be borne by farmers), and potential conflict with agricultural production. The decision tool identified significant economic and ecological efficiencies of scale, while also recognizing some important reasons for valuing certain small-scale projects for having smaller patches of restoration dispersed across the entire biome, creating a mosaic of large and small. These conclusions, and the methodology for reaching them, led to the application of this approach to the other five Brazilian biomes as part of the National Plan for Native Vegetation Recovery.

The argument in favor of larger scale occurred in a context where a national plan could connect areas regardless of ownership. It is also relevant for very large farms, ranches, haciendas, etc., to the extent that their owners are aware of the fertility losses that occur over time under industrial monocropping. Naturally, given the scale of the problem, such opportunities are appealing. Nevertheless, this should be balanced by a note about the realities facing small farmers in most of the world:

> Most regenerative agriculture can only happen at a small scale. This leads to impact investment opportunities that are also small-scale, which can make them risky and challenging to effectively underwrite for several reasons (eg – size of team, assets managed, concentration risk, whether it is worth it for an investment adviser to spend time evaluating a smaller opportunity, etc). I do not disagree with the advantages of large-scale projects, it can just take tremendous efforts to coordinate among many small-scale actors, which takes even more time and management effort (Jacob Israelow).[9]

Community restoration, from the bottom up

While the Brazilian example, and a few others mentioned earlier, have taken into account the impact on local people, and made some attempts to engage their cooperation by making known the benefits they could receive from restoration activities, virtually all of the approaches described in this paper have been top-down. Restoration of the world's damaged ecosystems is an enormous challenge; it will require enormous amounts of investment funds. With such big challenges, it is perhaps natural to look to big players. However a recent book by Gary Nabhan provides compelling examples of effective bottom-up actions by individuals or communities,

[9] Personal communication from Jacob Israelow, Founder and Managing Director, Dirt Capital Partners March 11, 2019.

depending mostly on their own efforts to provide the work that, in top-down projects, is contracted out by planners.

Tucson, Arizona, is a city that suffers from urban problems of poverty and job insufficiency, as well as a severe shortage of water. In 1998 a citizen of the city, Brad Lancaster, set out to capture rainwater from his street front, roof, and other hard surfaces, working up to the ability to store a hundred thousand gallons every year in containers in his yard. He also began to cut away curbs, allowing storm-water from the city's paved streets to flow into basins placed between the streets and the sidewalk, with the goal of planting there, and irrigating, trees bearing edible fruits. Initially the city proposed arresting those who cut curbs without permits, but by 2001 "Lancaster had worked with the city to legalize the process and get block grants to help one neighborhood after another transition their provision of moisture" to use harvested rainwater in support of fruit trees (Nabhan, 2018). The city caught on to the potential, and offered rebates to citizens who were using harvested rainwater instead of drawing on the scarce municipal supplies.

The results snowballed into a remarkable diversity of businesses, for example providing cisterns, gutters and drip irrigation systems, while Tucson became a center for the collection, sale, and free distribution of native, drought-tolerant varieties of edible plants. The Pima County library has created the largest free seed interlibrary loan program in the country. (The distribution from this and other seed libraries in the area count as "loans" because 60 percent of the people who checked out seed packets have returned some seeds after their harvests to replenish the stocks available to others the next season.)

> "By the summer of 2017, Tucsonians had local access to more than 2,020 named varieties of 130 cultivated annual food crops; 140 species of wild, native desert edible plants, and more than 200 varieties of domesticated fruit, nut, berry, and succulent edibles...Young entrepreneurs – from wild foragers and farmers to millers, bakers, tortilla makers, cactus syrup processers, distillers, brewers, *mole* makers, mixologists and fermenters – have begun to use these local ingredients to develop new products, most of them unique to Tucson or the border region" (Nabhan 2018, pp. 156–7).

The Tucson example is only one among many described in Nabhan's book, where local people confront the resource degradation problems in their own region, whether that be as large as a city or as small as one farm. Sometimes these stories start with one inspired leader, such as the "'harvest rain' guru", Brad Lancaster in Tucson. Sometimes they depend on a family's commitment, over multiple generations, to increasing the productivity of their land by building terraces or planting living hedgerows to prevent erosion. Sometimes they stem from knowledge and understanding about the land that is held by local groups, often including indigenous people and those who learn from them. Some of the cases described by Nabhan have benefitted from a little access to outside resources, whether of money or of knowledge; but more often the stories are of people using their own time and knowledge and muscle to bring back beauty, diversity, and productivity to the land on which they depend.

Conclusion

There are 7.7 billion people in the world today, with more than half living in urban settings. Who thinks about the health of soils in cities? Not very many people yet (an exception is HUMI, see below); however with the explosion of the field of soil science, knowledge is growing rapidly regarding the difference, for humans, between living in an environment of degraded soils and impoverished flora and fauna, vs. living in a place where the surfaces on which we humans travel allow water to flow or filter through; where native plant and animal species are encouraged to flourish; where anyone can find a genuine green space in just a ten minute walk from their home.[10] The EcoHealth Network[11] is one of several organizations now working to provide scientific analysis for the hypothesis of a strong link between soil health and human health. Another is the Healthy Urban Microbiome Initiative, "HUMI" in Adelaide, Australia.[12]

[10] This is the announced goal of The Trust for Public Lands.

[11] Based at the Missouri Botanical Gardens, in St Louis; the steering committee consists of James Aronson, Cristina Eisenberg, Laura Orlando and Neva Goodwin.

[12] HUMI focuses exclusively on urban areas, including the soils in built areas (e.g., rooftop gardens) as well as natural ones. Their hope is to engage 20 cities around the world in a carefully designed study protocol on the relationship of soil health and human health.

The evidence for benefits via psychological responses is extensive, but also requires firmer scientific underpinnings.

In rural areas, where nearly half the world's population still live, the need for continued advances in science is not about *whether* soil health is essential for ecological health, or whether ecological health is essential for human well-being; it is incontrovertible that the future well-being of our species will depend heavily on reversing the trend toward ever more degradation of the natural world. As reported by the United Nations in 2014, if farmers don't change their agricultural practices, "most of the soil they rely on to sustain their livelihoods will disappear within 60 years."[13] There are, however, many areas still requiring research on how to restore different areas most effectively and efficiently. The great variety of questions to be researched include how to assess soil health under a variety of circumstances, how to identify species of plants and animals that will be most beneficial in restoration, and then how to introduce or encourage these species. These are not simple matters, but they are doable. While the science is still maturing some actions will head in the wrong direction,, but usually significant ecological recovery can be achieved through a range of actions, some of them more "correct" than others.

---

[13] "Industrial-scale agricultural practices lean on polluting machinery and chemicals, which contaminate farmland. Insistence on monoculture–growing a single crop on the same patch of land–saps the soil of nutrients that more diverse crops deliver. And aggressive tilling breaks down soil structure and makes it harder for healthy land to regenerate." Quoted in https://www.fastcompany.com/90313818/general-mills-has-a-plan-to-regenerate-1-million-acres-of-farmland

This paper is based on the premise that a massive reversal of the interactions between humans and the rest of nature is essential – to turn away from spreading ecological destruction, towards restoration and renewal. The focus of the paper has been on where to find the financial resources required to make the necessary global investments in ecological restoration. It has considered private sources, public funds and government activities, public-private partnerships, the large pools of capital in insurance companies and pension funds, and the support that can come from philanthropic institutions and individuals. It has also touched briefly on the reality that, while local communities often have little access to outside funds, some manage extraordinarily well with their own resources.

It is tempting to try to quantify the size of the effort and the resources, and where they will come from, that are needed to make a strong turn away from global ecological degradation, towards vigorous renewal. To make such an assessment would require answering the following questions:

1. What is the potential for storing more carbon and other greenhouse gasses in soils and plants than is there now? (NB: this paper has not dealt at all with the great question of how much the storage capacity of oceans has been diminished by human actions, and how much it might be increased by changes in behaviour.)
2. In many cases the last 10–20% of a solution is much harder to achieve than the first 80–90%. Guessing that this is likely so in this case, and recognizing the

urgency to achieve a large impact as quickly as possible, how can we prioritize where to begin our efforts?

3. What are the different elements of the job to be done?
4. How can we calculate the cost in money and human effort that will be required to move from the current situation to where half of the potential greenhouse storage in ecosystems is achieved? 80%? 90%?
5. How can we assign the responsibility for this renewal among the different categories of players in this drama?

Others have made a variety of efforts at answering most of these questions, though not necessarily at a global scale. The "4 per thousand" initiative launched by France in 2015 proposes that "**An annual growth rate of 0.4% in the soil carbon stocks... per year, would halt the increase in the CO_2 concentration in the atmosphere related to human activities**" (4 per 1000, 2018). Although the writing on this initiative sometimes refers generally to soils – most broadly agroecology, agroforestry, conservation agriculture, landscape management, etc. – the focus is on agricultural land, specifying an intention "to demonstrate that agriculture, and in particular agricultural soils can play a crucial role where food security and climate change are concerned" (Ibid.). Thus the reasoning behind the French proposal is a good start on one approach to calculating the GHG storage that might be achieved on agricultural lands, but mostly ignores other lands, such as forests, wetlands, and cities.

This paper has shown just one example of a technology used for prioritization: the linear programing model with multicriteria spatial planning used by Strasburg et al. This approach, or modifications of it, or sometimes quite different, more qualitative approaches, will be required for different cases, at different scales. The IUCN has identified "global hotspots" of extraordinarily high diversity; this is an important value to be included in any prioritization scheme (as it was in Strasburg et al), but there are multiple other values. Other questions to be asked would include: "where is it easiest to mobilize resources?" (e.g., where there is already enthusiasm for ecological restoration), and "where is the biggest bang for the buck?" This could point to the areas closest to more intact lands, from whence natural seeding could occur (both above and below ground); in some cases it might point to the most degraded lands, in others, the least degraded. The scale under consideration will strongly affect the approach to prioritization: and the question of scale will depend on who are the prime movers – e.g., local farmers or forest dwellers, a municipality, a nation...?

The 4 per thousand initiative lists many elements of the job to be done, including the establishment of soil carbon monitoring systems; training programs for farmers and advisors on how to increase soil organic matter; establishment of appropriate public policies and a suitable regulatory environment for the transition of agricultural systems; and opening carbon markets to new sectors such as agriculture and agroforestry. In addition, there is need for a great deal of research to continue expanding our understanding of how the system works. This requires engaging the best of modern science along with serious attention to experiential knowledge, whether that is embodied in first, second or 10th generation workers on the land or in indigenous peoples. Both of these kinds of knowledge will be needed in programs of capacity building, to rapidly increase the number of professionals and students who can engage with restoration work on the ground. Not least is the need to inform the global public of the great importance of ecological restoration, for their own health and wellbeing and that of coming generations. People view electrification, for example, or access to modern transportation or communications media as critical to their wellbeing; global and local ecological restoration is no less important. This message needs to be conveyed, vigorously and credibly.

The answer to question #4 is probably no – we cannot really calculate what it will take to achieve any large fraction, let alone the whole, of the job to be done. And question #5 is equally hard to answer. We know that some portion of the responsibility rests on actors in the public purpose sector – governments and institutions working for the common good; and that governments will doubtless need to play a leading role. As the world is today, the private, for-profit sector must also take on much of the need; it is to be hoped that global cultures and economies will change enough so that they will feel ever more motivated to do so (Great Transition Initiative, 2010). And then there are the individuals and groups who are motivated not by government regulation or incentive, nor by the profit motive, but because they understand how intimately human wellbeing is connected with the health of the earth. Success, to the extent it is achieved, will come from combining local passion and knowledge with global and local science and knowledge, and with financial capital, in ways that work for immediate human needs combined with long-term movement toward ecological restoration.

All people – young and old, rich and poor, rural and urban – are affected by ecological degradation. To address these massive meta-externalities, scientists and planners, as well as those with large financial resources – businesses, governments, investors, and philanthropists – must work with, and support, the needs and the knowledge of the people and communities that are directly interacting with the state of the natural world. How much should each do? The only answer is: as much as they possibly can – which is probably more than they now realize they can.

References

4 per 1000 (2018) "Welcome to the '4 per 1000' Initiative" at https://www.4p1000.org/

Aronson, James and Sasha Alexander (2013) "Ecosystem Restoration is Now a Global Priority: Time to Roll up our Sleeves" *Restoration Ecology*, Vol. 21, No. 3, pp. 293–296

Aronson, James, Suzanne J. Milton and James N. Blignaut (Eds.) (2007) *Restoring Natural Capital: Science, Business and Practice*. Washington DC: Island Press.

Anzilotti, Eillie (2019) "General Mills has a plan to regenerate 1 million acres of farmland". FastCompany.
https://www.fastcompany.com/90313818/general-mills-has-a-plan-to-regenerate-1-million-acres-of-farmland

Callahan, David (2014) "Foundation Impact Investing Is Gaining Steam. But Can Proponents Answer These Four Questions? *Inside Philanthropy*: https://www.insidephilanthropy.com/home/2018/3/13/foundation-impact-investing-is-gaining-steam-but-can-proponents-answer-these-four-questions

Great Transition Initiative (2010) "When the World Rules Corporations: Pathway to a Global Corporate Charter." Tellus Institute. www.greattransition.org/archives/perspectives/Perspective_Global_Corporate_Charters.pdf

de Groot, Rudolf, Martin de Wit, Erica J. Brown Gaddis, Carolyn Kousky, William McGhee and Mike D. Young (2007) "Restoring natural capital: science, business and practice." In Aronson et al. *Restoring Natural Capital: Science, Business and Practice*. Washington DC: Island Press.

Hooke, Roger LeB., José F. Martín-Duque (2012) "Land transformation by humans: A review." *GSA Today*, Volume 22, Issue 12 (December).

Iftekhar, Md Sayed, Maksym Polyakov, Dean Ansell, Fiona Gibson, Geoffrey Kay (2017) "How economics can further the success of ecological restoration." *Conservation Biology*, Volume 31, Issue 2; April.

Mirowski, Phillip (1989) *More Heat than Light.* Cambridge: Cambridge University Press,

Nabhan, Gary (2018) *Food from the Radical Center.* Washington DC: Island Press.

Pires, Mark (2004) "Watershed protection for a world city: the case of New York." *Land Use Policy*, 21, pp. 161–175.

Rohr, Jason R., Emily S. Bernhardt, Marc W. Cadotte, and William H. Clements (2018) "The ecology and economics of restoration: when, what, where, and how to restore ecosystems." *Ecology and Society*, June.

Sekera, June (2016) *The Public Economy in Crisis: A call for a new public economics.* Springer.

Society for Ecological Restoration International Science & Policy Working Group (2004) "The SER International Primer on Ecological Restoration". www.ser.org & Tucson: Society for Ecological Restoration International.

Strassburg, Bernardo B. N., Hawthorne L. Beyer, Renato Crouzeilles, Alvaro Iribarrem, Felipe Barros, Marinez Ferreira de Siqueira, Andrea Sánchez-Tapia, Andrew Balmford, Jerônimo Boelsums Barreto Sansevero, Pedro Henrique Santin Brancalion, Eben North Broadbent, Robin L. Chazdon, Ary Oliveira Filho, Toby A. Gardner, Ascelin Gordon, Agnieszka Latawiec, Rafael Loyola, Jean Paul Metzger, Morena Mills, Hugh P. Possingham, Ricardo Ribeiro Rodrigues, Carlos Alberto de Mattos Scaramuzza, Fabio Rubio Scarano, Leandro Tambosi, and Maria Uriarte (2018) "Strategic approaches to restoring ecosystems can triple conservation gains and halve costs." *Nature Ecology & Evolution* **volume 3**, pages 62–70, published 17 December. https://www.nature.com/articles/s41559-018-0743-8?WT.feed_name=subjects_conservation

USDA (2015) Farm Service Agency Fact Sheet, December; retrieved 3-9-19 at https://www.fsa.usda.gov/Assets/USDA-FSA-Public/usdafiles/FactSheets/archived-fact-sheets/consv_reserve_program.pdf

WBCSD (World Business Council for Sustainable Development) (2018) "The Business Case for Investing in Soil Health." http://catalogue.unccd.int/1034_The_Business_Case_for_Investing_in_Soil_Health.pdf

Chapter 6

Degrowth: a theory of radical abundance

Jason Hickel

As the climate crisis worsens and the carbon budgets set out by the Paris Agreement shrink, climate scientists and ecologists have increasingly come to highlight economic growth as a matter of concern. Growth drives energy demand up and makes it significantly more difficult – and likely infeasible – for nations to transition to clean energy quickly enough to prevent potentially catastrophic levels of global warming. In recent years, IPCC scientists have argued that the only feasible way to meet the Paris Agreement targets is to actively scale down the material throughput of the global economy. Reducing material throughput reduces energy demand, which makes it easier to accomplish the transition to clean energy.

Ecological economists acknowledge that this approach, known as degrowth, is likely to entail reducing aggregate economic activity as presently measured by GDP. While such a turn might seem inimical to human development, and indeed threaten to trigger a range of negative social consequences, proponents of degrowth argue that a planned reduction of throughput can be accomplished in high-income nations while at the same time maintaining and even improving people's standards of living. Policy proposals focus on redistributing existing income, shortening the working week, and introducing a job guarantee and a living wage, while expanding access to public goods.

As debates unfold around what these policies might look like and how to implement them, here I step back to consider the deeper economic logic of degrowth theory. On the surface, degrowth sounds like an economics of scarcity, as many on both the right and left have been quick to allege. But in fact exactly the opposite is true. A long view of the history of capitalism reveals that growth has always depended on enclosure. The Lauderdale Paradox first articulated by James Maitland holds that an increase in "private riches" is achieved by choking off "public wealth". This is done not only in order to acquire free value from the commons but also, I argue, in order to create an "artificial scarcity" that generates pressures for competitive productivity.

Degrowth seeks to invert the Lauderdale Paradox. By calling for a fairer distribution of existing resources and the expansion of public goods, degrowth demands not scarcity but rather abundance (see Sahlins, 1976; Galbraith, 1998; Latouche, 2014; D'Alisa et al., 2014). I build on this insight to show that such an approach not only embodies an alternative to a growth-oriented economy, but in fact offers an antidote to the driving mechanism of growth itself, thus releasing both humans and ecosystems from its grip. By advancing a theory of abundance, degrowth provides a feasible political pathway toward an ecological economy fit for the Anthropocene.

The Paris Agreement and the degrowth imperative

In 2018 the UN Intergovernmental Panel on Climate Change (IPCC) published a special report outlining what it will take to prevent global warming of more than 1.5°C over preindustrial levels. The report concluded that global emissions must be halved by 2030 and reach net zero by the middle of the century. It is a dramatic trajectory, and requires a rapid reversal of direction for our civilization. There is at present no agreed plan for accomplishing this. The voluntary pledges made by signatories to the Paris climate agreement in 2015 entail no absolute reductions to global emissions, and set us on a path to 3.4 degrees of warming by the end of the century – significantly exceeding the 1.5 and 2°C limits established by the Paris agreement.

The primary reason for this problem is that economic growth is projected to drive energy demand up at a rate that outpaces the rollout of clean energy capacity (Raftery et al., 2017). This has already presented a problem in the 21st century. Today the world is producing 8 billion more megawatt hours of clean energy each year than in 2000, which is a significant increase. But over the same period, energy demand has grown by 48 billion megawatt hours. In other words, new clean energy capacity covers only 16% of new demand. It is of course technically possible to scale up clean energy output to cover total global energy demand (Jacobson and Delucchi, 2011). But the question is whether it is feasible to do so at a rate that is fast enough to respect the carbon budget for 1.5 or 2°C, while at the same time growing the global economy at the usual pace.

We can assess this question by looking at projected rates of decarbonization. If we assume that global GDP continues to grow at 3% per year (the average from 2010-2014), then decarbonization must occur at a rate of 10.5% per year for 1.5°C, or 7.3% per year for 2°C. If GDP slows down and grows at only 2.1% per year (as PWC predicts), then decarbonization must occur at 9.6% per year for 1.5°C, or 6.4% per year for 2°C. All of these targets are significantly beyond what existing empirical models indicate is feasible (see Hickel and Kallis, 2019). A few brief examples will serve to illustrate this point. Schandl et al. (2016) indicate that decarbonization can happen by at most 3% per year under highly optimistic policy conditions. The C-ROADS tool (developed by Climate Interactive and MIT Sloan) projects decarbonization of at most 4% per year under the most aggressive possible abatement policies: high subsidies for renewables and nuclear power, plus high taxes on oil, gas and coal. In a recent review of existing evidence, Holz et al. (2018) find that the rate of decarbonization required to meet the Paris targets is "well outside what is currently deemed achievable, based on historical evidence and standard modelling."

IPCC scientists and authors have been aware of this problem for some time. In the Fifth Integrated Assessment Report (AR5), they dealt with it by assuming the future existence of speculative "negative emissions" technologies. The theory is that while business-as-usual growth will cause emissions to exceed the carbon budget in the medium term, that is fine so long as we find a way to remove carbon from the atmosphere later in the century. The dominant proposal for achieving this is known as BECCS, or bioenergy with carbon capture and storage. BECCS entails developing large tree plantations around the world to absorb CO_2 from the atmosphere, harvesting the biomass, burning it for energy, capturing the emissions at source and storing the waste underground. In AR5, the vast majority of scenarios for 2°C (101 of the 116) rely on BECCS to the point of achieving negative emissions.

BECCS is highly controversial among scientists, however. There are a number of concerns. First, the viability of power generation with CCS has never been proven to be economically viable or scalable (Peters, 2017). Second, the scale of biomass assumed in the AR5 scenarios would require plantations covering land two to three times the size of India, which raises questions about land availability, competition with food production, carbon neutrality, and biodiversity loss (Smith et al., 2015; Heck et al., 2018). Third, the necessary CO_2 storage capacity may not exist (De Coninck and Benson, 2014; Global CCS Institute, 2015). Anderson and Peters (2016) conclude that "BECCS thus remains a highly speculative technology" and that relying on it is therefore "an unjust and high stakes gamble": if it is unsuccessful, "society will be locked into a high-temperature pathway". This conclusion is shared by a growing number of scientists (e.g., Fuss et al., 2014; Vaughan and Gough, 2016; Larkin et al., 2017; van Vuuren et al., 2017), and by the European Academies' Science Advisory Council (2018).

Responding to these concerns, the IPCC (2018) has for the first time published a scenario for reducing emissions in line with the Paris Agreement that does not rely on speculative negative emissions technologies. Developed by Grubler et al. (2018) and known as Low Energy Demand (LED), the scenario works by reducing global energy consumption by 40% by 2050, which makes it much more feasible to achieve a transition to 100% clean energy. The key feature of this scenario is that global material production and consumption declines significantly: "The aggregate total material output decreases by close to 20 per cent from today, one-third due to dematerialization, and two-thirds due to improvements in material efficiency." LED differentiates between the global North and South. Industrial production and consumption declines by 42% in the North and 12% in the South. Given improvements in energy efficiency, this translates into industrial energy demand declining by 57% in the North and 23% in the South.

The LED model represents a "degrowth" scenario – a planned reduction of the material and energy throughput of the global economy. Its inclusion in the IPCC report as the only scenario that does not rely on questionable negative emissions technologies suggests that degrowth may be the only feasible way to achieve the emissions reductions required by the Paris Agreement. This is a major milestone in climate mitigation theory. What is appealing about this approach is that it not only addresses emissions and climate change, but also reduces ecological impact across a range of other key indicators, including deforestation, chemical pollution, soil depletion, biodiversity loss, and so on (Rockstrom et al., 2009; Steffen et al., 2015).

There are a number of policies that would help to achieve reductions in material throughput in line with the LED scenario. One would be to legislate extended warranties on products, so that goods like washing machines and refrigerators last for 30 years instead of ten. Another is to ban planned obsolescence, and to introduce a "right to repair" so that products can be fixed cheaply and without proprietary parts. We could legislate reductions in food waste (as South Korea, France and Italy are doing), tax red meat to promote a shift to less resource-intensive foods, ban single-use plastics and disposable coffee cups, and end advertising in public spaces to reduce pressures for material consumption. Ultimately, however, to accomplish significant and sustained reductions will likely require imposing a cap on annual material use and tightening it year by year until it reaches what ecologists identify as sustainable levels (50 billion tons per year on a global scale, or 6-8 tons per capita; see Dittrich et al., 2012; Hoekstra & Wiedmann, 2014; UNEP IRP, 2014; Bringezu, 2015).

The degrowth hypothesis

The idea of degrowth was first articulated in the early 21st century by ecological economists and post-development theorists (e.g., Latouche, 2009; Victor, 2008; Jackson,

2009; Alier, 2009; Kallis, 2011; Kallis, 2018), and in recent years has captured public attention, even appearing in popular media outlets. The objective of degrowth is to scale down the material and energy throughput of the global economy, focusing on high-income nations with high levels of per capita consumption. The idea is to achieve this objective by reducing waste and shrinking sectors of economic activity that are ecologically destructive and offer little if any social benefit (such as marketing, and the production of commodities like McMansions, SUVs, beef, single-use plastics, fossil fuels, etc.).

Degrowth scholars acknowledge that reductions in aggregate throughput are likely to entail reductions in aggregate economic activity as measured by GDP, given the historically tight coupling between throughput and output (see Hickel and Kallis, 2019; Ward et al., 2016; UNEP, 2017). At first glance, this may seem a troubling prospect. Economists and policymakers have become accustomed to equating GDP growth with human progress and improvements in well-being, so it might seem sensible to conclude that a decline in GDP must necessarily entail a decline in well-being. After all, a reduction in GDP sounds like a recession, and recessions have a range of harmful social effects: firms lay off workers, unemployment rises, and as people lose their jobs they become unable to pay for access to housing, food, healthcare, education and other basic goods. Moreover, states, firms and households find themselves unable to pay debts, heightening risk of financial crisis.

A recession is categorically different to degrowth, however. A recession is a shrinkage of the existing economy (an economy that requires growth in order to remain stable), while degrowth calls for a shift to a different kind of economy altogether (an economy that does not require growth in the first place). The literature on degrowth argues that it is possible to reduce aggregate economic activity in high-income nations while at the same time maintaining and even improving indicators of human development and well-being. This can be accomplished with a series of integrated policy reforms. For instance, as dirty and socially unnecessary industries close down and aggregate economic activity contracts, unemployment can be prevented by shortening the working week and redistributing necessary labour (into cleaner, more socially useful sectors) with a job guarantee. Wage losses due to a reduction in working hours can be prevented by increasing hourly wages with a living wage policy. To protect small businesses that may find it difficult to pay significantly higher hourly wages, a universal basic income scheme could be introduced, with dividends funded by taxation on carbon, wealth, land value, resource extraction, and corporate profits. These policies have been successfully modelled in degrowth scenarios developed by D'Allessandro et al. (2018) and Victor (2019).

The core feature of degrowth economics is that it requires a progressive distribution of existing income. This inverts the usual political logic of growth. In their pursuit of improvements in human welfare, economists and policymakers often regard growth as a substitute for equality: it is politically easier to grow total income and expect that enough will trickle down to improve the lives of ordinary people than it is to distribute existing income more fairly, as this requires an attack on the interests of the dominant class. But if growth is a substitute for equality, then by the same logic equality can be a substitute for growth (Dietz and O'Neill, 2013). By distributing existing income more fairly we can improve human welfare and accomplish social objectives *without* growth – and therefore without additional material and energy throughput. A shorter working week plus a job guarantee and a living wage policy, as described above, are central mechanisms for accomplishing this. So too is investment in public services. By expanding access to high-quality, generous public healthcare, education, affordable housing, transportation, utilities and recreation facilities, it is possible to enable people to access the goods they need to live well without needing high levels of income to do so.

Existing empirical evidence demonstrates that it is possible to achieve high social indicators without high levels of GDP per capita. Past a certain point, the relationship between GDP per capita and social indicators begins to break down. Take life expectancy, for instance; while there is a general correlation between GDP per capita and longevity (countries with higher GDP per capita generally have better life expectancy), the relationship follows a saturation curve with sharply diminishing returns (Preston, 2007; Steinberger & Roberts, 2010). Longevity depends on other important variables besides GDP, such as investment in universal healthcare. For example, Costa Rica's healthcare system allows the country to match US life expectancy with only one-fifth of the US GDP per capita (Sánchez-Ancochea and Martínez Franzoni, 2016). Similarly, there is a tenuous relationship between GDP per capita and happiness, or well-being (see Easterlin, 1995; Easterlin et al., 2010). In the United States and the United Kingdom, for instance, happiness levels have remained unchanged since the early 1970s, despite significant growth in real GDP per capita. According to the Gallup World Poll, many countries (Germany, Austria, Sweden, Netherlands, Australia, Finland, Canada, Denmark, and most notably Costa Rica) have higher levels of well-being than the United States does, with less GDP per capita.

The same pattern applies to many other social indicators. The GDP per capita of Europe is 40% lower than that of the US, and yet Europe performs better in virtually every social category, as European countries tend to be more equal and more committed to public goods. But even European countries have significant room for improvement. Inequality in Europe has worsened significantly since 1980. From a degrowth perspective, this represents an opportunity: there is no *a priori* reason why Europe's social performance cannot be improved still further – without any additional growth – by distributing existing income more fairly and using progress taxation to expand public goods.

It is not just that GDP is not strongly correlated with human development after a point – it is also that GDP growth past a certain threshold tends to have a negative impact. Alternative metrics of economic progress, such as the Genuine Progress Indicator (GPI), make this effect visible. GPI starts with personal consumption expenditure (also the starting point for GDP) and adjusts using 24 different components, such as income distribution, environmental costs and pollution, while adding positive components left out of GDP, such as household work. Kubiszewski et al. (2013) find that in most countries GPI grows along with GDP until a particular threshold, after which GDP continues to grow while GPI flattens and in some cases declines. The authors draw on Max-Neef (1995) to interpret this threshold as the point at which the social and environmental costs of GDP growth become significant enough to cancel out consumption-related gains (Deaton, 2008; Inglehart, 1997).

Of course, one might argue that economic growth is necessary for mobilizing resources to invest in the technological change required to shift the world toward sustainability. But there is no evidence for the assumption that *aggregate* growth is necessary for achieving this. If the objective is to achieve specific kinds of technological innovation, it would make more sense to invest in those directly, or incentivize innovation with policy measures (e.g., caps on carbon and resource use), rather than to grow the whole economy indiscriminately (which would include growth of dirty and destructive industries) while blindly hoping for a specific outcome.

The scarcity machine

While the scholarship on degrowth has outlined the policy changes that would be necessary to achieve a safe and equitable transition to an ecological post-growth economy, the deep logic of such an economy remains undertheorized. Are the reforms that degrowth scholars propose in and of themselves sufficient to euthanize the capitalist growth imperative? Here I want to address this question by elaborating further on the argument that expanding public goods and services is central to a successful degrowth scenario. This argument is much deeper and more profound than it appears at first glance, and opens up a number of fruitful lines of inquiry.

Let us begin with an example that is close to my own experience. In London, house prices are astronomically high, to the point where a normal one-bedroom flat may cost £2,000 per month to rent, or £600,000 to buy. These prices are fictional; they are no indication of the actual cost of building a house, or even of land, but are rather largely a consequence of the rapid privatization of the public housing stock in Britain since 1980, as well as financial speculation, zero-interest rate policy and quantitative easing, which has driven asset prices up in the wake of the 2008 financial crisis to the extraordinary benefit of the rich. Meanwhile, wages in London have not kept pace with housing prices. In order to purchase housing, then, Londoners have to either increase their aggregate working hours or take out loans, which are effectively a claim on their future labour. In other words, people are required to work unnecessarily long hours to earn additional money simply in order to access shelter, which they were previously able to access with a fraction of the income. In the process, they produce additional goods and services that must find a market, thereby creating new pressures for consumption – pressures that manifest in the form of, for example, aggressive and increasingly insidious advertising schemes.

The fictionally high housing prices in London therefore ultimately compel everyone to contribute unnecessarily to the juggernaut of ever-expanding production and consumption, with all of the corresponding ecological consequences that this entails.

This is a problem that is as old as capitalism itself. And it has a name: enclosure. Ellen Eiksins Wood (1999) has argued that the origins of capitalism lay in the enclosure movement in England, during which wealthy elites – empowered by the Statute of Merton of 1235 – fenced off commons and systematically forced peasants off the land in a violent, centuries-long campaign of dispossession. This period saw the abolition of the ancient "right to habitation", once enshrined in the Charter of the Forest, which guaranteed ordinary people access to land, forests, game, fodder, waters, fish and other resources necessary for life. In the wake of enclosure, England's commoners found themselves subject to a new regime: in order to survive they had to compete with each other for leases to farm on newly privatized land. Leases were allocated on the basis of productivity, and were reassessed at regular intervals. In order to retain their leases, peasants had to find ways to intensify their production *vis-à-vis* their competitors (with whom they used to relate convivially and in cooperation as kin and neighbours), even if it was in surplus to their actual needs or even desires. Those who fell behind in the productivity race would lose their access to land and face starvation.

There are two things going on here with enclosure. The first is straightforward primitive accumulation, whereby the commons (land, natural resources, etc.) are acquired for free. This process is essential to the creation of capitalist surplus, or profit: capitalism always needs an outside, external to itself, from which it can draw uncompensated value. But there is also something else at stake here – something even more important, a more powerful and dynamic force. The emergence of the enormous productive capacity that characterizes capitalism depended in the first instance on subjecting humans to artificial scarcity. Scarcity – and the threat of hunger – created the impetus for competitive productivity and *served as the engine of growth*. The scarcity was *artificial* in the sense that there was no actual net depletion of resources: all of the same land and forests and waters remained, just as they always had, but people's access to them was restricted. Scarcity was, in this sense, created in the process of elite accumulation. And it was enforced by state violence: peasant uprisings against enclosures were repeatedly put down by force, and often by massacre (Fairlie, 2009).

Michael Perelman (2000) observes that the historical record is full of commentary by British landowners and elites celebrating enclosure as a tool for enhancing the "industry" of peasants whose access to abundant commons rendered them given to leisure and "insolence". It is worth reciting some of the more notable examples of this sentiment. The Quaker John Bellers (1695) wrote "Our forests and great commons make the poor that are upon them too much like the Indians, being a hindrance to industry, and are nurseries of idleness and insolence." The agriculturalist Arthur Young (1771) noted that "everyone but an idiot knows that the lower classes must be kept poor, or they will never be industrious". The Reverend Joseph Townsend (1786) emphasized that "it is only hunger which can spur and goad them on to labour," while pointing out that

> "legal constraint... is attended with too much trouble, violence, and noise... whereas hunger is not only a peaceable, silent, unremitted pressure, but as the most natural motive to industry, it calls forth the most powerful exertions... Hunger will tame the fiercest animals, it will teach decency and civility, obedience and subjugation to the most brutish, the most obstinate, and the most perverse."

Patrick Coquhoun, a powerful Scottish merchant, saw poverty as an essential precondition for industrialization:

> "Poverty is that state and condition in society where the individual has no surplus labour in store, or, in other words, no property or means of subsistence but what is derived from the constant exercise of industry in the various occupations of life. Poverty is therefore a most necessary and indispensable ingredient in society, without which nations and communities could not exist in a state of civilization. It is the lot of man. It is the source of wealth, since without poverty, there could be no labour; there could be no riches,

> no refinement, no comfort, and no benefit to those who may be possessed of wealth."

It was David Hume (1752), though, who elaborated an explicit theory of "scarcity": "Tis always observed, in years of scarcity, if it be not extreme, that the poor labour more, and really live better."

The same process – the production of scarcity for the sake of generating capitalist growth – unfolded around much of the rest of the world during the period of European colonization, often even more clearly. Across British Africa, colonizers faced what they called "The Labour Question": how to get Africans to work in mines and on plantations for low wages during a time when slavery was no longer an option. Colonizers discovered to their dismay that Africans were content with their subsistence lifestyles, where they had all the land and livestock they needed to live, and showed no inclination to do back-breaking work in European industries. Wages were not high enough to attract people voluntarily into the capitalist labour market. The solution that colonizers settled on was to either force people off their land (the Native Lands Act in South Africa being perhaps the best known example of this), or to force them to pay taxes in European currency. Either course of action left people with no option but to sell their labour for wages. The creation of artificial scarcity meant that in order to access the means of survival, people had to participate in the juggernaut of ever-increasing productivity (with, again, pressures for concomitant consumption, generally elsewhere in the world system).

In India, British colonizers sought to find ways to compel Indians to shift from subsistence farming to cash-crops for export (Davis, 2000; Patnaik, 2018). They found that people were unwilling to make this transition voluntarily as they already enjoyed sufficient livelihoods, and even during times of drought had robust systems of mutual aid to ensure their well-being. Colonial policy, beginning with the British East India Company and continuing under the Raj, was to systematically dismantle the support systems that people relied on: destroying communal granaries, privatizing communal irrigation systems, enclosing commons that people used for wood and fodder and game, and taxing peasants into debt. As during the British enclosures, the explicit purpose was to put people at the mercy of hunger, and thereby compel them not only to participate in cash-cropping, but to *compete* with one another in that industry. This approach increased agricultural productivity, but at the expense of people's lives: it left peasants so vulnerable to fluctuations in climate and markets that tens of millions died needlessly of famine under British rule, including up to 30 million who perished during the last decades of the 19th century, the heyday of the Victorian era.

The same process of enclosure and forced proletarianization played out repeatedly during the period of European colonization – not just under the British but under the Spanish, Portuguese, French and Dutch as well – with examples far too numerous to recite here. In all cases the creation of artificial scarcity was leveraged, purposefully, as the engine of capitalist expansion.

Today, in our now almost completely proletarianized world, people continue to feel the force of scarcity in the constant threat of unemployment. Workers must become ever-more disciplined and productive at work or else lose their jobs to someone who will be more productive still – usually someone poorer and more desperate. But there is a paradox: as productivity rises, less labour is needed in order to produce the same quantity of goods and services. As a result, workers get laid off and find themselves with no means of livelihood. The state, desperate to reduce unemployment and forefend against social and political crisis, must then find ways to grow the economy in order to create new jobs so that people can survive – cutting taxes and regulations on businesses, providing access to cheap energy and raw materials, enabling debt-fueled consumption, and so on. Aware of this dynamic, workers and unions join in the chorus calling for more growth, and tend to elect politicians who can promise it most credibly. Scarcity, then, creates recruits to the ideology of growth.

Even people who are concerned about ecological breakdown are forced to submit to this logic: if you care about human lives, then you must call for growth first and foremost, regardless of the ecological consequences; we can deal with the environment later, once everyone has enough. But there will be no later, because the problem of scarcity is never solved – there is never enough. Whenever scarcity is about to be solved, it is always quickly produced anew. In 1930, Keynes famously predicted that the economy would rapidly become so productive and replete that people would have to work for no more than 15 hours a week to satisfy all their material needs, thus freeing up more time for leisure. Productivity has long since surpassed the point of abundance that Keynes foresaw, and yet his prediction about work has never come true, because instead of translating productivity gains into shorter working hours, higher wages and guaranteed employment, capitalists have captured the benefits for themselves, increasing their profits while keeping wages low, and retaining the threat of unemployment in order to discipline labour.

In this way, capitalism transforms even the most spectacular productivity gains not into abundance and human freedom, but into new forms of artificial scarcity. It must, or else it risks shutting down the engine of accumulation itself – killing off the goose that lays the golden egg.

Here it becomes clear that inequality itself drives artificial scarcity, just as enclosure did in an earlier era. In the 1970s, the United States had a lower poverty rate, higher average real wages, and higher happiness levels than it does today, despite having less than half of today's per capita income. The difference has to do with distribution: in the 1970s, income was shared more fairly, leading to better social outcomes. Virtually all of the yields of growth since 1980 have been accumulated by the rich, leaving the rest of the society in a state of what can only be called artificial scarcity. The same process plays out in every nation that has seen rising inequality, and indeed on a global scale as well. Today 4.2 billion people in the world (60% of humanity) live on less than the equivalent of $7.40 per day, the minimum necessary for normal human life expectancy and basic nutrition. Since

1980, the incomes of the richest 1% have grown by 100 times more than those of the poorest 60%, and now stand at $18.7 trillion (World Inequality Report, 2018). This is three times more than it would take to cover the poverty gap and lift everyone in the world above $7.40/day. In other words, shifting a third of the income of the richest 1% to the poorest 4.2 billion people could end global poverty in a stroke, while still leaving the 1% with $175,000 per year.

We can also see the logic of artificial scarcity at work in the realm of consumption. Industrialists who fear that people's existing needs are too limited to absorb capitalism's immense productive output must seek to create new needs, or else the juggernaut will grind to a halt. This is accomplished by various means. One is to expand desires through sophisticated advertising campaigns – and to extend these campaigns into all public and private spaces – manipulating people's emotions and psychology to create new "needs" for products that promise to grant them a sense of self-esteem, status, identity, sexual prowess and so on that did not exist before and indeed do not have to exist. Another is to create products that are designed to break down quickly (like laptops and smart phones today) or become rapidly obsolete (as with the rise of throwaway fashion), and which therefore must be replaced more frequently than would otherwise be necessary. Another is to preclude the development of public goods in order to ensure that people have no choice but to purchase private alternatives: for instance, blocking the construction of effective public transportation systems in order to ensure a steady stream of demand for the automobile industry.

On top of this, a significant portion of consumption in highly-industrialized countries is driven by an artificial scarcity of time. As pressure on labour mounts, the structural compulsion to work unnecessarily long hours leaves people with so little time in the day that they must pay firms to do things that they would otherwise be able to do themselves: cook meals, clean their homes, watch their children, care for their elderly parents. Meanwhile, the stress of overwork creates needs for anti-depressants, sleep aids, alcohol, dieticians, gym memberships, therapy, marital counselling, expensive holidays, and other products that people would otherwise be less likely to feel they require. To pay for these products and services, people need to work yet more to increase their incomes, driving a vicious circle of unnecessary production and consumption.

All of this reveals an interesting contradiction. The ideology of capitalism is that it is a system that generates immense abundance (just think of all the products that one sees displayed on television and in shopfronts, which parade as an extraordinary cornucopia of stuff). But in reality it is a system that relies on the constant production of scarcity.

Resolving the Lauderdale Paradox

The pattern by which capitalist growth generates scarcity was first noticed in 1804 by James Maitland, the 8th Earl of Lauderdale, in his *Inquiry into the Nature and Origin of Public Wealth and into the Means and Causes of its Increase.* Maitland introduced what became known as the

"Lauderdale Paradox". He pointed out that there is an inverse correlation between "private riches" and "public wealth", such that an increase in the former can only come at the expense of the latter (see Foster and Clark, 2009).

"Public wealth," Maitland wrote, "may be accurately defined, – *to consist of all that man desires, as useful or delightful to him.*" In other words, public wealth comprises goods that have an intrinsic use value even in abundance, including air, water, and food. Private riches, on the other hand, consist "*of all that man desires as useful or delightful to him; which exists in a degree of scarcity.*" In other words, Maitland sought to explain how private riches depend on goods having an exchange value that increases in proportion to their scarcity. By way of example, he pointed out that if one were to enclose an abundant resource like water and establish a monopoly over it, one could charge people to access it and therefore increase one's private riches. This would also increase what Maitland called the "sum-total of individual riches" – what today we call GDP. But this increase, of private riches and GDP, can be accomplished only by curtailing access to what was once abundant and free.

Maitland recognized that this was happening during the process of European colonization. He stopped short of theorizing enclosure (unlike Henry George and Karl Marx later in the century), but he did point out that colonialists would often resort to burning down trees that produced fruits and nuts so that local inhabitants wouldn't be able to live off of the natural abundance of the earth, but would be compelled instead to work for wages and purchase food from Europeans in order to feed themselves. For the sake of private riches and GDP, what was once abundant and free had to be made scarce. The iconic example of this was of course the salt tax that the British Raj imposed on India. Salt was abundant and free all along India's coasts, but the British banned people from gathering it and taxed its consumption in order to create what became a significant revenue stream for the colonial government. The abolition of public wealth generated private riches.

We can see this same process happening today in the endless waves of privatization that have been unleashed all over the world since 1980, of education, healthcare, transportation, libraries, parks, swimming pools, water, even social security. At a time when globalization has run its course, trade protections have been dismantled around the world, wages are as low as they can reasonably get, and consumer markets are increasingly saturated, continued growth requires new rounds of what David Harvey (2003) has called accumulation by dispossession – the enclosure of the remaining stock of public wealth. Social goods everywhere are under attack – they must be made scarce for the sake of growing the GDP. People must be made to pay in order to acquire goods that they used to access for free. And in order to pay, they will of course have to work more, placing them once again under pressure to compete with one another to be ever-more productive – a pressure justified, again, for the sake of growing the GDP. Indeed, our society's obsession with GDP growth as the primary public policy objective reveals the entrenchment of the Lauderdale Paradox as political common sense, the ultimate triumph of enclosure: the growth of "private riches" has come to stand in for Progress itself. Meanwhile, conveniently – and tellingly – there is no indicator that charts the concomitant collapse of public wealth.

This logic reaches its apogee in the contemporary vision of austerity, which was rolled out across Europe in the wake of the 2008 financial crisis. What is austerity, really? It is a desperate attempt to re-start the engines of growth by slashing public investment in social goods and welfare protections – everything from elderly heating allowances to unemployment benefits to public sector wages – chopping away at what remains of the commons so that people deemed too "comfortable" or "lazy" are placed once again under threat of hunger, and forced to increase their productivity if they want to survive. This logic is overt, just as it was in the writings of people like John Bellers and David Hume. During the government of British Prime Minister David Cameron and his Finance Minister George Osborne, welfare cuts were conducted explicitly in order to get "shirkers" to work harder and to be more productive ("workfare", they called it). Scarcity must be induced for the sake of more growth, as Maitland pointed out. In the logic of austerity, scarcity and growth emerge as two sides of the same coin, just as during the enclosures.

Today there is a new element added to this dynamic, where the Lauderdale Paradox now reveals itself also in the process of ecological breakdown that is unfolding around us on a planetary scale. Since the 1950s there has been an extraordinary increase in global GDP (often referred to as the "Great Acceleration"), but this growth in "private riches" has come at the expense of an extraordinary depletion of the living world, given the tight coupling between GDP and material and energy throughput. The majority of the planet's tropical forests have been destroyed, agricultural soils are largely degraded, rates of species extinction are now 1,000 times faster than the background rate prior to the Industrial Revolution, while CO_2 emissions have caused climate change and ocean acidification, destabilizing terrestrial and marine ecosystems and threatening food chains. This is the ultimate price of the longstanding plunder of "free" value from nature. And by destabilizing the biosphere on which human life depends, it becomes clear that the greatest public wealth of all – the integrity of the planetary biosphere – has been sacrificed for the sake of private riches.

So what will happen? How will capitalism resolve this crisis? This brings us to an important point. In response to the threat of ecological breakdown, one might say that all we need to do is cap emissions and material use and scale these down to sustainable levels, much like the IPCC's degrowth scenario has proposed (as I described above). Some insist that once we have done this, there is no reason that GDP cannot then continue to grow indefinitely, while the biosphere recovers. But once emissions are outlawed and material use is capped at a low level, where will capitalism acquire its free inputs, if not from energy-dense fossil fuels and from nature? It will have to turn to the other primary source of value, namely, human labour. We can expect, then, that in a state of ecological emergency, capitalism will seek growth by finding new ways to squeeze workers.

Some progressive economists, like Dean Baker (2018), insist that continued growth needn't be so rapacious. Baker insists that we can scale down material and energy throughput *and* protect labour rights (effectively placing limits on both of the sources of value that capitalism draws upon), and *still* have growth. There is no reason that the new value cannot be purely immaterial, he says.

There are good reasons to believe that Baker is incorrect in this assumption. As capitalist growth has for all of its history been tightly coupled to material and energy throughput (even during the transition to services in the global North), to imagine that GDP can continue to grow while throughput declines is to go against all available evidence, and indeed to imagine a completely different kind of economy – one that has never before existed. If we are going to imagine a new economy altogether, why not imagine one without growth? This brings us to the key point. It is not growing throughput that is ultimately the problem: the problem is the growth imperative itself. To illustrate, one can imagine that in an economy where growth *must* happen despite a cap on throughput, and where all new value therefore has to be immaterial, capital would seek to enclose immaterial commons that are presently abundant and free (knowledge, songs, green spaces, maybe even parenting, physical touch, love, and perhaps even the air itself) and sell them back to people for money. Subject to these new waves of artificial scarcity, people would find themselves compelled to work and earn wages in the new immaterial industries simply in order to acquire immaterial goods that used to be freely available. This may be an ecological economy, but it is not an economy that makes any sense, or one that anyone would actually want to live in.

The point of this imaginary exercise is to illustrate that while capping throughput might create the conditions for an ecological economy and indeed cause material and energy throughput to decline, it does not neutralize the deeper violence of the juggernaut, which is the logic of growth itself. Such a move might be adequate in a pragmatic sense, but it is intellectually unsatisfying. The only way to resolve the Lauderdale Paradox is to reverse it: to reorganize the economy around generating an abundance of public wealth even if doing so comes at the expense of private riches. This would liberate humans from the pressures generated by artificial scarcity, thus neutralizing the juggernaut and releasing the living world from its grip.

A theory of radical abundance

What would this look like? Let's return to the example with which we began: the London housing market. Imagine if we were to even just partially decommoditize London's housing stock; for example, imagine the government was to cap the price of housing at half its present level. Prices would still be outrageously high, but Londoners would suddenly be able to work and earn significantly less than they presently do without any loss to their quality of life. Indeed, they would gain in terms of time they could spend with their friends and family, doing things they love, improvements to their health and mental well-being, and so on. And by needing to work less they would contribute to less overproduction, and therefore ease concomitant pressures for unnecessary consumption.

The same thought experiment can be applied to all social goods that have either been made to be artificially scarce or that would otherwise be simple to manage as commons. And here I have in mind not only healthcare and education, which are already generally well-recognized as public goods by most social democracies, but also other key goods that are essential to people's well-being, like internet, housing and public transportation, as in the vision of Universal Basic Services outlined by academics at University College London (IGP, 2017). On top of this, new "utilities" like Uber and AirBnb could be taken into public ownership, or public alternatives could be created, thus enabling the emergence of "platform commons" which would allow people to exchange their material resources (cars, homes) without having to pay exorbitant and unnecessary fees to private monopolies. Employment too could be considered a common good – and indeed this would be crucial: a shorter working week with a job guarantee and a living wage, plus legislation to ensure that all productivity gains are delivered back to workers in the form of higher wages and shorter hours. And by banning advertising in public spaces we could reclaim our streets (and attention) as commons and liberate people from the sense of scarcity that advertising induces.

By de-enclosing and expanding the commons, and by redistributing existing income more fairly, we can enable people to access the goods that they need to live well without needing high levels of income (and therefore additional growth) in order to do so. People would be able to work less without any loss to their quality of life, thus producing less unnecessary stuff and therefore generating less pressure for unnecessary consumption. Meanwhile, with more free time people would be able to have fun, enjoy conviviality with loved ones, cooperate with neighbors, care for friends and relatives, cook healthy food, exercise and enjoy nature, thus rendering unnecessary the patterns of consumption that are driven by time scarcity. And opportunities to learn and develop new skills such as music, maintenance, growing food and crafting furniture would contribute to local self-sufficiency (Alexander and Gleeson, 2019).

Liberated from the pressures of artificial scarcity, the compulsion for people to compete for ever-increasing productivity would wither away. We would not have to feed our time and energy into the juggernaut of ever-increasing production, consumption and ecological destruction. The economy would produce less as a result, yes – but it would also *need* much less. It would be smaller and yet nonetheless much more abundant. In such an economy private riches (or GDP) may shrink, as Maitland pointed out, reducing the incomes of corporations and the very rich, but public wealth would increase, significantly improving the lives of everyone else. Suddenly a new paradox emerges: abundance is revealed to be the antidote to growth.

If austerity represents the apogee of the Lauderdale Paradox, where public wealth is sacrificed for the sake of generating private riches, what becomes clear from the above is that degrowth is the very opposite. This is an important point. Some have attempted to smear degrowth as a new version of austerity, this time promoted by the left rather than the right – an extreme manifestation of old-school environmentalists who want to force everyone to live miserable lives. But exactly the opposite is true. While austerity calls for scarcity in order to generate more growth, degrowth calls for abundance *in order to render growth unnecessary.* Abundance, then, is the solution to our ecological crisis. If we are to avert climate breakdown, the environmentalism of the 21st century must articulate a new demand: a demand for radical abundance.

References

Anderson, K., & Peters, G. (2016) "The trouble with negative emissions." *Science*, 354(6309), pp. 182-183.

Baker, Dean (2018) "Will degrowthing save the planet?" *Beat the Press.*

Bringezu, S. et al. (2015) "Possible target corridor for sustainable use of global material resources." *Resources*, 4, pp. 25-54.

D'Alisa, G., Demaria, F., & Kallis, G. (2014) "From austerity to depense." In *Degrowth: a vocabulary for a new era*. Routledge.

D'Allessandro et al. (2018) "EUROGREEN Model of job creation in a post-growth economy." The Greens and EFA in the European Parliament.

Davis, M. (2002) *Late Victorian holocausts: El Niño famines and the making of the third world.* Verso Books.

De Coninck, H., and Benson, S. M. (2014) "Carbon dioxide capture and storage: issues and prospects." *Annual Review of Environment Resources*, 39, pp.243-270.

Dietz, R., & O'Neill, D. (2013) *Enough is enough: Building a sustainable economy in a world of finite resources.* San Francisco/London: Berrett-Koehler/Routledge.

Dittrich, M., Giljum, S., Lutter, S., & Polzin, C. (2012) "Green economies around the world." *Implications of resource use for development and the environment.* Vienna: SERI.

European Academies Science Advisory Council (2018) "Negative Emission Technologies: What role in meeting Paris Agreement targets?" EASAC Policy Report 35.

Fairlie, S. (2009) "A short history of enclosure in Britain." *The Land*, 7, pp. 16-31.

Foster, J. B. and Clark, B. (2009) *Monthly Review* (01 Nov) https://monthlyreview.org/2009/11/01/the-paradox-of-wealth-capitalism-and-ecological-destruction/

Fuss, S. et al. (2014) "Betting on negative emissions." *Nature Climate Change,* 4(10), pp. 850-853.

Galbraith, J. K. (1998). *The affluent society.* Boston, MA: Houghton Mifflin Harcourt.

Global CCS Institute. (2015) *Global Status of CCS 2015: Summary Report.*

Grubler, Arnulf et al. (2018) "A low energy demand scenario for meeting the 1.5C target and sustainable development goals without negative emissions technologies." *Nature Energy,* 3, pp. 515-527.

Harvey, David (2003) *The New Imperialism.* Oxford University Press.

Heck, V. et al. (2018) "Biomass-based negative emissions difficult to reconcile with planetary boundaries." *Nature Climate Change,* 8.

Hickel, Jason and Giorgos Kallis (2019) "Is Green Growth Possible?" *New Political Economy.*

Hoekstra, A. Y. & Wiedmann, T. O. (2014) "Humanity's unsustainable environmental footprint." *Science,* 344, pp. 1114-1117.

Holz, C., Siegel, L. S., Johnston, E., Jones, A. P., & Sterman, J. (2018) "Ratcheting ambition to limit warming to 1.5°C–trade-offs between emission reductions and carbon dioxide removal." *Environmental Research Letters*, 13(6), 064028.

IGP (Institute for Global Prosperity) (2017) *Social prosperity for the future: a proposal for Universal Basic Services.* London: University College London.

IPCC. (2018) *Global Warming of 1.5°C – Summary for Policymakers.* Switzerland: IPCC.

Jackson, T. (2009) *Prosperity without growth: Economics for a finite planet.* Routledge.

Jacobson, M. Z., & Delucchi, M. A. (2011) "Providing all global energy with wind, water, and solar power, Part I: Technologies, energy resources, quantities and areas of infrastructure, and materials." *Energy policy*, 39(3), pp. 1154-1169.

Kallis, G. (2011) "In defense of degrowth." *Ecological Economics*, 70(5), pp.873-880.

Kallis, G. (2018) *Degrowth.* Newcastle-upon-Tyne: Agenda Publishing.

Larkin, A. et al. (2017) "What if negative emissions technologies fail at scale?" *Climate Policy.*

Latouche, S. (2009) *Farewell to growth.* Polity.

Latouche, S. (2014) "Essays on Frugal Abundance." *Degrowth: Misinterpretations and Controversies, Part 1.* Republished by the Simplicity Institute.

Patnaik, U. (2018) *Agrarian and Other Histories.* Columbia University Press.

Perelman, M. (2000) *The invention of capitalism.* Duke University Press.

Peters, G. (2017) "Does the carbon budget mean the end of fossil fuels?" *Climate News.*

Raftery et al. (2017) "Less than 2 °C warming by 2100 unlikely." *Nature Climate Change.*

Rockström, J. et al. (2009) "Planetary boundaries: exploring the safe operating space for humanity." *Ecology and Society*, 14, no. 2.

Sahlins, M. (1976) "The original affluent society." In *Culture and Practical Reason*. University of Chicago Press.

Schandl, H. et al. (2016) "Decoupling global environmental pressure and economic growth." *Journal of Cleaner Production*, 132, 45-56.

Smith, P. et al. (2016) "Biophysical and economic limits to negative CO_2 emissions." *Nature Climate Change*, 6(1), pp. 42-50.

Steffen, W., et al. (2015) "Planetary boundaries: Guiding human development on a changing planet." *Science*, 347, no. 6223.

United Nations Environment Programme (UNEP) (2017). *Assessing Global Resource Use.*

Van Vuuren, D. P. et al. (2017) "Open discussion of negative emissions is urgently needed." *Nature Energy*, 2, pp. 902-904.

Vaughan, N.E., and Gough, C. (2016) "Expert assessment concludes negative emissions scenarios may not deliver." *Environmental Research Letters*, 11.

Victor, P. (2008) *Managing without Growth: Slower by Design, Not Disaster*. Edward Elgar Publishing.

Ward, J. et al. (2016) "Is decoupling GDP growth from environmental impact possible?" *PloS one* 11(10).

Wood, E. M. (2002) *The Origin of Capitalism: A Longer View*. Verso.

World Inequality Report (2018) Published by the World Inequality Database, https://wir2018.wid.world/.

Chapter 7

Environmental financialization: what could go wrong?[1]

Eric Kemp-Benedict & Sivan Kartha

1. Introduction

In 2012, Food and Water Watch (2012) introduced a new term of art, "the financialization of nature". It was quickly taken up (e.g., Silvertown, 2015; Kill, 2015). Studies that use the term warn against a range of related policy proposals: monetization of ecosystem services and natural capital; pursuit of "green" growth; and creation of financial markets for environmental assets. Defenders of natural capital accounts and ecosystem service valuation respond that they are simply providing information (e.g., Potschin et al., 2016; Schröter and van Oudenhoven, 2016) and that the information is critical if ecosystems are to be protected (Daily et al., 2009; Costanza et al., 2014). In this paper we address one particular aspect of the financialization of nature: the claim that tradable financial instruments and their

[1] The authors would like to thank two SEI reviewers for their helpful and constructive comments. The work presented in this paper was funded in part by the Swedish International Development Cooperation Agency (Sida).

exchange on financial markets can be effectively enlisted to protect nature. We argue that the abstractions needed to make financial assets usable also makes them unsuitable for protecting ecosystems.

The environmental motivation for financializing nature is clear. When natural capital is not explicitly integrated into economic decision-making, services derived from nature are uncounted benefits (positive externalities), and the harms done to it uncounted costs (negative externalities), in financial and economic transactions. As natural environments are being degraded and destroyed at an alarming rate, whatever value people might place on nature is clearly not translating into conservation and care. A range of voices, from influential environmental economists (e.g., Chichilnisky and Heal, 1996; 2000) to environmental activists (e.g., Krupp 2008) to corporate leaders (Carbon Pricing Leadership Coalition, 2016; e.g., The B Team 2016), advocate the creation and market exchange of financial "environmental assets".

We question the conclusion that financial markets – specifically, trading in ecosystem-backed securities – is an effective strategy for preserving natural capital. For a financial asset to be usable, it must be fungible; that is, it must be comparable to, and exchangeable with, other assets. From the investor's point of view, different financial assets (or portfolios of assets) are distinguished along a narrow set of relevant dimensions – price, expected rate of return, and risk – that abstract from the underlying real assets. In contrast, ecosystems are not exchangeable and differ along manifold and complex dimensions. We argue that this complexity, and the challenges inherent in abstracting beyond it, pose a threat to natural capital and ecosystem services when they are "securitized" to back tradeable financial assets.

A financial asset is a right to a stream of income payments. It therefore has a value that is only weakly tied to the particularities of the underlying physical asset. While the fundamental value of the financial asset – the income stream – depends ultimately on the viability of the underlying physical asset, its market value is determined in trades of the financial asset, and can diverge from the fundamental value. An extreme example of the consequences of such divergence is the financial crisis precipitated by the collapse of the sub-prime mortgage market.

Our claim is that securitizing ecosystem services will lead to simplification and degradation of ecosystems. They are examples of Karl Polanyi's (2001, chapter 6) "fictitious commodities". Like land, labor and money, treating ecosystem services as commodities, contrary to fact, leads to perverse outcomes. Securitization creates strong incentives to maximize output of commodifiable services at the expense of broader ecosystem function and non-commodifiable services – as well as the provision of local ecosystem-based livelihoods. We will argue that ecosystems have suffered from this dynamic in the past, and will continue to do so, despite the somewhat expanded list of ecosystem services that we recognize today, and despite good-faith efforts to protect the ecosystems behind environmental financial assets.

For natural capital, the question is how well the characteristics of financial assets reflect the value of the underlying natural asset. If preserving natural capital is the goal, as proponents of ecosystem services valuation claim, then the paramount interest is to maintain ecosystem function. When instead the investor's desire for high and reliable returns is given priority – as financial markets do, by design – it creates incentives to maximize the provision of the ecosystem services that yield the largest and most stable payments. This will typically come at the expense of other ecosystem flows, both those that are valued by people (but not commodified) and those that are not valued by people. In parallel, growing economies tend to place increasing pressures on ecosystems for their regulating services, such as waste management and carbon dioxide absorption. This process is sketched in Figure 1.

Figure 1 Enhancing commodified ecosystem services (ES) and their associated financial assets tends to degrade ecosystem function

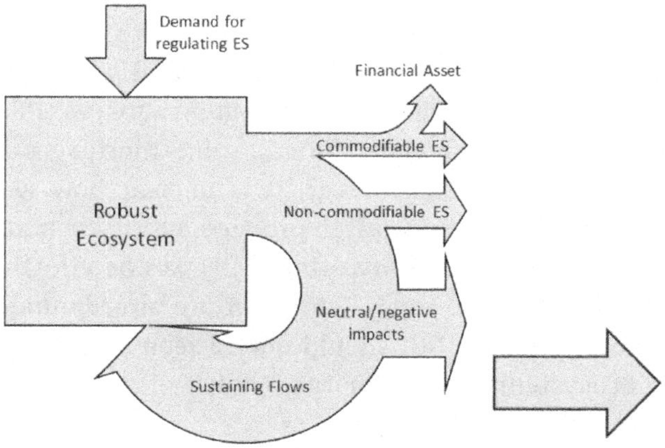

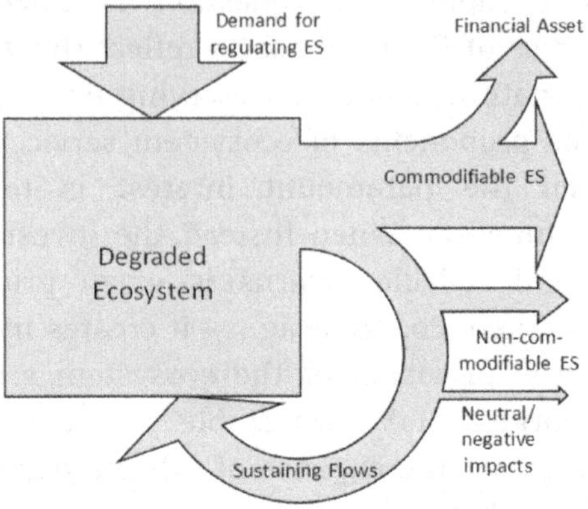

We will identify three partial exceptions to this broad claim. First, when a permit to place a non-specific pressure on ecosystems is traded, in principle it can help protect ecosystem function. For example, trading a restricted supply of permits to emit SO_x or CO_2 from industrial plants reduces demand for gas regulation across a broad range of ecosystems. Whether permits are effective in practice is open to debate; regardless, the same partial exception does not apply to ecosystem-based carbon sequestration, where pressure is being placed upon the gas regulating services of a specific ecosystem. Second, payment for ecosystem services (PES) schemes could, in principle, function like mortgages in PES-backed securities. However, it is unclear how well such an approach would work in practice. Third, we argue that socially-responsible investing (SRI) can be effective in mitigating harm to ecosystems that are already under production. However, SRI should not be seen as a way to protect ecosystem function more broadly.

2. The process of environmental financialization

In environmental financialization, the financial asset represents a stream of payments that is associated with commodified ecosystem services that have had some monetary value placed on them, and that are ultimately tied to some underlying natural capital. Ecosystems – the natural capital – produce flows of goods and services of value to people – the ecosystem services (see Costanza and Daly, 1992). They also produce flows of no or negative value to people while contributing to ecosystem function (such as floods, predators and fire). Although the idea goes back centuries, if not millennia (Gómez-Baggethun et al., 2010), the term "natural capital" was introduced in the 1970s by Schumacher (1973) as a metaphor to frame conservation in business terms. The concept of ecosystem services – that is, the flow of various benefits arising from natural capital – also appeared in the 1970s (Westman, 1977), and was later popularized by the Millennium Ecosystem Assessment (MA, 2005). Only after the 1970s were these concepts taken literally: natural capital becoming a stock that can be assigned a monetary value so that nature can be properly reflected in commercial transactions (e.g., Pearce et al., 1989), with the monetary value based on valuations of the ecosystem services provided. The twin concepts of natural capital and ecosystem services are motivated by a desire to account for the benefits that humans derive from nature, while markets for ecosystem services aim to bring those values explicitly into economic decision-making.

In a foundational paper, Costanza et al. (1997) provided a list of ecosystem services: 1) gas regulation (such as CO_2, SO_x, O_3); 2) climate regulation; 3) disturbance regulation (e.g., storm protection); 4) water regulation; 5) water supply; 6) erosion control and sediment retention; 7) soil formation; 8) nutrient cycling; 9) waste treatment; 10) pollination; 11) biological control (e.g., pest or weed control); 12) refugia (such as habitats for migratory species); 13) food production; 14) raw materials; 15) genetic resources; 16) recreation; 17) cultural. Any given ecosystem might provide some or all of these services, which people value directly, while carrying out many other functions, which they value only to the extent that a healthy ecosystem continues to provide the preceding list of services. But not all of the items in the list can be commodified to provide a stream of payments.

Much of the work on ecosystem services focuses on valuation, or assigning monetary values to those services (Farber et al., 2002). The values do not have to correspond to actual exchanges of money, and they often do not – we discuss this at length in the next section. The purpose of valuation is to compare the normally invisible ecosystem services to other goods and services using a well-recognized unit of value. In this way, for example, Costanza et al. (1997) found the total value of their list of ecosystem services to exceed global gross national product. Valuation thus serves a communicative role. Defenders of the ecosystem services concept emphasize this aspect of valuation, arguing that it should not be equated to monetization and market exchange (e.g., Schröter and van Oudenhoven, 2016).

Nonetheless, to underpin a financial asset, an ecosystem service must produce a flow of actual payments. And to create a market in that financial asset, different instances of the flow of payments must be comparable. Thus, the payment streams underlying a security must, to some degree, act as commodities – that is, one payment stream should be effectively indistinguishable from another. If that is not the case, then the process of securitization will hide potentially crucial information about their differences.

The difficulty of commodifying ecosystem services can be seen in a list of "proxy commodities" proposed by Landell-Mills and Porras (2002) for forest-based ecological services. Of the twelve instruments for biodiversity conservation in their review, only three have the potential to provide a steady stream of payments that could back a broadly exchangeable financial instrument: biodiversity credits/offsets, bioprospecting rights, and equity shares in businesses that market themselves as "biodiversity-friendly" (although others would question this claim; see Apostolopoulou and Adams, 2017). The others relate to ecosystem services that are not broadly interchangeable: locally-specific contractual arrangements (easements, conservation concessions, development rights, managerial contracts, and research permits), non-commodity goods and services (from biodiversity-friendly companies), one-time transactions (debt-for-nature swaps and land acquisition), or policy measures (protected areas) distinct from financial markets.

Finally, the commodified payment stream must be securitized. This process establishes a tradable right to an income derived from that stream of payments. The whole process is shown in Figure 2: first, assigning an economic value for the ecosystem service (valuation); second, establishing a fungible proxy for the ecosystem service, in that quantities of equal economic value are also physically equivalent (commodification); third, creating tradable rights to an income stream associated with the ecosystem or service (securitization).

Figure 2 Levels of abstraction and simplification involved in financializing environmental goods.

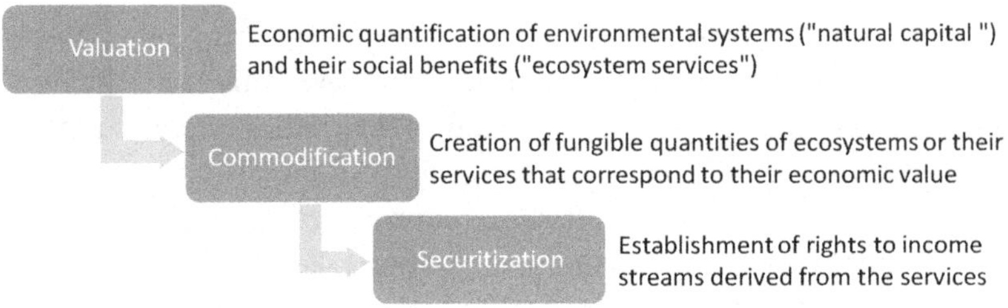

We now examine these three successive steps – valuation, commodification, and securitization – involved in environmental financialization. In particular, we look at them as a progressive process of simplification and abstraction.

2.1 Valuation: defining an economic value

Farber et al. (2002) distinguish between value systems, value, and valuation in the introductory paper to a special issue on ecosystem services. *Value systems* are normative and moral frameworks that guide action. Within their value systems, people assign *value* to actions or objects in the degree to which they meet user-specified goals, objectives or conditions. *Valuation* is then the process by which values are assigned to actions and objects. As Farber et al. point out, because valuation ultimately rests on the particular value system, whether the economic value of an ecosystem service can be defined in a meaningful manner has not been settled. Until it is, we are blocked at the first step in Figure 2.

However, in case economic notions of value prevail, Farber et al. considered the concept of "value" in economic history: the ancient distinction between "use value" and "exchange value"; the classical labor theory of value; and marginalism, which is the basis of nearly all economic policy analysis today.[2]

The argument for valuing natural capital and ecosystem services is that nature has a very high use value, a proposition to which the classical economists would likely have agreed (Gómez-Baggethun et al., 2010). In contrast, they believed that the exchange value of goods and services derives from the cost of labor used to produce them; natural capital and ecosystem services are

---

[2] The literature on ecosystem services contrasts "use value" with "option value" and "non-use value" (de Groot et al., 2010). The sum is Total Economic Value (TEV). Option value is the value of deferring use to an uncertain future. Non-use value is the value of leaving something alone. We recognize the validity of the concepts, and note that they are inherently friendly to the notion of sustainability. We question, however, the identification of option value or non-use value as uniquely "economic" notions. After all, they can be enforced through exclusion, *e.g.* through the expansion or establishment of a protected area, without any recourse to a market or market prices.

provided by nature at no cost, so their exchange value is zero. When the services provided by nature can be appropriated, the one holding title can extract a rent, but the services are prior to the economy proper.

In marginalist theory, people reach an optimal allocation of resources when the marginal utility they derive from application of the resource is equal across different ends. The great theoretical advantage is that value in exchange derives from value in use (Farber et al., 2002): marginal utility dictates the price. In an economy with multiple agents all seeking to maximize their utility by exchange with one another, the end result is (in theory) a set of prices that ensures demand is equal to supply. According to the welfare theorems of neoclassical economics, the final result is "optimal", in the special sense that it is Pareto efficient. The second welfare theorem says that any Pareto efficient outcome is the result of a competitive equilibrium, thus justifying the construction of markets in order to achieve social goals.

The equivalence of classical use value and exchange value in marginalist theory means that one of the two can be dropped from the lexicon. However, we note some well documented flaws in the marginalist story. The requirements for a market to actually lead to efficient outcomes (even in the narrowly defined sense of Pareto efficiency) are very stringent, and not seen in reality (Gowdy and Erickson, 2005). Markets are generally not competitive, there are substantial barriers to entering and leaving markets, and information is far from perfect. It might be thought that it is enough that the Paretian conditions hold approximately, but that expectation is dashed by the general theorem of the second best (Lipsey and Lancaster, 1956). If conditions depart from the ideals of perfect competition, free entry and exit, and perfect information, then adding or subtracting a constraint may raise welfare, lower it, or leave it the same, and the new optimum need not look anything like the one calculated without the constraints.

In the case of ecosystems, the relevant constraints required to preserve the ecosystem may be unknowable. Ecosystems are complex systems in constant flux. The concept of ecosystem services is a useful idealization that provides heuristic guides to manage the complexity, but those idealizations do not translate well into concrete policy (Evans 2018). Attempts to apply ecosystem services assessment in practice reveals its deficiencies, leading practitioners to extend and "patch" the framework in an attempt to better reflect reality. Each extension and patch corresponds to a new constraint in a welfare analysis, so each is likely to change the optimum solution. If any constraints are left out – as they almost certainly will be from the standpoint of preserving ecosystem function – then a constructed market cannot be expected to yield an optimal outcome.

2.2 Commodification: ecosystem services as fungible assets

The source of exchange value for ecosystem services in marginalist theory is the same as in classical economics: the service must be appropriated before anyone will pay for it. Marginalist theory introduces the useful concept of a "shadow price", which is the value derived from a marginal change in the availability of some unpriced good or service, but unrealized shadow prices cannot underlie a financial asset. Neither can other forms of indirect valuation, such as contingent valuation, in which people are asked how much they would pay to accept or give up a good. Enforceable property rights are central to the design of markets for ecosystem services in either classical or marginalist theory.

Even in cases where an economic value may be sensibly assigned to a given ecosystem service in a given locale, it may not be sensible to then construct a corresponding financial asset with enforceable property rights that can be widely traded on a market. This would require those ecosystem services to be fungible across the domain of that market, *i.e.*, for there to be a common currency that allows for faithful comparison of their value across the full range of contexts that the market is intended to encompass. Without this fungibility, a common market through which they are exchanged at a common market price cannot be constructed.

By way of example, Driesen (2005) writes

> "suppose that the Army Corps of Engineers allows a developer to fill in a 100 acre wetland, but requires it to purchase 100 acres of restored wetlands in order to make up for it. This sounds like an environmentally responsible trade. But is it? We cannot tell without a lot more information. Some wetlands play a critical role in preventing floods. Others are less important

> in that regard. Some wetlands provide critical habitat for endangered species; others do not. Wetlands also vary in their value in providing water filtration."

The non-fungibility of a given acreage of wetlands, and indeed the absence of any simple valuation metric that can handle the diversity of wetlands makes a tradeable financial security in wetlands preservations of questionable usefulness.

Even if specific ecosystem services lend themselves to commodification as a fungible financial asset, it does not follow that it holds for the ecosystem as a whole. For example, a mangrove may provide climate regulation (by storing carbon), disturbance regulation, water regulation, nutrient cycling, waste treatment, pollination, refugia, genetic resources, and cultural services. Moreover, as Redford and Adams (2009) point out, some ecosystem services are not benign; mangroves are home to carnivores and disease vectors. Thus, even if one service may be treatable as a commodity, the ecosystem itself – the natural capital – cannot be reliably valued through market mechanisms. Aside from heavily managed landscapes, such as forest plantations or cropland, these non-commodity flows will predominate. Indeed, the dynamic illustrated in Figure 1 can be seen in the global transformation of natural forests and grasslands into forest plantations and monoculture croplands.

In Costanza et al.'s (1997) list of 17 ecosystem services, each of the items can be valued more or less reliably – that was the point of their paper – although refugia and cultural amenities are particularly challenging. But only four are suitable candidates for creating fungible financial securities that represent the underlying ecosystem services: gas regulation; climate regulation; food production; raw materials; and genetic resources. The output of two of these items – food production and raw materials – are already traded on commodity markets, albeit without the explicit goal of ecosystem protection. Genetic resources also have become economic assets. While one genetic resource is not substitutable for another, the patenting of genome sequences and entire genomes (Jensen and Murray, 2005) shows that exclusive rights can be established, even if it is not entirely clear what the object of those rights actually is (Calvert, 2007). The rights are sought in order to secure a future stream of income, which can then provide the basis for a financial asset.

Gas regulation itself – the ecosystem service – has not been put on markets, but pollution permits have; the same can be said for climate regulation. In both cases, markets for pollutant emissions permits have been implemented. The commodity here is not the ecosystem service itself, which cannot be appropriated, but rather the permit to place a burden on the shared natural system that provides that ecosystem service. Markets have been established for gases that disperse over global (ozone-depleting compounds and greenhouse gases including CO_2) or large regional (SO_x) scales.

Climate regulation poses unique problems. The global climate is shared, and measures that mitigate drivers of climate change are welcome. However, ecosystem-based measures for climate regulation are problematic for the reason alluded to above: it is only one of many services provided by any particular ecosystem. Thus, the Amazon may be a vast store of carbon, but it is also a home to many indigenous and non-indigenous people, a watershed, a biodiversity reserve, a source of non-timber forest products, a potential source of land for commercial agriculture or forestry, and the site of underground mineral resources, as well as a climate regulator through means other than serving as a carbon sink, such as albedo control and effects on the global hydrological cycle.

One might argue that any failure to identify commodities among ecosystem services is a failure of imagination. After all, the gas regulating services of nature are not themselves amenable to commodification; pollution permits are an indirect but effective "proxy" commodity for pricing the pressure on ecosystems. However, as illustrated by the list compiled by Landell-Mills and Porras (2002) that was introduced earlier, it is not easy to identify proxies that are truly commodities. Moreover, even when ecosystem services can plausibly be treated as commodities (individual agricultural and forest products and mineral resources), they are invariably coupled with others that cannot, which means the ecosystem itself cannot be properly valued by markets.

2.3 Securitization, financialization and the environment

The final step in the financialization process is to generate tradable rights to an income stream derived from a commodifiable unit of natural capital or ecosystem service.

In some cases, rights to natural capital or ecosystem services permit firms to produce goods. For example, operating a paper mill requires rights to timber, which the mill's owners ensure either by owning forest land or by buying timber futures contracts for later delivery. To take another example, an emissions permit allows a firm to operate its marginally higher-emitting equipment in order to produce goods and services, while it rewards the seller of the permit for low-emissions operations. Once goods and services are sold, firms receive income from the sale, so holding the rights to the ecosystem service is valuable.

The connection between the ecosystem service and the financial asset need not be so direct. For example, a publicly-traded firm may seek to expand its market and boost its stock price by appealing to green consumers and investors. By establishing a reputation for good stewardship, a firm can secure a segment of the market in which it operates. That can reassure conventional investors of a reliable, if narrow, profit stream, and convince green investors to purchase stock. Another indirect instrument is certified emissions reductions (in contrast to tradable emissions permits). In this case, a firm pays someone else to reduce emissions so the firm can operate its marginally higher-emitting equipment. As we discuss below, in each of these cases the abstraction inherent in commodification and securitization makes these approaches particularly problematic.

In this paper we have used "securitization" to mean any process for creating a security – that is, a tradeable financial asset. In a narrow sense, securitization is used to mean the bundling together of illiquid financial assets (ones that are hard to buy and sell, such as a mortgage) as collateral on liquid securities (ones that are easy to buy and sell, such as a mortgage-backed security or collateralized debt obligation). To the extent that the values of the bundled assets move independently of one another, the risk of a loss from the diversified bundle is lower than the risk of a loss from any one of the underlying assets. The expectation of acceptable risk, combined with greater liquidity, invites a broad range of investors to finance the underlying assets (Schwarcz, 1994).

It is theoretically possible for a payment for ecosystem services (PES) scheme to back a securitized instrument, because it includes a flow of monetary payments.[3] Such schemes are entirely location-specific (Jack et al., 2008), and so do not naturally lend themselves to commodification. However, the community providing the ecosystem services could sell rights to a portion of the payments received under the scheme. If the community has access to credit – not at all certain in many countries

[3] A classic example of PES is when a comparatively wealthy city is downstream from a relatively poor rural area. The economic incentives for the upstream rural communities may be to conserve their vegetation and topsoil, but it may also be to deforest and grow crops, resulting in larger variation in river flow and greater pollutant run-off and silt loads. To maintain the quality and reliability of their water supply, the city can contract to pay the upstream communities to maintain vegetation cover. It is not straightforward to implement and sustain such schemes, and it is not always clear what counts as PES (Schomers and Matzdorf, 2013). Nevertheless, under some conditions they can be a useful tool in the local policy toolkit (Salzman, 2005).

– it could use the payment stream as security on a loan, either to get the scheme itself started or for some other purpose. The promise of payment from the downstream users would be collateral on the loan. The risks to the bank issuing the loan would be that community providing the service failed to uphold its conservation responsibilities or the downstream users refused to pay. The bank could assess the severity of the risk if it was familiar with the parties involved. The resulting loan would be illiquid, but the bank could securitize it by bundling it with similar loans.

We could not find any examples of this practice. Securitization presumes that the loans are offered, which already presents a substantial barrier. Access to credit is often limited for the communities where PES is practiced, and there is very limited evidence that PES schemes improve access. In a study of farmers in Kenya, participation in extension and ecosystem services (EES) was associated with only a small and statistically insignificant reduction in interest rates (Benjamin et al., 2016). The large bulk of the extensive literature on PES does not address this question at all. Given the rarity and potential difficulties of securitizing PES schemes, we note its possibility but do not consider it further.

3. Forms of environmental financialization and their problems

From the preceding discussion, we can identify four characteristic ways in which ecosystems and other forms of natural capital, or ecosystem services, can back securities traded in financial markets:

1. The ecosystem service lends itself to commodification (e.g., agricultural, mineral or forest products);
2. A publicly traded firm establishes a reputation for treating ecosystems well;
3. A tradable permit to a share of the pressure placed on an ecosystem is established (e.g., a pollution permit);
4. A certified credit is issued for reducing pressure on ecosystems (e.g., through the Clean Development Mechanism or biodiversity credits)

Our central claim is that each of these mechanisms suffers to a greater or lesser degree from the cascade of abstractions in the production of natural financial assets, particularly in the process of commodification and securitization. A financial asset that is backed by natural capital necessarily reduces the multifaceted processes of viable ecosystems to a few of their services. In this way, ecosystem services obscure ecosystem function (Peterson et al., 2010).

3.1 Traditional natural resource commodities and ecosystems

For about 10,000 years, people have transformed plants and ecosystems to produce specific crops in abundance (Evans, 1996), and for thousands of years people have substantially deforested and altered temperate woodlands (Ellis, 2011). Markets for commodities are also ancient, established in some form several thousand years ago in China and Mesopotamia.

Production and exchange of agricultural commodities was not intended to protect ecosystem function, but it was certainly in the interest of agricultural societies to do so. And, to some degree, the value placed on agricultural commodities did encourage conservation. While managed forests and cropland are very different from the natural ecosystems they replace, they can persist (with significant human intervention) as ecosystems producing a stream of income that is worth protecting. Nevertheless, that protection is not absolute. Deforestation and erosion arguably contributed to the collapse of classical Mediterranean civilization (Hughes and Thirgood, 1982). More generally, agricultural expansion and deforestation both supported civilization and made it vulnerable to collapse when the climate changed (Tainter 1988; Wilkinson, 1997; Weiss and Bradley, 2001). When ancient civilizations modified natural environments for the optimal production of commodities, it compromised other ecosystem function.

These ancient patterns persist, and land-use pressures are growing in scale and scope (Foley et al. 2005). Highly altered and potentially vulnerable managed ecosystems are protected because they provide a stream of value; but as ecosystems are reoriented towards optimal production of specific commodities, they become vulnerable to stresses that they might have survived in the past. In the short run, this may not be evident. Intensively managed tree plantations are less susceptible than natural forest to pests, diseases and physical disturbance. Exotic species perform best, because they are less vulnerable than native plants to pests and diseases (Gadgil and Bain, 1999). The same could be said of intensively managed croplands planted to non-native monocultures. Indeed, reliable and high yields are the main reason to grow intensively-managed single-species crops. However, in the longer run and on larger scales, ecosystems populated by native species and with high functional diversity (Walker 1995) are more resilient and adaptive (Folke et al., 2004; Fischer et al., 2006; Thompson et al., 2009).

Moreover, even the relatively compromised function of agricultural ecosystems is not ensured by agricultural markets. As urban areas expand, agricultural land is displaced (Azadi et al., 2011), despite the existence of markets for agricultural products. This is also true of environmental markets. If carbon can be sequestered at lower cost on plantations than in natural forest, a carbon market may compete with plans to increase biodiversity (Hunt, 2008). Demand for bioethanol in the US displaces land for crops, which are then likely to be produced elsewhere, resulting in further land conversion and indirect carbon emissions (Chakravorty et al., 2014).

While these problems do not arise from financial transactions *per se* (in this case, in commodity exchanges), they arise from commodification, which is a necessary precursor to creating a financial security. Thus, in the case of agricultural commodities, it is not the way trading is done that is creating incentives to maximize short- and medium-term productivity, but rather the incentives to maximize productivity that shape the way trading is done.

3.2 Buying shares in ecologically-friendly firms

It is difficult to imagine a future in which commerce does not play a major role. Indeed, it is likely impossible to support the current global population without a connected and technologically advanced global economy, although it could conceivably differ significantly from our present one. If there is to be commerce, then it is better that it be ecosystem-friendly, and if we value nature, then let us show it through our investments. That, in a nutshell, is the argument behind buying shares in ecosystem-friendly firms as a way to place a value on ecosystem services.

Equity shares in companies are commodities distinguished by their historical and anticipated yield, as well as their risk profile. Socially-responsible investing (SRI) is no different in principle, but SRI funds restrict their equity shares to firms that meet certain social and environmental standards. They may passively monitor firm performance, aiming to stay alert to "green-washing" or changes in firm culture. Alternatively, they may actively use their influence as shareholders to encourage firms in their portfolio to maintain or improve adherence to standards. The extent to which they succeed relies on their motivation and capacity to faithfully monitor the firm's impacts across a potentially broad array of domains, and make financial decisions depending on the results.

The record to date on SRI is mixed. There is evidence that SRI has grown to the point that it is influencing corporate decision-making (Sparkes and Cowton, 2004), although early enthusiasm for "triple win" investment strategies has faded. Recognizing that there really are tradeoffs between protecting the environment and firm profits in today's markets, SRI is increasingly dependent on the prospect of future regulation to make a business case to firms (Richardson, 2009; Harmes, 2011). Moreover, SRI is more attractive to private and institutional investors than it is to fund managers, who have a fiduciary duty to the effective management of the funds entrusted to them. Fund managers are more focused on delivering an acceptable return at minimal risk than they are to social and environmental goals (Jansson and Biel, 2011).

As a commodification strategy, SRI works to the extent that a large number of assets passes through the investor's filter. Finer and more specialized filters are costlier to monitor, so investors usually rely on third-party organizations, such as the Global Reporting Initiative and Carbon Disclosure Project, which collect information from firms and present the results in a standard format for investors, civil society actors, and other interested parties.

Porter (1996) uses the phrase "mechanical objectivity" to describe the practice of impersonal comparison through rules-based assessment, particularly using standardized quantitative indicators or checklists. When applied to the multiple and complex social benefits of environmental systems, mechanical objectivity is problematic, because local and context-specific information, including unarticulated tacit knowledge, is essential for making value judgements. Establishing incentives on the basis of a particular set of indicators, however large that set may be, tends to shape the underlying system in the image of the indicators (Scott 1998). Hence, the business adage, "You get what you measure." When applied to ecosystems, rule-based systems do not fully capture the underlying ecosystem services, given the complexity of socio-ecological systems and the institutions tasked with managing them (Robertson, 2006).

One could counter that what is important in the process of financialization is that it creates a sufficient incentive for effective protection of the ecosystem service. We offer two objections to this position. First, valuing a particular ecosystem service means treating it as a commodity, which returns us to the first type of environmental financialization. It is possible that a system can be created that sustainably maximizes the production of a commodified ecosystem service. But that is not what an ecologically-friendly firm is promising. Rather, they promise to protect additional, non-commodifiable functions of a socio-ecological system. That brings us to the second point. Without knowing what might compromise the system, it is not possible, even in theory (Lipsey and Lancaster, 1956), to construct a "second-best" solution that accurately captures the value of protecting ecosystem function.

The commodity corresponding to an equity share in an ecologically-friendly firm is thus a share in a firm that has passed a filter for ecological friendliness. While superficially identical, these are different things. The filter is an abstraction that pools firms with quite different corporate cultures and approaches to ecosystem management operating in substantially different environments. While SRI represents a broader conception of nature and society than investment in traditional natural resource commodities, and is therefore worth pursuing, its effectiveness is limited. It cannot ensure that firms take account of the specificity, richness, and function of any particular ecosystem. Rather, it encourages better conservation practice in already-managed ecosystems.

3.3 Constraining pressure through tradable permits

Emissions permits are proxies for ecosystem services. They aim to preserve a range of ecosystem services by constraining the ability to impose pressure on ecosystems under an overall cap. Constraining pressure values leaving the ecosystem alone, without regard to the details of how the ecosystem functions. In contrast to the concept of "non-use value", this might be thought of as "restrained use value". Unlike non-use value, restrained use has a well-defined interpretation in marginal pricing theory: the price of a permit to place a pressure should equal the shadow price associated with loosening or tightening the limit on that pressure by a small amount. While shadow prices are theoretical constructs with no direct bearing on economic decision-making, they are made manifest when the permit to loosen the constraint becomes a tradeable asset.

Perhaps surprisingly, the theory of emissions permit pricing follows a different argument, while applying marginalist principles (Tol, 2011). In that theory, cost-benefit analysis reveals the social value of the viability of the ecosystem, and the social cost of an additional amount of pressure. We should therefore, as a society, be willing to pay for a reduction in the pressure in an amount equal to the marginal social cost. However, quite aside from the daunting and perhaps prohibitive conceptual and practical difficulties of calculating marginal social cost (Ackerman and Stanton, 2015, p. 29), society as a whole does not pay for the reduction; business do, and they typically pass on the cost through their markup. While welfare economics says that the marginal social cost should equal the shadow price encountered by firms, as we discussed earlier, the theoretical basis for that belief is weak.

The motivation for using cost-benefit analysis and social cost is that it purports to answer the question "How much is enough?" But there are other bases for that answer. Global negotiations on climate mitigation under the auspices of the United Nations Framework Convention on Climate Change (UNFCCC) have adopted a number of trading mechanisms (Schneider et al., 2016), and have also adopted an agreed goal of keeping warming below either 1.5°C or 2.0°C above pre-industrial levels. A global temperature rise can be interpreted as an emissions constraint, and converted into global emissions trajectories, which can then be apportioned between emitters. If emissions permit trading is efficient, then different firms' emissions should reach the point that they all face the same marginal cost of reducing emissions. This principle can be generalized: a limit is set through a biophysical determination of the level of ecosystem pressure consistent with social goals, and then a market in permits to apply pressure within the limit is implemented.

Permits to apply pressure can be converted into financial assets if the source of pressure can be commodified. That is, one firm's or household's pressure on the environment must be directly exchangeable with any other's. That is the case for emissions of any given greenhouse and ozone-depleting gases. These are globally dispersed gases linked to ecosystem damage through reasonably well-understood causal chains. However, value-laden questions arise when considering gases that have differential impact over time (e.g., relating to intergenerational justice), or distance (distributional fairness). Further problems arise with efforts to aggregate across different gases that act across different timescales (Shine, 2009; Neubauer and Megonigal, 2015). Perhaps most problematic is the fact that different measures for reducing pollution, even when they are equivalent in terms of the commodified unit of pressure reduction (e.g., tonnes of carbon dioxide), may have very different ecosystem impacts. For example, switching from a more carbon-intensive fuel to a less carbon intensive one (e.g., from coal to gas) based on an existing, widely disseminated conventional energy technology may directly provide the same quantity of pressure reduction as deploying a certain amount of a new, emerging, zero-carbon, renewable energy technology, but the latter may also provide learning-by-doing benefits, help achieve greater economies of scale, shift perceptions of technological riskiness, weaken socio-institutional carbon lock-in, and otherwise help induce further innovation, deployment, and thus emission reductions.

In addition to these challenges, permit trading schemes have been notoriously vulnerable to political interference on the part of vested interests, manifested in the form of set-asides, free allocations, exceptions, and other consequences of rent-seeking behavior (Markussen and Svendsen, 2005; Spash, 2010).

3.4 Reducing pressure with credits and offsets

In practice, one way in which firms and other entities are allowed to meet their obligations with regard to imposing environmental pressure is to pay someone else to do it on their behalf. In contrast to markets in permits, which convey a permit to apply a pressure by an entity that is under an overall cap, markets in credits enable an entity not under the cap that has achieved a reduction in pressure to sell that reduction to an entity under the cap. As with permit trading, the economic principle is that, through trade, mitigation targets can be met through the least cost means.

On its face, this seems like a modest extension of the idea of trading permits, and indeed it faces the same challenges outlined above. Yet, it also faces additional constraints. To take a highly relevant and concrete example, the Clean Development Mechanism (CDM) allowed countries under the cap (the Annex I countries) to meet their obligations in part by purchasing credits from parties outside the cap (the non-Annex I countries). Under the Paris agreements, a similar mechanism may evolve for trading "internationally transferred mitigation outcomes" (ITMOs). Recognizing the danger of generating mitigation credits at the expense of local development, both CDM projects and ITMOs are required to achieve goals other than mitigation, such as promoting sustainable development. However, it is only the mitigation outcome that is valued and commodified, and experience with the CDM has shown the hoped-for sustainable development benefits do not materialize (Olsen, 2007; Schneider, 2007; Sutter and Parreño, 2007).

The situation is similar to that of an SRI fund. If the actual mitigation outcome can be verified and quantified, then it can be made into a tradable commodity. However, to meet additional goals, such as contributing to sustainable development, achieving transparency, or ensuring environmental integrity, then it must also pass a filter. Applying the filter raises the problems of rule-based "mechanical objectivity" that plague SRI funds; the situations on the ground are too disparate to be effectively protected by creating tradable credits that have passed through a filter.

These challenges are exacerbated when the credits are derived from ecosystem-based projects, which entails a complex and location-specific ecosystem being reduced to a single ecosystem service. When firms trade pollution permits, they are exchanging reasonably fungible commodities, but that does not hold for ecosystem-based credits. This can be seen in the case of biodiversity offset banking, in which biodiversity credits are bought and sold (McKenney and Kiesecker, 2010; Bull et al., 2013), but implementation is problematic because conditions are location-specific and "biodiversity" does not have a uniform measure (Burgin, 2008). The same can be said of the wetlands example given by (Driesen, 2005) and discussed above.

A prominent example of ecosystem-based climate mitigation is actions for reducing emissions from deforestation and forest degradation (REDD), combined with sustainable management of forests, conservation of forest carbon stocks and enhancement of forest carbon stocks (REDD+). The REDD+ mechanism was explicitly included at the Cancún climate conference and has been widely promoted as a way to achieve emissions reductions while contributing to sustainable development in low-income countries. However, early enthusiasm gave way to more modest expectations as experience accumulated. Economic analysis has focused on capturing opportunity costs that might hinder adoption (Lubowski and Rose, 2013; Irawan et al., 2013), but even at the time of the Cancún conference, there were known problems of political economy in reforestation projects. For the Asia-Pacific region, Barr and Sayer (2012) noted that previous reforestation projects had displaced local communities, exacerbated inequalities, and accelerated biodiversity loss, among other perverse outcomes. As anticipated, issues of political economy (among others) have emerged in REDD+ projects (Corbera and Schroeder, 2017).

Credit mechanisms raise further difficulties. Credits are meant to measure reductions additional to any that would have taken place otherwise. However, it has long been recognized that ascertaining this "additionality" is epistemically fraught (Grubb et al., 1997). Certifying that reductions as measurable, reportable, and verifiable (MRV) is impossible if the world against which those reductions are measured is an unknowable counterfactual. If the uncertainty were minor and the errors random, the difficulties would be manageable. However, this is not the case, and there is large scope for overestimating reductions (Bernow et al., 2001). This expectation appears to be borne out (Schneider, 2009; Schiermeier, 2011).

The market in credits is structurally biased, as both the buyer and seller have an economic incentive to define their credits relative to as generous a counterfactual as possible, while certifying entities have an incentive to retain project developers as satisfied clients. Such a transaction is thus quite different from the normal market dynamic. In the most perverse cases, where revenues from credits significantly exceed mitigation costs, sellers have an incentive to *increase* emissions to reap the rewards of partial abatement, and this has been observed in practice (Schneider and Kollmuss, 2015).

In effect, the generation of credits results in the loosening of the cap, since entities under the cap that purchase the credits are free to emit more than the cap would otherwise allow. If all credits were indeed additional, the effect of the credit mechanism on environmental outcomes would be neutral. To the extent that non-additional credits are certified, environmental outcomes are worsened by the existence of the credit mechanism.

4. Discussion: the non-magical market

There are, broadly, two arguments for financializing nature. The first is a widely-held belief, inspired by neoclassical theory, that markets are the most efficient way to allocate scarce resources. However, the conditions needed for the neoclassical argument to hold are never met in practice and the information requirements for achieving a "second-best" outcome are highly unlikely to be met in the case of natural capital. The second is that because public finance is on the wane, we must turn to private finance to meet social and environmental goals (e.g., Rubino, 2000; Jenkins et al., 2004).

These arguments are connected. Aggressive promotion of market mechanisms as an all-purpose tool led ultimately to political acceptance (Burgin, 2012). Ronald Reagan memorably captured the spirit of this belief as the "magic of the market". This political reorientation has contributed, over time, to the erosion of non-market policy instruments and declining public funding for the environment. These trends powerfully motivate a search for ways to protect the environment through financial markets. Yet, while we support the sentiment, we believe the conclusion to be misguided.

We have argued that trading ecosystem-backed securities on financial markets will very likely undermine ecosystem function. The most important exception may be those few cases in which a tradable permit to broadly apply a pressure to multiple ecosystems can be established, for which a biophysical assessment can determine the total allowable pressure consistent with social and environmental goals, for which neglected pressures on ecosystem services and other socio-economic impacts are not, in fact, dominant, and where political economic factors do not allow rent-seeking behavior to undermine environmental integrity. This is an extremely narrow set of criteria that may leave only tradable emissions permits for widely-dispersed gases, though this has yet to be demonstrated. Two partial exceptions also apply. First, it is conceivable, although likely difficult, to create locally traded securities backed by payments for ecosystem services (PES). Second, if an ecosystem is already being used for production, then socially-responsible investing (SRI) might help by favoring firms that exhibit greater social and environmental responsibility.

This critique is relevant to climate finance, as well as development aid generally, where there is significant interest in garnering private sector investment to produce ecosystem benefits. The bulk of private adaptation financing hopes to produce ecosystem "goods". In contrast, our arguments suggest that the focus should rather be on avoided harm. For most conservation, mitigation, and adaptation challenges, local analysis and local solutions are needed, in which a broad array of policy approaches is put on the table. Some policy options might involve locally specific financial arrangements, such as PES, but many, such as establishing a protected area, will not. This is not to say that protected areas are always and everywhere preferable to PES. Indeed, the point is that given the complex reality of ecosystems and societies a solution that works in one place may very well fail in another. Nor do we mean to imply that problems of political economy vanish when decisions are made locally. In fact, they are present in any case. The difference is that there is not also a global market attracting relatively powerful actors to the detriment of less powerful ones.

Our critique is specific to financial markets. Valuation is, in itself, neutral. In principle it can be used to convey important insights, such as that the value of ecosystem services exceeds global GNP (Costanza et al., 1997) or to demonstrate the extent to which we are losing ecosystem services by degrading natural capital (Costanza et al., 2014). Yet, once nature is put in monetary terms, and the values are shown to be very large, there is an understandable tendency on the part of financial actors to try to capture some part of that stream of value. Even though the underlying natural assets are not commodities, financial actors can propose financial instruments that put an explicit monetary value on a subset of ecosystem services and treat them as commodities. The predictable result is that those services are amplified, while broader ecosystem function is impaired.

This leads us to the following recommendations. First, reserve valuation for high-level numbers, not for explicit policy guidance – at fine resolution, valuation can hide more than it reveals. Second, apply filters and mechanical objectivity (as in SRI) to influence economic activity only in cases where ecosystems are already under pressure due to the specific activities of firms, governments, or other entities. Third, restrict any new instruments of environmental financialization to the narrow case where it is possible to trade a permit to place non-specific pressure on ecosystems, unrelated to the specific activities of a particular entity.

5. Conclusion

Ecosystems are under pressure from economic activity. This has led to efforts to place a monetary value on nature (valuation) in order to highlight what is being lost when ecosystems are degraded. Valuation, in turn, arguably encourages the creation of environmentally-based commodities and tradable financial assets that reflect the value of ecosystem services. Yet, this is rarely appropriate. Nearly all ecosystem services are locally-specific and so non-fungible. Creating a tradable financial asset encourages the amplification of a narrow set of ecosystem services at the expense of broader ecosystem function.

We identify one case in which a new financial instrument might help protect ecosystems. That is when a tradable permit to place non-specific pressure on ecosystems can be established, where the pressure is not due to the direct operations of the firm, government, or other entity. The only examples we could identify are tradable permits in emissions of widely-dispersed gases, such as sulfur oxides (SO_x) and carbon dioxide (CO_2). Tradable biodiversity or climate mitigation credits do not fit under this heading. Socially responsible investment (SRI) can help to reduce harm in cases where ecosystems are already being exploited.

Valuation has a useful role to play in highlighting the importance of ecosystems in contexts where monetary values carry substantial weight. However, those presentations must emphasize that the streams of value are diffuse and in nearly all cases cannot be appropriated. When they are appropriated and sold, as with natural resource extraction, there is an incentive to maximize the provision of income-generating services at the expense of broader ecosystem function.

References

Ackerman, F. and Stanton, E. A., eds. (2015) *Climate Economics: The State of the Art*. New York: Routledge.

Apostolopoulou, E. and Adams, W. M. (2017) "Biodiversity offsetting and conservation: reframing nature to save it." *Oryx*, 51(1), pp. 23–31. DOI:10.1017/S0030605315000782.

Azadi, H., Ho, P. and Hasfiati, L. (2011) "Agricultural land conversion drivers: A comparison between less developed, developing and developed countries." *Land Degradation & Development*, 22(6), pp. 596–604. DOI:10.1002/ldr.1037.

Barr, C. M. and Sayer, J. A. (2012) "The political economy of reforestation and forest restoration in Asia–Pacific: Critical issues for REDD+." *Biological Conservation*, 154, pp. 9–19. DOI:10.1016/j.biocon.2012.03.020.

Benjamin, E. O., Blum, M. and Punt, M. (2016) "The impact of extension and ecosystem services on smallholder's credit constraint." *The Journal of Developing Areas*, 50(1), pp. 333–50. DOI:10.1353/jda.2016.0020.

Bernow, S., Kartha, S., Lazarus, M. and Page, T. (2001) "Cleaner generation, free-riders, and environmental integrity: clean development mechanism and the power sector." *Climate Policy*, 1(2), pp. 229–49. DOI:10.1016/S1469-3062(01)00006-7.

Bull, J. W., Suttle, K. B., Gordon, A., Singh, N. J. and Milner-Gulland, E. J. (2013) "Biodiversity offsets in theory and practice." *Oryx*, 47(3), pp. 369–80. DOI:10.1017/S003060531200172X.

Burgin, A. (2012) *The Great Persuasion: Reinventing Free Markets since the Depression*. Cambridge, MA: Harvard University Press.

Burgin, S. (2008) "BioBanking: an environmental scientist's view of the role of biodiversity banking offsets in conservation." *Biodiversity and Conservation*, 17(4), pp. 807–16. DOI:10.1007/s10531-008-9319-2.

Calvert, J. (2007) "Patenting genomic objects: genes, genomes, function and information." *Science as Culture*, 16(2), pp. 207–23. DOI:10.1080/09505430701387953.

Carbon Pricing Leadership Coalition (2016) Carbon Pricing Leadership Coalition. https://www.carbonpricingleadership.org/.

Chakravorty, U., Hubert, M.-H. and Moreaux, M. (2014) "Land allocation between food and energy." *Frontiers of Economics in China*, 9(1), pp. 52–69. DOI:10.3868/s060-003-014-0004-8.

Chichilnisky, G. and Heal, G. (2000) "Securitizing the Biosphere." In *Environmental Markets: Equity and Efficiency*. Columbia University Press.

Chichilnisky, G. and Heal, G. M. (1996) *Catastrophe Futures: Financial Markets and Changing Climate Risks*. ID 1625. Social Science Research Network, Rochester, NY. https://papers.ssrn.com/abstract=1625. SSRN Scholarly Paper.

Corbera, E. and Schroeder, H. (2017) "REDD+ crossroads post Paris: politics, lessons and interplays." *Forests*, 8(12), p. 508. DOI:10.3390/f8120508.

Costanza, R., d'Arge, R., Groot, R. de, Farber, S., Grasso, M., et al. (1997) "The value of the world's ecosystem services and natural capital." *Nature*, 387(6630), p. 253. DOI:10.1038/387253a0.

Costanza, R. and Daly, H. E. (1992) "Natural capital and sustainable development." *Conservation Biology*, 6(1), pp. 37–46. http://www.jstor.org/stable/2385849.

Costanza, R., de Groot, R., Sutton, P., van der Ploeg, S., Anderson, S. J., Kubiszewski, I., Farber, S. and Turner, R. K. (2014) "Changes in the global value of ecosystem services." *Global Environmental Change*, 26(Supplement C), pp. 152–58. DOI:10.1016/j.gloenvcha.2014.04.002.

Daily, G. C., Polasky, S., Goldstein, J., Kareiva, P. M., Mooney, H. A., Pejchar, L., Ricketts, T. H., Salzman, J. and Shallenberger, R. (2009) "Ecosystem services in decision making: time to deliver." *Frontiers in Ecology and the Environment*, 7(1), pp. 21–28. DOI:10.1890/080025.

de Groot, R. S., Alkemade, R., Braat, L., Hein, L. and Willemen, L. (2010) "Challenges in integrating the concept of ecosystem services and values in landscape planning, management and decision making." *Ecological Complexity*, 7(3), pp. 260–72. DOI:10.1016/j.ecocom.2009.10.006.

Driesen, D. M. (2005) "Trading and Its Limits." *Penn State Environmental Law Review*, 14, pp. 169–76.

Ellis, E. C. (2011) "Anthropogenic transformation of the terrestrial biosphere." *Philosophical Transactions of the Royal Society A: Mathematical, Physical and Engineering Sciences*, 369(1938). 1010–35. DOI:10.1098/rsta.2010.0331.

Evans, L. T. (1996). *Crop Evolution, Adaptation and Yield.* Cambridge University Press.

Evans, N. M. (2018) "Ecosystem services: on idealization and understanding complexity." *Ecological Economics.* DOI:10.1016/j.ecolecon.2018.10.014.

Farber, S. C., Costanza, R. and Wilson, M. A. (2002) "Economic and ecological concepts for valuing ecosystem services." *Ecological Economics*, 41(3), pp. 375–92. DOI:10.1016/S0921-8009(02)00088-5.

Fischer, J., Lindenmayer, D. B. and Manning, A. D. (2006) "Biodiversity, ecosystem function, and resilience: ten guiding principles for commodity production landscapes." *Frontiers in*

Ecology and the Environment, 4(2), pp. 80–86. DOI:10.1890/1540-9295(2006)004[0080:BEFART]2.0.CO;2.

Foley, J. A., DeFries, R., Asner, G. P., Barford, C., Bonan, G., et al. (2005) "Global consequences of land use." *Science*, 309(5734), pp. 570–74. DOI:10.1126/science.1111772.

Folke, C., Carpenter, S., Walker, B., Scheffer, M., Elmqvist, T., Gunderson, L. and Holling, C. S. (2004) "Regime shifts, resilience, and biodiversity in ecosystem management." *Annual Review of Ecology, Evolution, and Systematics*, 35(1), pp. 557–81. DOI:10.1146/annurev.ecolsys.35.021103.105711.

Food & Water Watch (2012). *Don't Bet on Wall Street: The Financialization of Nature and the Risk to Our Common Resources*. Food & Water Watch, Washington, DC, US. Fact Sheet.

Gadgil, P. D. and Bain, J. (1999). "Vulnerability of planted forests to biotic and abiotic disturbances." *New Forests*, 17(1–3), pp. 227–38. DOI:10.1023/A:1006546917313.

Gómez-Baggethun, E., de Groot, R., Lomas, P. L. and Montes, C. (2010) "The history of ecosystem services in economic theory and practice: From early notions to markets and payment schemes." *Ecological Economics*, 69(6), pp. 1209–18. DOI:10.1016/j.ecolecon.2009.11.007.

Gowdy, J. and Erickson, J. D. (2005) "The approach of ecological economics." *Cambridge Journal of Economics*, 29(2), pp. 207–22. DOI:10.1093/cje/bei033.

Grubb, M., Vrolijk, C. and Brack, D. (1997) *The Kyoto Protocol: A Guide and Assessment*. Royal Institute of International Affairs Energy and Environmental Programme, London. http://discovery.ucl.ac.uk/1503013/.

Harmes, A. (2011) "The limits of carbon disclosure: Theorizing the business case for investor environmentalism." *Global Environmental Politics*, 11(2). 98–119. DOI:10.1162/GLEP_a_00057.

Hughes, J. D. and Thirgood, J. V. (1982). Deforestation, erosion, and forest management in ancient Greece and Rome. *Journal of Forest History*, 26(2), pp. 60–75. DOI:10.2307/4004530.

Hunt, C. (2008) "Economy and ecology of emerging markets and credits for bio-sequestered carbon on private land in tropical Australia." *Ecological Economics*, 66(2), pp. 309–18. DOI:10.1016/j.ecolecon.2007.09.012.

Irawan, S., Tacconi, L. and Ring, I. (2013) "Stakeholders' incentives for land-use change and REDD+: The case of Indonesia." *Ecological Economics*, 87, pp. 75–83. DOI:10.1016/j.ecolecon.2012.12.018.

Jack, B. K., Kousky, C. and Sims, K. R. E. (2008) "Designing payments for ecosystem services: Lessons from previous experience with incentive-based mechanisms." *Proceedings of the National Academy of Sciences*, 105(28), pp. 9465–70. DOI:10.1073/pnas.0705503104.

Jansson, M. and Biel, A. (2011) "Motives to engage in sustainable investment: A comparison between institutional and private investors." *Sustainable Development*, 19(2), pp. 135–42. DOI:10.1002/sd.512.

Jenkins, M., Scherr, S. J. and Inbar, M. (2004) "Markets for biodiversity services: potential roles and challenges." *Environment: Science and Policy for Sustainable Development*, 46(6), pp. 32–42. DOI:10.1080/00139157.2004.10545160.

Jensen, K. and Murray, F. (2005) "Intellectual property landscape of the human genome." *Science*, 310(5746), pp. 239–40. DOI:10.1126/science.1120014.

Kill, J. (2015) *Financialization of Nature: Creating a New Definition of Nature*. Friends of the Earth International, Amsterdam, the Netherlands. Brochure.

Krupp, F. (2008) "The Making of a Market-Minded Environmentalist." *strategy+business*(51). https://www.strategy-business.com/article/08201?gko=97ea9.

Landell-Mills, N. and Porras, I. T. (2002) *Silver Bullet or Fools' Gold?: A Global Review of Markets for Forest Environmental Services and Their Impact on the Poor*. London: IIED.

Lipsey, R. G. and Lancaster, K. (1956) "The general theory of second best." *The Review of Economic Studies*, 24(1), pp. 11–32. DOI:10.2307/2296233.

Lubowski, R. N. and Rose, S. K. (2013) "The Potential for REDD+: Key Economic Modeling Insights and Issues." *Review of Environmental Economics and Policy*, 7(1), pp. 67–90. DOI:10.1093/reep/res024.

MA (2005). *Millennium Ecosystem Assessment, 2005. Ecosystems and Human Well-Being: Synthesis.* Washington, DC: Island Press.

Markussen, P. and Svendsen, G. T. (2005) "Industry lobbying and the political economy of GHG trade in the European Union." *Energy Policy*, 33(2), pp. 245–55. DOI:10.1016/S0301-4215(03)00238-6.

McKenney, B. A. and Kiesecker, J. M. (2010). Policy development for biodiversity offsets: a review of offset frameworks. *Environmental Management*, 45(1). 165–76. DOI:10.1007/s00267-009-9396-3.

Neubauer, S. C. and Megonigal, J. P. (2015) "Moving Beyond Global Warming Potentials to Quantify the Climatic Role of Ecosystems." *Ecosystems*, 18(6), pp. 1000–1013. DOI:10.1007/s10021-015-9879-4.
Olsen, K. H. (2007) "The clean development mechanism's contribution to sustainable development: a review of the literature." *Climatic Change*, 84(1), pp. 59–73.
DOI:10.1007/s10584-007-9267-y.

Pearce, D. W., Markandya, A. and Barbier, E. (1989) *Blueprint for a Green Economy*. Earthscan.

Peterson, M. J., Hall, D. M., Feldpausch-Parker, A. M. and Peterson, T. R. (2010) "Obscuring ecosystem function with application of the ecosystem services concept." *Conservation Biology*, 24(1), pp. 113–19. DOI:10.1111/j.1523-1739.2009.01305.x.

Polanyi, K. (2001) *The Great Transformation: The Political and Economic Origins of Our Time*. Beacon Press.

Porter, T. M. (1996) *Trust in Numbers: The Pursuit of Objectivity in Science and Public Life*. Princeton University Press.

Potschin, M. B., Primmer, E., Furman, E. and Haines-Young, R. H. (2016) "Have ecosystem services been oversold? A response to Silvertown." *Trends in Ecology & Evolution*, 31(5), pp. 334–35. DOI:10.1016/j.tree.2016.03.008.

Redford, K. H. and Adams, W. M. (2009) "Payment for ecosystem services and the challenge of saving nature." *Conservation Biology*, 23(4), pp. 785–87. DOI:10.1111/j.1523-1739.2009.01271.x.

Richardson, B. J. (2009) "Climate finance and its governance: Moving to a low carbon economy through socially responsible financing." *The International and Comparative Law Quarterly*, 58(3), pp. 597–626. http://www.jstor.org/stable/25622228.

Robertson, M. M. (2006) "The Nature That Capital Can See: Science, State, and Market in the Commodification of Ecosystem Services." *Environment and Planning D: Society and Space*, 24(3), pp. 367–87. DOI:10.1068/d3304.

Rubino, M. C. (2000) "Biodiversity finance." *International Affairs*, 76(2), pp. 223–40. DOI:10.1111/1468-2346.00131.

Salzman, J. (2005) "Creating markets for ecosystem services: notes from the field." *New York University Law Review*, 80, pp. 870–961.

Schiermeier, Q. (2011) "Clean-energy credits tarnished." *Nature*, 477(7366), pp. 517–18. DOI:10.1038/477517a.

Schneider, L. (2007) *Is the CDM Fulfilling Its Environmental and Sustainable Development Objec- Tives? An Evaluation of the CDM and Options for Improvement*. Oko Institut, Berlin, Germany.

Schneider, L. (2009) "Assessing the additionality of CDM projects: practical experiences and lessons learned." *Climate Policy*, 9(3), pp. 242–54. DOI:10.3763/cpol.2008.0533.

Schneider, L., Broekhoff, D., Cames, M., Healy, S., Fuessler, J. and La Hoz Theuer, S. (2016) *Market Mechanisms in the Paris Agreement – Differences and Commonalities with Kyoto Mechanisms*. German Environment Agency (UBA), Berlin. https://www.dehst.de/SharedDocs/Downloads/EN/JI-CDM/Differences_and_commonalities_paris_agreement_discussion_paper.html. Discussion Paper.

Schneider, L. and Kollmuss, A. (2015) "Perverse effects of carbon markets on HFC-23 and SF_6 abatement projects in Russia." *Nature Climate Change*, 5(12), pp. 1061–63. DOI:10.1038/nclimate2772.

Schomers, S. and Matzdorf, B. (2013) "Payments for ecosystem services: A review and comparison of developing and industrialized countries." *Ecosystem Services*, 6(Supplement C), pp. 16–30. DOI:10.1016/j.ecoser.2013.01.002.

Schröter, M. and van Oudenhoven, A. P. E. (2016) "Ecosystem services go beyond money and markets: reply to Silvertown." *Trends in Ecology & Evolution*, 31(5), pp. 333–34. DOI:10.1016/j.tree.2016.03.001.

Schumacher, E. F. (1973) *Small Is Beautiful; Economics as If People Mattered*. New York: Harper & Row.

Schwarcz, S. L. (1994) "The alchemy of asset securitization." *Stanford Journal of Law, Business & Finance*, 1, pp. 133–54.

Scott, J. C. (1998) *Seeing like a State: How Certain Schemes to Improve the Human Condition Have Failed*. New Haven: Yale University Press.

Shine, K. P. (2009) "The global warming potential—the need for an interdisciplinary retrial." *Climatic Change*, 96(4), pp. 467–72. DOI:10.1007/s10584-009-9647-6.

Silvertown, J. (2015) "Have ecosystem services been oversold?" *Trends in Ecology & Evolution*, 30(11), pp. 641–48. DOI:10.1016/j.tree.2015.08.007.

Sparkes, R. and Cowton, C. J. (2004) "The maturing of socially responsible investment: A review of the developing link with corporate social responsibility." *Journal of Business Ethics*, 52(1), pp. 45–57. http://www.jstor.org/stable/25075231.

Spash, C. L. (2010) "The brave new world of carbon trading." *New Political Economy*, 15(2), pp. 169–95. DOI:10.1080/13563460903556049.

Sutter, C. and Parreño, J. C. (2007) "Does the current Clean Development Mechanism (CDM) deliver its sustainable development claim? An analysis of officially registered CDM projects." *Climatic Change*, 84(1), pp. 75–90. DOI:10.1007/s10584-007-9269-9.

Tainter, J. A. (1988) *The Collapse of Complex Societies*. Cambridge University Press.

The B Team (2016) "Towards a Plan B for Business: The B Team Progress Report, June 2013-June 2016." The B Team. New York, NY. http://www.bteam.org/towards-a-plan-b-for-business/.

Thompson, I., Mackey, B., McNulty, S. and Mosseler, A. (2009). *Forest Resilience, Biodiversity, and Climate Change*. CBD Technical Series, 43. Secretariat of the Convention on Biodiversity, Montreal, Quebec, Canada. Technical Report.

Tol, R. S. J. (2011) "The social cost of carbon." *Annual Review of Resource Economics*, 3(1), pp. 419–43. DOI:10.1146/annurev-resource-083110-120028.

Walker, B. (1995) "Conserving biological diversity through ecosystem resilience." *Conservation Biology*, 9(4), pp. 747–52. http://www.jstor.org/stable/2386983.

Weiss, H. and Bradley, R. S. (2001) "What drives societal collapse?" *Science*, 291(5504), pp. 609–10. DOI:10.1126/science.1058775.

Westman, W. E. (1977) "How much are nature's services worth?" *Science*, 197(4307), pp. 960–64. DOI:10.1126/science.197.4307.960.

Wilkinson, T. J. (1997) "Environmental fluctuations, agricultural production and collapse: a view from bronze age upper Mesopotamia." In *Third Millennium BC Climate Change and Old World Collapse*. NATO ASI Series. Springer, Berlin, Heidelberg. pp. 67–106. DOI:10.1007/978-3-642-60616-8_4.

Chapter 8

Elements of a political economy of the postgrowth era

Max Koch

Introduction

Thresholds for biophysical processes such as climate, biodiversity and the nitrogen cycle are being approached or crossed (Steffen et al., 2018). In relation to climate change, the Intergovernmental Panel on Climate Change (IPCC, 2014) highlights that concentrations of CO_2 and other greenhouse gases in the atmosphere have risen to levels that are unprecedented in at least the last 800,000 years, with the burning of fossil fuels being the main reason behind the 40% increase in CO_2 concentrations since the Industrial Revolution. By the end of the 21st century, the IPCC projects the global surface temperature increase to exceed 1.5°C relative to the period 1850–1900 in all but the lowest and most optimistic scenario considered. Exceeding this threshold, beyond which uncontrollable climate change with frequent droughts, floods and storms plus largely unpredictable climate feedback effects is expected, is increasingly likely. Other scenarios predict global temperatures to rise by as much as 4.8°C. The higher end of this range – and in particular the unprecedented speed of the temperature rise – is far outside the experience of human civilization and would expose at least 70% of the world's population to deadly heat stress (Ramanathan et al., 2017). Warming will continue beyond 2100.

More recently, the IPCC specified that we have just 12 years for global warming to be kept to a maximum of 1.5°C.[1] The mainstream policy response, which has been actively promoted by the Organisation of Economic Cooperation and Development, the World Bank, the United Nations Environment Programme as well as the European Union, is the promotion of "green growth" (Dale et al., 2016) or "ecological modernisation". The obvious advantage of the green-growth path is that it does not make any enemies. Not accidently, there is broad socio-economic and political support for this policy course, ranging from green to liberal parties and from trade unions to employers' organisations. The belief that climate change mitigation is compatible with a growth-oriented, largely uncoordinated and finance-driven capitalism is also reflected in the "market-oriented" mitigation policies adopted (Koch, 2014). Central to any evaluation of the feasibility of green-growth strategies is the distinction between "absolute" and "relative" decoupling of Gross Domestic Product (GDP) growth from carbon emissions and resource use. While resource impacts have declined relative to GDP in a range of rich countries, they have either not done so in absolute terms at all or not to the extent that climate scientists regard as

[1] The *Guardian*, 8 October 2018; https://www.theguardian.com/environment/2018/oct/08/global-warming-must-not-exceed-15c-warns-landmark-un-report.

necessary to avoid catastrophic climate change (Antal and Van Den Bergh, 2014). Not only have improvements in energy efficiency hitherto been offset by increases in the overall scale of economic activity, but the prospects of achieving this in the future to the required extent are very low indeed. In contrast, comparative research (Fritz and Koch, 2016; O'Neill et al., 2018) continues to indicate a strong link between the level of economic development measured in GDP per capita, on the one hand, and carbon emissions and ecological footprints of production and consumption, on the other.

Given the lack of evidence for absolute decoupling of GDP growth, material resource use and carbon emissions a range of approaches have been tabled, the common denominator of which is the questioning of what Daly and Farley (2011) call the "growth imperative".[2] The point of departure of this literature is that the ecological crisis is a structural feature of global capitalism, while the common goal is the reembedding of production and consumption norms into planetary boundaries – through a decrease of matter and energy throughputs, particularly in the rich countries. Also, authors closer to the mainstream, such as Gordon (2012) or Streeck et al. (2016), now regard the issue how economic systems may function under these circumstances and in the absence of economic growth as a crucial research theme within economics and beyond. To continue and deepen the debate on an economics oriented towards sustainable provision of the goods and services necessary for the

[2] See for recent overviews Büchs and Koch (2017); Cosme et al. (2017); Pichler et al. (2017); Weiss and Cattaneo (2017); Kallis et al. (2018) and Barca et al. (2019).

welfare of all human beings, now and in future, and within planetary boundaries, this paper outlines some of the elements and analytical steps that may turn out to be useful for formulating a political economy of the postgrowth era. My point of departure is the ecological critique of neoclassic economics. Subsequently, I will revisit Marx's Critique of Political Economy and its potential capability of unifying the monetary (or exchange value) with the matter and energy (or use value) aspects of production and consumption patterns. The following section considers institutional variations of capitalism and the example of the regulation approach. This approach has originally been tabled for the institutional analysis of different growth strategies within capitalist development, but may nevertheless prove useful for an understanding of economics without growth. Using the analytical toolbox developed in the previous sections, the last section outlines some general features of a "global steady-state" economy highlighting the centrality of the provision of sustainable needs satisfiers and the role of one particular institutional form in the transition from a growth to a postgrowth economy: that of the state.

The ecological critique of neoclassical economics

Modern economic theory often proposes a circular flow of exchange, a repetitive cycle linking money and commodities as well as households and companies. It is understood as being circular and reversible: a "return to capital" basically means that the original capital spent, augmented by a surplus, returns to its owner, and the process of capital valorisation starts over again on a greater scale. Yet the circular monetary value aspect of economics is coupled with a physical flow and throughput of matter and energy, which is ultimately linear (Daly, 1985). In the neoclassical perspective, especially, the production of goods and services is analysed from the standpoint of the growth of monetary value. The latter is seen as indefinite, while the roles played by energy and natural resources are often not mentioned. If the issue of resources is discussed at all, it is assumed that "substitution" processes will sort out the problem of depletion. Indeed, in the case of "perfect competition", an essential requirement of neoclassical growth models, "the price system will see that 'correct' substitutions are made at the right times" (Miernyk, 1999, p. 75).

Ecological economists emphasise that the sidelining or removal of biophysical processes from reasoning in economics is a problematic trend. Rather than exclusively focusing on the movement of values, ecological economics deals with both the extraction of raw materials and their elimination in the form of waste as the "first and last phases of all economic activity" (Deléage, 1994, p. 40). Hence, if the overall scale of production increases, deposit sites will also grow. Since nature is unable to recycle all the new waste, the social cost of waste management increases as a result and becomes a socially contested issue (Corvellec, 2018). Building on the pioneering work of Nicholas Georgescu-Roegen (1971), ecological economics stresses that in production, transport, communication and consumption, processes of irreversible material and energy transformation take place (Spash, 2017). The overall increase in entropy resulting from production processes is greater that its local decrease arising from the production of a concrete good. The portion of free and unbound energy declines in comparison with that of bound and dissipated energy that can no longer be converted into work (in the physical sense). Ecological economists also emphasise that environmental sources and sinks of energy and raw materials are finite, that is, they can be used only once. Indeed, "if the entropic process were not irrevocable, i.e., if the energy of a piece of coal or of uranium could be used over and over again ad infinitum, scarcity would hardly exist in man's life" (Georgescu-Roegen, 1971, p. 6). Though entropy as such is inevitable, since matter continues to dissipate and disperse into less ordered and less useful forms, the entropy rate is subject to historical specific economic and social conditions. Daly, for example, does not assume that entropy remains constant in a "steady-state economy", but instead advocates the minimisation of the matter-energy throughput that such an economy would enable. The advantage of Daly's economic and ecological model over neoclassical growth models is, therefore, not that it is characterised by "constant throughput", but that it provides a "strategy for good stewardship, for maintaining our spaceship and permitting it to die of old age rather than the cancer of growthmania" (Daly, 1974, p. 16).

Both entropy rate and throughputs of matter and energy vary over time and with regard to the different historical forms of organising economies. Georgescu-Roegen identified different stages in human development with associated dominant technological paradigms, which he called "Promethean" technologies. Through the qualitative conversion of energy, these produce an irreversible change in the relationship between economic agents and nature and cause profound modifications in natural ecosystems and human societies. The three Promethean transitions are animal husbandry, fire, and heat engines, each marking an evolutionary step within human economic development. Georgescu-Roegen further observes that any Promethean technology requires a particular set of natural resources: fire requires wood, steam engines require coal etc. (see Beard and Lozada, 1999). Anthropogenic climate change did not emerge during the era of hunting and gathering or in agricultural societies but instead began with the Industrial Revolution, which itself was accompanied by an unprecedented increase in the depletion and burning of fossil fuels, which accelerated the entropy process as well as the greenhouse effect. Neither could the Industrial Revolution have proceeded without concomitant political and socio-economic upheavals. Hence, Georgescu-Roegen made an important step beyond an exclusively natural scientific view on entropy, environmental degradation and climate change. By embedding these into historically diverse socio-economic circumstances, he criticises the most recent economic stage for its reckless consumption of fossil fuels. His famous dictum that "matter matters, too" serves as a powerful reminder that the depletion of resources is not only a physical and ecological, but also a political, social and ethical issue. Yet his analysis, though providing a historically comparative perspective, largely remains at the level of technological – that is, Promethean – innovations. As a result, Georgescu-Roegen and some of his followers lack the analytic tools to provide a theory on the specific socio-economic relations that would explain the current lack of moderation in fossil fuel depletion. Such historically specific configurations are always associated with economic categories, social relations (including power relations) and corresponding modes of consciousness within which the actors make sense of their economic interaction. I will next argue that Karl Marx, in his Critique of Political Economy, provided central elements for such an analysis for the capitalist mode of production.

Capitalism and nature in Marx's Critique of Political Economy

As far as the general work process is concerned Marx (1973, p. 320) does not deviate from Georgescu-Roegen and more recent ecological economists in that "all epochs of production have certain common traits, common characteristics". No production is possible "without an instrument of production, even if this instrument is only the hand". Throughout the centuries and even going back to "simple exchange or barter" (Marx, 1973, pp. 267-8), the purpose of this operation has always been the production of use values. Closely associated with production at this most abstract level is the general concept of labour and the work process, which is itself directly linked to nature: "a process by which man, though his own actions, mediates, regulates, and controls the metabolism between himself and nature" (Marx, 1990, pp. 283). For Marx, therefore, labour is the connecting link between nature and human beings, who, in order to survive, must interact with nature and transfer natural raw materials into use values. While Marx and Georgescu-Roegen could easily have agreed on the fact that this "metabolic" relation constitutes the universal condition upon which human life is sustained, the former departs from the latter by stressing that the notion of "production in general" is no more than "a rational abstraction in so far as it really brings out and fixes the common element and thus saves us repetition..." (Marx, 1973, p. 320). For Marx, real production processes and the associated relationship between economic agents and nature take place in specific social forms, and it is the particular features this relationship takes in the capitalist mode of production, during which the current ecological crisis emerged.

Far from disregarding natural laws, Marx made the pivot (*Springpunkt* in the German original) of his Critique of Political Economy the dual nature of commodities as constituting both exchange value and use value and of work as producing both abstract value and concrete products through the transformation of raw materials and energy. Understanding this "double character" provides insight not only into further economic categories, associated social relations and modes of consciousness but also into the corresponding tensions between the capitalist economy and the ecological system that, among other things, amplify the greenhouse effect (Koch, 2012; 2018). Exchange value refers to the commodity's monetary value for the seller, while use value is concerned with the material and/or symbolical usefulness for the purchaser. However, under the imperative of valorisation, the concrete and material aspect of labour, which is reflected in use values of commodities, is subordinated to abstract labour and exchange value and, hence, somewhat sidelined. Use values, matter and energy are not of primary interest on capitalist markets but instead their form as values, that is, repositories of abstract, socially necessary labour. The societal handling of ecological goals – such as sustainable land-use practices, the preservation of species diversity, clean air and water as well as non-congested transport networks – have to respect the priority of valorisation.

A further tension between the monetary form of values and the principles of natural reproduction is that the former is completely divisible from monetary quota, while the natural world, of which the work process is composed, represents "highly interconnected and interdependent material, biological and thermodynamic systems of varying entropy levels" (Burkett, 2005, p. 144). Furthermore, monetary claims on wealth in the forms of bank accounts, stocks or bonds are highly mobile, and this often contradicts the locational fixedness and specificities of ecosystems. Finally, while money, valorisation and GDP growth are quantitatively unlimited and hence reversible, low-entropy matter and energy are not. The Earth's stock of fossil fuels, in particular, is confined, and the existing stock can only be burnt once. In other words, it is irreversible. There is a structural tension between the value and money form of societal wealth and its material and energy substance.

Profit production is possible due to the fact that a commodity is available for sale that has the use value of creating exchange value and which can be used longer than that which represents the cost of its own reproduction: labour power. In capitalism, producers – as wage-earners – are largely separated from their means of subsistence and production and have no alternative but to offer the only commodity at their disposal on "labour markets". The other "factors of production" – land, raw materials, fuels, auxiliaries etc. – can be purchased on separate markets, and it is only through the intermediation of employers, who hold the necessary capital, that the former comes in contact with the latter. This implies that capitalism's reproduction requirements are distinct from the material and ecological preconditions for the reproduction of labour power and the other factors of production. For capitalist production, all that matters is that these factors and the ingredients of material production are separately available for purchase, and in forms that can be combined in the production process of capital. Given this precondition, capitalist reproduction tends to disrespect the imperatives of natural reproduction such as the preservation of the fossil fuel stocks due to its inherent tendency to expand the scale of production.

In the chapters of *Capital* on cooperation, the manufactory and "machinery and modern industry", Marx (1990, Part IV) discusses the advancement of the division of labour and how the work process became independent from the subjective limitations of individual workers through the systematic application of natural forces and the natural sciences. The Industrial Revolution introduced tools and machinery that reduced the role of many individual workers to that of an "appendage". When the work process had an industrial foundation, then the subjugation of nature under capital became more complete. Now nature was

> "for the first time... purely an object for humankind, purely matter of utility; ceases to be recognized as a power to itself; and the theoretical discovery of its autonomous laws appears merely as a ruse so as to subjugate it under human needs, whether as an object of consumption or as a means of production" (Marx, 1973, p. 409).

Marx also demonstrates that expanding scales of production normally coincide with greater amounts of throughput of raw materials and auxiliary substances, especially in the form of fossil fuels, and available energy. All other things being equal, an increase in productivity means that a given work force processes a larger quantity of raw materials and consumes more energy. Rising demand for raw materials and available energy normally leads to rising prices, for example, for crude oil, creating incentives for individual companies to recycle and to use a given quantity of materials or fuels in more efficient ways. Marx (2006, Chapter 5) described this as a long-term trend towards a greater "economy in the employment of constant capital". Yet progress in the efficiency of raw and auxiliary materials does not fundamentally alter the link between the expansion of the scale of production and the increase in the material and energy throughput, a phenomenon that had already been observed by Jevons in the 1860s (Jevons, 1865). Greater efficiency in the use of a fossil energy source such as coal or oil leads to an *increase* in demand – not to a decrease. On the contrary, this increase becomes the precondition for further capital expansion and economic growth.

In summary, an analysis of the ecological crisis, and climate change in particular, that takes the double nature of the commodity and work as both value in motion and a concrete stock of invested time- and place-specific assets of matter and energy as point of departure is able to demonstrate that the capitalist mode of production is oriented towards unlimited and short-term valorisation, quantitative and geographic expansion, circularity and reversibility, while the principles that guide the ecological system involve stable and sustainable matter and energy transformations and throughputs as well as irreversibility (Koch, 2012). Capital's "expansionism" tends to be accompanied by the degradation of the environmental conditions of production and especially reductions in their ability to act as sources and sinks for the permanently increasing flow and throughput of matter and energy (Clark and York, 2005). When these sources and sinks cease to function, their decelerating impact on the greenhouse effect is nullified, thus increasing the risk of positive feedback mechanisms within the climate system.

Beyond the "mode of production": the regulation approach

Capitalist development proceeds in not only socially, but also in ecologically contingent forms. Though it does not get rid of the use-value element and the corresponding matter and energy side of production altogether, it nevertheless tends to negate and dispel them as much as possible. However, it is sometimes forgotten that the structural tensions and contradictions between the capitalist growth economy and the ecological system discussed above are located at a relatively high level of abstraction: Marx's capitalist mode of production, where abstraction is made from institutional regulation and individual actors are reduced to their roles as "character masks", that is, to the roles they play in the production and circulation processes of capital. Yet the tensions inherent to the capitalist mode of production take on different forms in concrete social formations and historical periods of capitalist development, presenting themselves as continuous development or as rupture, depending on diverse types of institutional regulation (Becker and Raza, 2000; Paterson and Laberge, 2018). An institutional attempt to link analyses of production patterns to those of consumption has been proposed by the regulation approach, which – at least in the Parisian version – continues key insights from Marx's Critique of Political Economy ("modes of production" and "social formations"), and complements them with "intermediary concepts" ("accumulation regime" and "mode of regulation"). These concepts express the largely non-variable conditions of agents involved in the relations of production and exchange, as well as the historical changes that these relations undergo during different phases of capitalist growth (Boyer and Saillard, 2002). While the abstract features of capitalism are seen as largely trans-historical, both crises in the accumulation process and phases of expanded production are addressed in the context of their institutional embedding. Hence, the focus on "regulation" allows for going beyond the "mode of production" and for analysing concrete periods of accumulation and growth, both in historical and comparative perspective (Brand and Wissen, 2015).

The regulatory settings required for continued and expanded capital accumulation are socially, culturally and politically constructed and contested within a myriad of societal struggles, in which the relations both within and between social classes play a prominent role. The notion of "intermediary concepts", in particular, highlights that the articulation of a given social formation in time and space corresponds to particular structural features and institutional forms. "Accumulation regimes" are associated with certain historical phases and development paths or growth strategies, which take the form of compatible commodity streams of production and consumption, reproduced over a long period of time. They differ historically, for example, as to whether intensive or extensive, export-oriented or import-oriented forms dominate, or vice versa. All accumulation regimes are associated with a specific industrial paradigm, a dominant principle of division of labour and a corresponding "mode of consumption". In contrast to neoclassical economics, the regulation approach does not view consumption as an isolated or behavioural phenomenon – the result of autonomous individual choices – but within its social genesis and context. Aglietta (1987, p. 154) conceptualises consumption as "an organized set of activities, which – while predominantly private – became subject to a general logic of the reconstitution of energies expended in social practices and the preservation of abilities and attitudes implied by the social relations". What and how much we buy and consume is of the greatest relevance for ecological issues such as the carbon cycle, since these decisions are normally bound to matter and energy transformations that more often than not necessitate the burning of fossil fuels. The regulation approach insists – contradicting the predominant notion of *homo economicus* in the neoclassical perspective – that purchase decisions or the "demand side" of economics are neither "formally rational" nor "autonomous", but instead are greatly influenced by structural factors such as income inequality and corporate sales strategies. This resonates with sociological and anthropological consumption research, which points out that purchases of things are not in the first place about the goods themselves but rather about the symbolic messages that purchase acts express and mediate (Bourdieu, 1982; see Boyer, 2008).

A "mode of regulation" comprises an ensemble of social networks as well as rules, norms, and conventions, which together facilitate the seamless reproduction of an accumulation regime. The term "mode of regulation" stresses the fact that capitalism does not reproduce itself only upon the basis of the immanent logic dealt with in Marx's *Capital*, but that its stabilisation also requires "institutional forms": these comprise the wage relation or "wage-labour nexus" (Bertrand, 2002); the enterprise form; the nature of money (Guttmann, 2002); the state (Jessop, 2002); and international regimes (Aglietta, 2002). Subsequent analyses have added another institutional form: that of "energy regimes" (Koch, 2012; Cahen-Fourot and Durand, 2016), which consider, among other things, the environmental impacts of historical growth strategies such as Fordism or finance-driven capitalism. Critical geographers have complemented these institutional forms with a notion of geographic scales, which determine the main spatial boundaries within which structural coherence is sought (Brenner, 2004). Regulationists view the institutional forms that help stabilise capitalist development during particular growth periods as the hard-won products of social struggles and diverse and often contradictory interests. Regulation in its concrete forms is, hence, not simply the product of the strategies of the dominant classes, which themselves are divided by different competitive interests, but always reflect a degree of a compromise with dominated groups. Modes of regulation and patterns of governance vary considerably depending on the nature of such institutional compromises.

Among the institutional forms, the state plays a crucial role within the regulation of capitalist growth. Materialist state theory (Gramsci, 1999; Poulantzas, 1978; Bourdieu, 2015) constructs the state as a relatively autonomous political sphere, where social classes and groups represent their interests in indirect and mediated ways. As political parties and interest groups raise variable issues such as religion, age, and the environment, these interests and issues are sometimes in the focus of government action, only to be superseded by others at later points in time. As a corollary, state policies cannot be reduced to the strategic interests of single actors, but rather develop as a result of the heterogeneity and changing dynamic of social forces that influence state institutions. Once such a coalition of relatively powerful actors has been formed and managed to influence the general direction of state policies, however, it takes on the character of a relatively homogenous social force and appears to "act" as if it were a single actor: the more socially coherent the coalition of forces that influences the state, the lesser the contradictions across its policies. Hence, according to the mentioned state theorists – and provided the necessary bottom-up mobilisation (Buch-Hansen, 2018) – the existing state apparatus could be used to initiate a social and ecological transition beyond the growth imperative.

Features of a postgrowth economy within planetary limits

Previous sections have demonstrated that any re-embedding of the economy into planetary boundaries is unlikely as long as the top priority of economic growth in policymaking is upheld. Yet it is far from obvious what a sustainable postgrowth economy could look like. Though outlining some of its general features is difficult and speculative, I would nevertheless argue that such somewhat utopian thinking is required to encourage debate and eventually social change (Levitas, 2013). Using recent empirical results from comparative studies into the links of GDP/capita, environmental sustainability, social inclusion and individual wellbeing as well the analytical tools introduced in previous sections – especially steady-state economy, use value/exchange value, rescaling and the state as an institutional form within socio-economic regulation – I sketch out some of the key features of what may be called a "global steady-state economy".

Herman Daly's "steady-state economy" (Daly, 1974) is the most cited case of an economic system that functions within ecological boundaries. It is a model of an economy that does not grow in the sense that it keeps the level of throughput (extraction of raw materials from nature and their return to nature as waste) as low as possible and ideally within the regenerative and assimilative capacities of the ecosystem. However, the original concept of a steady-state economy was not developed at the global level. Yet environmental threats such as climate change are global issues, because for the atmosphere it does not matter from which part of the globe greenhouse gases are emitted. Accordingly, the ecological footprint and the associated matter and energy throughput of the whole planet would need to shrink if the world's mode of production and consumption were to respect ecological limits. However, due to massive differences in economic development and unprecedented socio-economic global inequality (Piketty, 2014) such a re-embedding of the world's production and consumption patterns would imply different challenges for different regions and nations. Recent comparative research demonstrates that not only nations' social inclusion, wellbeing and democracy scores largely increase with GDP per capita but also their ecological footprints and carbon emissions. According to Fritz and Koch (2016), who divided 138 countries into five clusters of economic development measured as GDP / capita ("poor", "developing", "emerging", "rich" and "overdeveloped" countries), it is only the poorest group of countries that could currently be seen as environmentally sustainable.

Such a global comparison of national eco-social performances has repercussions for the internal structure of a global steady-state economy as countries at different levels of economic development would need to undertake different measures to achieve a maximum in wellbeing within ecological limits. The policy challenge for "poor" countries would be to enhance the quality of life and social inclusion while maintaining low ecological footprints and carbon emissions; "developing" and "emerging" countries face the double challenge of combining individual wellbeing with social welfare, while preserving relatively low amounts of matter and energy throughput as well as carbon emissions; "rich" and especially the "overdeveloped" countries would need to produce and consume differently so that lesser amounts of material resources and fossil energy are used and to make production and consumption processes more environmentally sustainable (Fritz and Koch, 2016, p. 48). Hence, the socio-ecological transitions required for setting up a global steady-state economy would involve issues of redistribution of wealth, labour, time and natural resources both within and between countries. It goes without saying that these issues are contested and that addressing them would involve questioning the material interests of the currently rich and powerful. Future growth-critical research would therefore be well advised to consider features of the global class structure more systematically (Leonardi, 2019).

In a situation where the production and consumption patterns of the vast majority of countries are beyond the Earth's carrying capacities, it is for the time being difficult to see how a re-embedding of the global economy into planetary boundaries could, in terms of welfare provision, mean much more than the satisfaction of basic human needs (Koch and Mont, 2016). The question how this can be done for all human beings and across generations should hence be prioritised not only in degrowth research (Koch et al., 2017; Büchs and Koch,

2019) but also in corresponding efforts within economics. Needs differ from wants and preferences in that they are non-negotiable and universalisable (Gough, 2017). They do not vary over time and across cultures but according to the ways in which a specific culture at a particular point in time attempts to satisfy them. In a corresponding transformational change strategy, critical thresholds for the universal provision of human needs or for a "minimally decent life" would constantly need to be (re-)defined in light of the advances of scientific and practical knowledge. Economic systems would be assessed according to their ability to produce a critical minimum of appropriate need satisfiers (Max-Neef, 1991; Guillén-Royo, 2016).

The centrality of human needs in degrowth research and that of the provision of sustainable needs satisfiers in economics presupposes probably the most significant shift from a growth to a postgrowth economy: that from monetary growth or exchange value orientation to an understanding and steering of the economy in biophysical or use-value terms. The second necessary shift relates to the fact that a re-embedding of global production and consumption patterns into planetary limits would imply different challenges for different regions and nations. If already the "developing" countries assembled in Fritz and Koch's second poorest cluster (Fritz and Koch, 2016) work and live, so to speak, beyond their ecological means, this has repercussions for the scales that socio-economic and environmental regulation target – from the national towards global but also local levels. In global governance networks, thresholds for matter and energy throughput would need to be determined in accordance with natural science expertise. These limits would at the same time delineate the leeway within which national and local economies may evolve. A new division of labour between the various regulatory levels is envisioned by Kothari (2018, p. 254) who proposes assigning "a minimal set of matters" to the global level, while the bulk of decision-making would "go to the most local level feasible", where he assumes that diverse approaches to meeting collective goals are most "accepted and encouraged".

The priority of use value and biophysical parameters as well as the new division of labour across scales would in all likelihood mean a lesser role and a stricter regulation of market forces than currently. Though the allocative efficiency of markets is generally accepted in most steady-state concepts, these would need to operate in much narrower limits given the primacy of global sustainability, needs satisfaction and intergenerational justice. Instead, a "steering state" would at various levels and scales be *primus inter pares* in a mixed economy and a governance network of public, collective, communal and private actors. New combinations of state and common ownership may be developed in relation to the governance of socio-natural resources such as energy and water. This downscaling of regulatory power from national welfare and environmental institutions to local levels is addressed by several contributors to steady-state economics, degrowth and social enterprises (Johanisova et al., 2013). These highlight the need to replace the current global production and trade systems with economies based on cooperative principles and oriented towards local production and consumption cycles. Some local and voluntary grassroots initiatives have proven quite efficient in environmental terms, even though they often face difficulties in sustaining themselves over time (Howell, 2012).

In relation to the national level of state regulation, Buch-Hansen (2014) argues that present institutional diversity is likely to affect degrowth trajectories as well as the concrete shaping of national steady-state economies and corresponding state apparatuses. Just as contemporary capitalist societies are diverse, so would steady-state economies take many different forms in different countries. This variety also relates to the capability of institutional reflexivity, of learning processes from "best-practice" countries. Comparative research into wellbeing, prosperity and environmental performance of existing countries relative to GDP/capita (Fritz and Koch, 2016) suggests that there are better than average performing countries in each part of the world (for example, Switzerland in Europe, Costa Rica and Uruguay in Latin America) that could be singled out for in-depth institutional analysis.

State eco-social policies in the rich countries

Within any analysis of a transition of the global economy towards a steady state, special emphasis would need to be placed on the currently rich countries. These do not only have the bulk of historical responsibility for the ecological crisis, but also continue to consume an amount of environmental resources that cannot be generalised to the rest of the planet without further crossing planetary boundaries. For these countries especially economic growth as top policy priority would need to be deprioritised and replaced by biophysical parameters as well as by a general policy orientation on basic needs satisfaction. In a corresponding transformation, public "eco-social policies" could start from the "double injustice" (Walker, 2012) according to which the poorest countries, and, within the rich countries, the poorest household groups, who are least responsible for environmental damages such as climate change are in the worst position to cope, and to afford mitigation and adaptation (Büchs et al., 2011). For example, ecological investment into retrofitting houses only has a chance of being perceived as legitimate if it is accompanied by social policies that, among other things, assist homeowners in paying for ecologically useful measures. Conversely, the recent protests of the so-called "yellow vests" in France indicate that the pursuit of ambitious climate policies with a simultaneous reduction in the tax burden of the rich has little chance of proving particularly popular with the electorate.

Policy initiatives that tackle social inequality and the environmental crisis at the same time could be informed and guided by need theories. Here, Ian Gough's "dual strategy" may provide a collective and critical way of distinguishing needs from luxuries in a particular national or local setting (Gough, 2017, p. 169). Accordingly, citizens, experts and government representatives would work together in democratic forums to identify the goods and services necessary to satisfy a given need, and the level of this satisfaction within particular social and cultural contexts. One example would be "social tariffs" that could adjust energy tariffs in line with energy need. This would require energy companies to "operate a 'rising block tariff', with lower tariffs for initial units of electricity or gas consumed, and higher tariffs for successive units" (Gough, 2017, p. 140). Beyond the energy sector, public policy initiatives can stimulate a recomposition of consumption. En route to a global steady-state economy, Western consumption rates would need to decrease disproportionately so that citizens of other parts of the world and future inhabitants of the planet could enjoy an improvement in their material standard of living. Conspicuous consumption would then not be regarded as a symbol of distinction and legitimacy but as a "negative externality". Consumers would be required to pay for the negative impacts this imposes upon others. Governments in the rich countries can encourage certain ways of consumption (for example, vegetarian diets, local holidays, use of public transport and cycling) and discourage others (for example, meat consumption, holidaying in distant locations, car and plane use). Such policy initiatives may be facilitated by a growing dissatisfaction of the public with the consumerist lifestyle. According to Soper (2016), citizens are increasingly disenchanted with consumer culture because of its negative side effects such as time scarcity, high levels of stress, traffic congestion, and due to the increasing displacement of other pleasures of life and well-being by shopping mall culture.

Both ecological economists and critical political economists regard a redistribution of wealth and income – both within and across countries and in an intergenerational perspective – as a crucial element of a wider ecological and social transition. In degrowth circles, especially, maximum limits on income and wealth are seen as critical to maintaining global warming within the 1.5°C range. After reaching the maximum income, people would be incentivised to devote their further energies to non-economic pursuits. However, there is a lack of concrete proposals as to how a maximum limit on income and wealth could be implemented (see Buch-Hansen and Koch, 2019 for an overview from a degrowth perspective). At the other end of the scale, structural inequality could be addressed through either minimum or basic income schemes, which could be co-financed from general revenues, an increasingly progressive income tax, eco-taxes and/or from depletion and emissions certificate auctions. Andersson (2009, p. 3), for example, assumes equivalence between basic income financed by green taxes and the distribution of equal and transferable rights to use scarce environmental resources and to emit a given quantity of greenhouse gases. A range of authors therefore postulate a new architecture of taxation to finance a postgrowth economy. If the tax base were linked to the throughput of finite resources, external costs, which private enterprises currently enjoy as "free gifts" from nature, would be internalised and considered in their cost calculations. Further eco-social policy suggestions involve carbon rationing, alternative ways of housing (Nelson and Schneider, 2018) and local currencies (Hornborg, 2017). Overall, there is no lack of more or less developed policy suggestions to which activists may turn. The problem continues to be that these are fragmented and in need of being unified in a coherent strategy for the social and ecological transformation of the rich countries.

Conclusion

Planetary boundaries are either approached or already crossed, and there is no evidence of an absolute decoupling of GDP growth, material resource use and carbon emissions in the near future. The IPCC is not the only one warning that significant and qualitative changes in the structures of the global economy would need to be initiated very soon to have a realistic chance of keeping global warming near 1.5°C by the end of the century. Economics as independent academic discipline can make an enormous contribution towards an understanding of this social and ecological transition. In this paper, I have addressed some of the themes that appear to be of special relevance in this regard.

I started from the critique of the neoclassical perspective launched by ecological economists. While the former tend to reduce economic action to monetary flows, the latter emphasise matter and energy transformations in the economic process without paying too much interest to the social relations and power asymmetries that are nevertheless inherent to capitalist economies. I suggested that a possible way of unifying standard and ecological economics is a re-discovery of Marx's Critique of Political Economy, and here especially his point of departure: the double character of commodity and labour of embodying both use value and exchange value. In abstract form, this already covers many of the contradictions of an economy that is geared to monetary growth (exchange value moment), but cannot get rid of the accompanying transformations in matter and energy (use value moment), which accelerate climate change and related ecological threats. Hence, an economics in keeping with the times would address both the use value and exchange value moments of current production and consumption patterns without losing sight of the power relations inherent in the current global political economy. Overcoming these would allow the economy to be understood as biophysical and social process, and to reorientate economics towards the goal of providing adequate amounts of environmentally sustainable needs satisfiers for all people, now and in future.

A consideration and reinterpretation of institutional economics such as the regulation approach may help make this general reorientation of the economic process more concrete. This approach has distinguished several institutional forms for the regulation of "growth" strategies, which could be used as a point of departure for specific research projects into postgrowth economics: which would be the central features of the wage relation, the enterprise form, the form and function of money and of the international political regime of the postgrowth era? What would be an operational division of scale in a corresponding mode of regulation? In the present paper, I have mentioned the potentially crucial role of the state as an institutional form in a "great transition" that leading sustainability scientists regard as necessary to avoid the potentially catastrophic consequences of climate change and biodiversity loss (Rockstöm, 2015). Particularly Poulantzas' concept of "condensation" of wider societal struggles within the state indicates that the political actions of the state are far from independent from what goes on beyond it. If mobilisation by socio-ecological and growth-critical groups reached a critical momentum, the existing state apparatus could be used to initiate the required ecological and social transition.

In a postgrowth context, the state's policy priority of achieving economic growth would be replaced by the goal of reembedding production and consumption patterns into planetary limits. Public economic, social and environmental policies would be oriented at minimising the matter and energy throughput and the provision of a sufficient amount of sustainable need satisfiers. While state capacity to act in the environmental domain would increase significantly, if the growth proviso were replaced by a sustainability proviso, state power would be used to build transnational networks and to act as *primus inter pares* together with various private, semi-private and non-profit actors to ensure the respect of ecological limits in global production and consumption patterns.

References

Andersson J.O. (2009) "Basic income from an ecological perspective", *Basic Income Studies* 4 (2), pp. 1-8.

Aglietta, M. (1987) *A Theory of Capitalist Regulation: The US Experience*. London: Verso.

Aglietta, M. (2002) "The International Monetary System." In R. Boyer and Y. Saillard (eds), *Régulation Theory. The State of the Art*. London: Taylor and Francis, pp. 64–72.

Antal M. and Van Den Bergh J. (2014) "Green Growth and Climate Change: Conceptual and Empirical Considerations." *Climate Policy*, 16 (2), pp. 165-177.

Barca, S., Chertkovskaya, E. and Paulsson, A. (eds) (2019) *Towards a Political Economy of Degrowth*. London: Rowman and Littlefield.

Beard, T.R. and Lozada, G.A. (1999) *Economics, Entropy and the Environment. The Extraordinary Economics of Nicholas Georgescu-Roegen*. Cheltenham: Edward Elgar.

Becker, J. and Raza, W. (2000) "Theory of Regulation and Political Ecology: An Inevitable Separation?" *Ambiente & Sociedade*, 5 (11), pp. 5-17.

Bertrand, H. (2002) "The Wage-labour Nexus and the Employment System." In R. Boyer and Y. Saillard (eds), *Régulation Theory. The State of the Art*. London: Taylor and Francis, pp. 80–86.

Bourdieu, P. (1984) *Distinction: A Social Critique of Judgement and Taste*. Cambridge: Harvard University Press.

Bourdieu, P. (2015) *On the State. Lectures at the Collège de France 1989-1992*. Cambridge: Polity Press.

Boyer, R. (2008) "Pierre Bourdieu, a Theoretician of Change? The View from Regulation Theory." In A. Ebner and N. Beck (eds), *The Institutions of the Market: Organizations, Social Systems, and Governance*. Oxford: Oxford University Press, pp. 348–98.

Boyer, R. and Y. Saillard (eds) (2002) *Régulation Theory. The State of the Art*. London: Taylor & Francis.

Brand, U. and Wissen, M. (2015) "Stategies of a Green Economy, Contours of a Green Capitalism." In: K. van der Pijl (ed.) *The International Political Economy of Production*. Celtenham: Edward Elgar, pp. 508-523.

Brenner, N. (2004) "Urban Governance and Production of New State Spaces in Western Europe, 1960–2000." *Review of International Political Economy*, 11 (3), pp. 447–88.

Buch-Hansen, H. (2014) "Capitalist Diversity and De-growth Trajectories to Steady-state Economies." *Ecological Economics*, 106, pp. 173-179.

Buch-Hansen, H. (2018) "The Prerequisites for a Degrowth Paradigm Shift: Insights from Critical Political Economy." *Ecological Economics*, 146, pp. 157-163.

Buch-Hansen, H. and Koch, M. (2019) "Degrowth through Income and Wealth Caps?" *Ecological Economics*, in press.

Büchs, M., Duwe, S. and Bardsley, N. (2011) "Who Bears the Brunt? Distributional Effects of Climate Change Mitigation Policies." *Critical Social Polic* 31 (2), pp. 285-307.

Büchs, M. and Koch, M. (2017) *Postgrowth and Wellbeing: Challenges to Sustainable Welfare*. London: Palgrave Macmillan.

Büchs, M. and Koch, M. (2019) "Challenges for the Degrowth Transition: The Debate about Wellbeing." *Futures*, 105, pp. 155-165.

Burkett, P. (2005) "Entropy in Ecological Economics: A Marxist Intervention." *Historical Materialism*, 13 (1), pp. 117-52.

Cahen-Fourot, L. and Durand, C. (2016) "La Transformation de la Relation Sociale à l'énergie du Fordisme au Capitalisme Néolibéral: Une Exploration Empirique et Macro-économique Comparée dans les Pays Riches (1950-2010)." *Revue de la Régulation: Capitalisme, Institutions, Pouvoirs.* 20 (2), available online: https://journals.openedition.org/regulation/12015.

Clark, B. and York, R. (2005) "Carbon Metabolism: Global Capitalism, Climate Change, and the Biospheric Rift." *Theory and Society*, 34 (4), pp. 391-428.

Corvellec, H. (2018) "Waste as Scats: For an Organizational Engagement with Waste." *Organization*, available online: http://dx.doi.org/10.1177/1350508418808235.

Cosme, I., dos Santos, R.F. and O'Neill, D. (2017) "Assesssing the Degrowth Discourse: A Review and Analysis of Academic Degrowth Proposals." *Journal of Cleaner Production*, 149, pp. 321-334.

Dale, G., Mathai, M.V. and de Oliveira, J.P. (Eds) (2016) *Green Growth: Ideology, Political Economy and the Alternatives*. London: Zed Books.

Daly, H. (1974) "The Economics of the Steady State." *American Economic Review*, 64 (2), pp. 15-21.

Daly, H. (1985) "The Circular Flow of Exchange Value and the Linear Throughput of Matter-Energy: A Case of Misplaced Concreteness." *Review of Social Economy*, 43 (3), pp. 279-97.

Daly, H. and Farley, J. (2011) *Ecological Economics. Principles and Applications*. Washington, Island Press.

Deléage, J.P. (1994) "Eco-Marxist Critique of Political Economy." In M. O'Connor (Ed) *Is Capitalism Sustainable? Political Economy and the Politics of Ecology*. New York / London: The Guildford Press, pp. 37-52.

Fritz, M. and Koch, M. (2016) "Economic Development and Prosperity Patterns around the World: Structural Challenges for a Global Steady-state Economy." *Global Environmental Change*, 38, pp. 41-48.

Georgescu-Roegen, N. (1971) *The Entropy Law and the Economic Process*. Cambridge and London: Harvard University Press.

Gordon, R. (2012) "Is US Economic Growth Over? Faltering Innovation Confronts the Six Headwinds." Working Paper 18315. Cambridge: National Bureau of Economic Research.

Gough I. (2017) *Heat, Greed and Human Need: Climate Change, Capitalism and Sustainable Wellbeing*. Cheltenham: Edward Elgar.

Gramsci, A. (1999) *Selections from the Prison Notebooks*. London: The Electric Book Company Ltd.

Guillén-Royo, M. (2016) *Sustainability and Wellbeing: Human Scale Development in Practice*. London: Routledge.

Guttmann, R. (2002) "Money and Credit in Regulation Theory." R. Boyer and Y. Saillard (Eds), *Régulation Theory. The State of the Art*. London: Taylor and Francis, pp. 57–63.

Hornborg, A. (2017) "How do Turn an Ocean Liner: A Proposal for Voluntary Degrowth by Redesigning Money for

Sustainability, Justice and Resilience." *Journal of Political Ecology*, 24 (1), pp. 623-632.

Howell, R.A. (2012) "Living with a Carbon Allowance: The Experiences of Carbon Rationing Action Groups and Implications for Policy." *Energy Policy*, 41, pp. 250-258.

Intergovernmental Panel on Climate Change (IPCC) (2014) *Climate Change 2014: Synthesis Report. Summary for Policymakers*. Geneva, http://www.ipcc.ch/pdf/assessment-report/ar5/syr/SYR_AR5_SPMcorr2.pdf.

Jessop, B. (2002) *The Future of the Capitalist State*. Cambridge: Polity Press.

Jevons, W.S. (1865) *The Coal Question. An Inquiry Concerning the Progress of the Nation, and the Probable Exhaustion of Our Coal-Mines*. London: Macmillan and Co.

Johanisova, N., Crabtree, T. and Frankova, E. (2013) "Social Enterprises and Non-market Capitals: A Path to Degrowth?" *Journal of Cleaner Production*, 38, pp. 7-16.

Kallis, G., Kostakis, V., Lange, S., Muraca, B., Paulson, S. and Schmelzer, M. (2018) "Research on Degrowth." *Annual Review of Environment and Resources*, 43, pp. 291-316.

Koch, M. (2012) *Capitalism and Climate Change. Theoretical Discussion, Historical Development and Policy Responses*. London: Palgrave Macmillan.

Koch M. (2014) "Climate Change, Carbon Trading and Societal Self-defense." *Real-world Economics Review*, 67, pp. 52-66.

Koch, M. (2018) "The Naturalisation of Growth: Marx, the Regulation Approach and Bourdieu." *Environmental Values*, 27 (1), pp. 9-27.

Koch, M. and Mont, O. (Eds) (2016) *Sustainability and the Political Economy of Welfare*. London: Routledge.

Koch, M., Buch-Hansen, H. and Fritz, M. (2017) "Shifting Priorities in Degrowth Research: An Argument for the Centrality of Human Needs." *Ecological Economics*, 138, pp. 74-81.

Kothari, A. (2018) "Towards Radical Alternatives to Development." In H. Rosa and C. Henning (Eds), *The Good Life Beyond Growth: New Perspectives*. London: Routledge, pp. 251-262.

Leonardi, E. (2019) "Bringing Class Analysis Back In: Assessing the Transformation of the Value-Nature Nexus to Strengthen the Connection Between Degrowth and Environmental Justice." *Ecological Economics*, 156, pp. 83-90.

Levitas, R. (2013) *Utopia as Method: The Imaginary Reconstitution of Society*. London: Palgrave Macmillan.

Marx, K. (1973) *Grundrisse: Foundations of the Critique of Political Economy*. Harmondsworth: Penguin.

Marx, K. (1990) *Capital: A Critique of Political Economy*. Vol. 1. London: Penguin Classics.

Marx, K. (2006) *Capital: A Critique of Political Economy*. Vol. 3. London: Penguin Classics.

Max-Neef, M. (1991) *Human Scale Development. Conception, Application and Further Reflections*. New York: The Apex Press.

Miernyk, W.H. (1999) "Economic Growth Theory and Georgescu-Roegen Paradigm." In K. Mayumi, and J.M. Gowdy (Eds), *Bioeconomics and Sustainability. Essays in Honor of Georgescu-Roegen*. Cheltenham: Edward Elgar, pp. 69-81.

Nelson, A. and Schneider, F. (Eds) (2018) *Housing for Degrowth: Principles, Models, Challenges and Opportunities*. London: Routledge.

O'Neill, D., Fanning, A.L., Lamb, W.F. and Steinberger, J.K. (2018) "A Good Life for All Within Planetary Boundaries." *Nature Sustainability* 1, pp. 88-95.

Paterson, M. and Laberge, X. P. (2018) "Political Economies of Climate Change." *WIREs Climate* Change, 9 (2), p. 506.

Pichler, M, Schaffartzik, A., Haberl, H. and Görg, C. (2017) "Drivers of Society-nature Relations in the Anthropocene and their Implications for Sustainability Transformations." *Current Opinion in Environmental Sustainability*, 26, pp. 32-36.

Piketty, T. (2014) *Capital in the Twenty-first Century*, Cambridge. Mass. / London: The Belknap Press.

Poulantzas, N. (1978) *State, Power and Socialism*. London: NLB.

Ramanathan V., Molina L., Zaelke D. et al. (2017) "Well under 2 Degrees Celsius: Fast Action Policies to Protect People and the Planet from Extreme Climate Change."
www.igsd.org/wp-content/uploads/2017/09/Well-Under-2-Degrees-Celsius-Report-2017.pdf

Rockström, J. (2015) "Bounding the Planetary Future: Why we Need a Great Transition." *Great Transition Initiative*, 9, pp. 1-13.

Soper, K. (2016) "The Interaction of Policy and Experience: An 'Alternative Hedonist' Optic." In M. Koch and O. Mont (Eds), *Sustainability and the Political Economy of Welfare*. London: Routledge, pp. 186-200.

Spash, C.L. (Ed) (2017) *Routledge Handbook of Ecological Economics: Nature and Society*. London: Routledge.

Steffen, W., Rockström, J. et al. (2018) "Trajectories of the Earth System in the Anthropocene." *Proceedings of the National Academy of Science of the United States of America*, 115 (33), pp. 8252-8259.

Streeck, W., Calhoun, C., Toynbee, P. and Etzioni, A. (2016) "Does Capitalism have a Future?" *Socio-economic Review*, 14 (1), pp. 163-183.

Walker G. (2012) *Environmental Justice: Concepts, Evidence and Politics*. London, Routledge.

Weiss, M. and Cattaneo, C. (2017) "Degrowth – Taking Stock and Reviewing an Emerging Academic Paradigm." *Ecological Economics*, 137, pp. 220-230.

Chapter 9

Victim of success: civilisation is at risk

Peter McManners

Introduction

Our current civilisation is the grandest and most ambitious in all of human history. Humans have never been more powerful, more capable, or richer. This should be the best of times as we enjoy peak civilisation; but it could be short-lived. If history is our guide, we can be sure that civilisation will collapse. Every civilisation which has existed in all of history, no matter how powerful it becomes, ends up falling apart (Ehrlich and Ehrlich, 2013). There is little reason to suppose that our current civilisation will be different. When collapse comes, there will be accusations of incompetence as people wonder why we did not see it coming. We don't see it because we do not open our eyes to the reality of our situation. Using logical analysis (rather than wishful thinking) it is clear to see that collapse is inevitable on the path we are on. Stopping the collapse may no longer be possible, as we have moved so far away from a sustainable track, but we should at least try to soften the landing. It may even be possible to prevent collapse entirely if we take bold action without delay. As we consider what action to take, it is important to understand that to keep civilisation on the rails in the coming decades will require switching track.

I believe that we should approach this challenge with optimism to engage people with an upbeat message which motivates and encourages positive action. However, I admit that a decade ago, I descended into a dark negative mindset. My mental nose-dive was due to disturbing insights from my research, leading me to really understand the glorious stupidity of the current generation of world leaders. They are wilfully blocking progress in shifting the economy onto a safe track. My response was to write the book *Victim of Success: Civilisation at Risk* (McManners, 2009). One person confided in me, that after reading the book they didn't sleep for a week. What kept them awake was my explanation of the mechanism of the collapse of civilization. This comes from simple logical analysis and was entirely believable when I wrote it a decade ago, and remains entirely believable today. The pending collapse of civilisation filled just the first few chapters of the book; most of the book was about how the future could be different. I explained that if we changed our ways, civilisation could be shifted onto a safer track. The ideas in my book did not diffuse out to a wide audience; perhaps people were so completely fixated by continuing with a successful economic formula that they do not want to have to consider changing it. A decade later, I write this essay knowing that we are exactly and precisely where I predicted we would be, on the road to ruin. Collapse now seems ever more certain; but the choice is still available to change direction. My focus in here is to outline how to make the adjustment of mind-set required to make it possible to contemplate a major shift to a different economic model which can intercept our decline before it happens and secure a safe future for humanity.

A global civilisation

The great civilisations of the past have been regional in extent. The ancient Egyptians of the Middle East, the Incas of South America and the Romans in Europe. When their civilisations collapsed, the fallout was isolated to a region; the world was not at risk. Other civilisations with roots in other regions could arise and become the next great civilisation. The British Empire was perhaps the first civilisation to reach into almost every corner of the globe. It too disintegrated of course, because that is the transitory nature of human success. The current global civilisation is not regionally based, or based on any particular cultural identity, but is connected through economic globalisation. Our great civilisation draws resources from everywhere with the potential to provide people everywhere with what they need. Whilst there are ample resources, economic globalisation seems to be the economics of success. This civilisation is not "Roman" or "Inka"; "European" or "American" in nature, but "economic". Focussing intently on economic outcomes, and using economics as the glue which holds global society together, has delivered increasing output, increasing wealth and makes us materially better off. This approach is based on the assumption that this is good for society.

The prime measures of success used by politicians and governments are economic metrics such as growth, wealth, income and consumption. The policies of free-trade and open markets, maximise efficiency and work to expand the economy. A commonly used metaphor to explain the benefits of a global market, is that a rising tide lifts all boats. To extend the metaphor into the future, what happens when the tide goes out? Who will be safe and who will be left stranded? We also welcome that some people become incredible wealthy (even when it seems excessive) because the benefits will "trickle down" to lower levels of society. How about when resources run short? The rich will no doubt ensure they secure their supplies; what will be left for the less wealthy? Because we measure success by economic metrics, it should come as no surprise that the needs of society can be side-lined in the narrow pursuit of keeping the economy motoring along.

It is not just society that comes under pressure from economic myopia. The ecosystem is also put at risk. In the words of economics, the ecosystem is an "externality" so does not generally enter as a factor into the economic analysis. This leaves it vulnerable, to be used, abused, and if necessary, sacrificed for the greater economic good. The consequences for a civilisation based on economic principles is that society is steamrollered and the ecosystem degraded. It is a logical certainty that such a civilisation is time limited.

The ecosystem

Civilisations come and go, whilst the ecosystem continues as the backdrop on which history plays out. It is the long-term foundation on which society depends. We have assumed that it will always be there for us to support how we live and enable us to thrive. We now understand that the planet is not an unlimited resource but has boundaries which must be respected if it is not to be damaged (Rockström et al., 2009). Human civilisation used to live within the constraints of the ecosystem, not by design, but simply because our impact was small compared with what seemed like a vast planet of limitless resources. Now our advanced capabilities have become so powerful and operate at such vast scale that we are capable of destroying planet Earth. We are slow to realise that our new-found capability is so profound that we need to take responsibility to use it wisely. This means working out how to reconcile economic aspirations with the fundamental constraint of a planet of finite resources.

If the challenge was simply resources, the outcome of hitting resource limits would be relatively straightforward. As resources run out, we would find alternatives and take greater care in reclaiming and reusing materials. It is not that simple; it is not just resources constraints we should be worried about. We are totally reliant on the ecosystem to be able to live. Operating society and the economy in ways which retain the integrity of our life support system should be fundamental to any system of human governance. In the past we did not have the capability and the scale to do lasting damage, so we did not have the means of suicide.

Now we are capable of turning off our life support. I don't think any sane person would take deliberate action to alter the ecosystem so that it no longer supports us. However, neglecting to protect it can have the same outcome. To advance to the next level of human progress we need to learn to be stewards of the ecosystem. The economic principles which have been used to drive policy are out of date, need to be rethought and replaced. If economists can't do this, then people outside economics will have to step in and insist that principles are set which transcend economics and frame the formulation of economic policy. It will take careful analysis and deliberate decisions to overrule the narrow pursuit of economic efficiency. Instead of a maximised global economy operating without constraint, the world needs a civilised economy which supports human aspirations to live well into the long future.

Human impacts on the ecosystem in the era of globalisation are not careful manipulations to improve how it supports society, but are more akin to acts of vandalism. Conventional economics regards the ecosystem as there to be exploited, and within the economic model there are no intrinsic safeguards. Almost any impact is acceptable provided sufficient benefit shows up in the economic figures. Pushing back against the guardians of economic orthodoxy meets resistance. It is exceedingly hard to convince mainstream economists that the ecosystem should be protected and conserved using higher order principles to frame the economic analysis. Environmentalists can be accused of being unworldly dreamers, when in fact it is the economists who continue working on the assumption that the ecosystem will remain intact by default, who are in cloud-cuckoo land. It is a fact that natural systems are being overexploited; we need to understand that this is the inevitable consequence of economic globalisation. Environmental regulations cannot provide the solution unless we first fix the flawed economic model.

Reconciling economics and ecology

We have allowed an economic system to evolve, and become deeply rooted in the current version of civilisation, which is in conflict with the ecosystem. At first, this was not obvious. The original architects of economics, such as Adam Smith and David Ricardo, proposed economic concepts which still underpin modern economics. They developed the ideas of economic efficiency and free-trade in a world in which human activities did not have the scale to destroy the planet. If these great economists were alive today, and could sit down with us to discuss the challenges of the 21st Century, I believe they would engage with the debate and support making changes to economics. Adam Smith, I feel sure would be particularly concerned. In addition to writing the seminal economics book "An Inquiry into the Nature and Causes of the Wealth of Nations" (Smith, 1776), Adam Smith also wrote an equally powerful book "The Theory of the Moral Sentiments" (Smith, 1759). From this we can be reasonably sure that Adam Smith would be concerned at the impact of the current globalised economy; and horrified that the economic theory he developed had not be been reframed to fix the problem. He would fully expect that, more than two centuries later, economics would have advanced to address the emerging challenges of environmental consequences, not by adding yet more detail and ever more exotic theory, but rethinking the fundamental basis of economic theory. Society needs first and foremost policy to secure social cohesion within a safe-guarded ecosystem. Economic policy has to sit in support of such aspirations. Social and environmental policy should frame economic policy (McManners, 2014). This is blindingly obvious when you pause and think deeply, but pushing back against two centuries of economic "wisdom" is not easy. It is particularly hard when, as I argue, economic efficiency becomes a secondary objective applying to the implementation of policy, and not to setting policy objectives. We don't *need* growth and expansion; but we do need social cohesion and ecosystem stability. Abandoning the growth objective leaves economists struggling to know what to do. We need nothing less than reframing economics for the 21st Century.

Ideas for how economics should change are outlined in other essays in this special issue of the *Real-World Economics Review*. Not all will pass close scrutiny, but all should be considered. Policy adjustments have been proposed; such as trading carbon hoping it might help to reduce emissions; and offsetting ecosystem destruction in one place with eco-conservation in another. These are well-intentioned first efforts but fall well short of resolving the situation we face. Stating clearly the challenge is perhaps a good place to start.

> "The challenge is to reframe economics to support cohesive society living on a finite planet in ways which safeguard the ecosystem into the long future."

To my mind, this is fundamental. The key words are "cohesive", "finite" and "safeguard". It is worth reflecting on current economics and the extent to which social cohesion, planetary limits and ecosystem attributes such as biodiversity, are included in economic modelling and frame economic policy. The answer is very little. Social policy and environmental policy are seen as separate to, and in some people's minds, inferior to economic policy. No wonder the world is in such a dire predicament.

It may seem that to solve challenges such as climate change, biodiversity loss and social disintegration, we need social and environmental policy. The reality is that the way economics is framed must change, to have any chance of making progress. The new economic policy has to start with principles, which sit above the economic analysis, and frame the development of economic policy. This shift of mindset is huge. If we liken the progress of civilisation to a train running on a railway network, it is like switching the points to switch onto a different track. Trying to make adjustments to the controls in the cab on the train will make no difference except to slow or speed up the train. The only way to arrive in a better place and secure the future is onto a different track. The switch required is a total reframing of how society and the economy operates.

Reframing economics fulfils two purposes. The first is to bring economics back under the umbrella of higher-level objectives and aspirations. The second is to set up economics as the enabler of high-level policy. Whilst economic principles are used to drive high-level policy we are seduced by the apparent success; but this success has consequences. There is a lot to like; many people are better off, both at the top of the wealth pyramid and at the bottom were people are being lifted out of absolute poverty. We have achieved all this through exploiting the planet's resources at an accelerating rate without constraining the economy to respect the capacity of the ecosystem. This has brought us to a civilisation which is already consuming more than the planet's capacity, with consumption increasing apace (WWF, 2018). Something has to give; and that is likely to be civilisation itself; unless we do something bold and dramatic to change direction.

We are as rich as we are, because we are not constrained by resource limits or the need to take care of the planet. We are rich at the expense of other people being exceedingly poor in the future. Poor in terms of resources and environment – which is what really matters. It would be irrelevant to use economic numbers to argue that a planet stripped bare and covered in concrete has greater economic value. This situation is deeply wrong. The people who will suffer are our children and grandchildren. This disregard for them is perhaps the first sign of social cohesion falling apart. As resource limits are reached, game theory research by the economist Petros Sekeris shows that conflict between nations is the expected consequence (2014). Combined with accelerating environmental degradation and an economy which hits the buffers these will be dangerous times. As the younger generation starts to allocate blame to those currently in power it is highly likely that social cohesion will unravel. This is not a prediction which I enjoy making, but on the current track, this outcome seems certain.

To be able to set up economics as the enabler of policy requires total clarity of the current situation. It is now absolutely clear that adopting economic principles as high-level policy drivers could only ever be a short-term blast. This opens up the debate about the future unconstrained by the baggage of old economics to work out the economics to apply going forward.

In this essay, my objective has been to alter the policy frame and steer policy makers' mindsets to think differently. If I have made a persuasive case for fundamental change, which can win people over to be willing to be part of the change process, then I will have succeeded.

Assuming it will be possible to win the argument that 20th century economics needs to be overruled, this will have no substantive impact until the new economic toolbox is developed to replace the old. I hesitate to go too far in explaining what I believe should comprise such an economic toolbox, because this is a debate in which I do not have all the answers; and a debate in which it is quite right that there should be many voices. However, I think it is worthwhile to explain one tool, which I propose should make it into the new economic toolbox.

This tool is the principle of Proximization (explained in detail below). I do not insist here, in this essay, that it is accepted and adopted. I use it to illustrate that reframing economic policy and adopting a different mind-set does indeed lead to developing different economic tools.

Developing the tools for the new tool box of economics will include new and innovative ideas. It will also include established economic tools, but applied in different ways according to principles which become bedded in the new economy. This requires people inside and outside economics to work out the detail. One trap to avoid is capture by the edifice of conventional economics. Long-established experts who have been responsible for developing and expanding 20th century economics, dominate the peer-review process for leading economic journals. They can get very defensive when new ideas threaten the old. Peer-review has a valuable role to maintain quality and rigour of published academic research. It can weed out bad research and papers based on spurious logic. The process should not be used to stifle debate in building the economics needed to deal with 21st century challenges.

The development of the new economic toolbox must be based on a fundamental shift in focus from economics as the policy driver, to economics as the policy enabler; from master to servant. It will not be easy to gain acceptance for this change of focus. I have experienced first-hand "expert" economists who will not accept that their discipline will be less powerful and less influential, with regard to high-level policy decisions, than it has been in the past. I also risk accusations of economic incompetence when I argue that there are decisions and choices which come above, and are superior to, pure economic efficiency. Such accusations would be odd because it is plainly obvious that decisions within the family, within the community, and by national governments are driven by people's aspirations and what they believe to be morally and ethically right. Such parameters sit alongside precise constraints such as environmental limits and the availability of resources. Applying the old 20th century economics involves crafting an economic and business case. Once completed, the case is subsequently subjected to social and environmental impact assessments. This sequential approach reinforces the dominance of economics and ensures that long-term overexploitation of resources and environmental damage is almost inevitable. Positioning the social and environmental analysis at the front of the process, sends the economic analysis down a different track.

As more and more people accept the demotion of economics to an enabling function, the easier it will become to build the new economic toolbox. This is a task for many people, both inside and outside economics. A multitude of ideas are needed to enter the debate, to be scrutinised, criticised, adjusted and then applied to real world situations. Nothing should be ruled out by those of a fixed "economic" mindset in the search for good policy derived from high-level choices. As a consensus emerges of what comprises "good" policy, this can guide and frame the design of appropriate economic policy.

One tool – proximization

Proximization is an example of a policy frame designed to ensure economics is an enabler rather than driver. This shows that the shift of mind-set proposed in this essay is not just a subtle alteration of how economics is framed but leads to significantly different policy. I don't lay claim to ground-breaking novel ingenious economic theory. Economics has always been an enabler, but the overreach of economics causes the problem. Reining back allows economics to regain its standing as a respected and useful means of analysis. I use proximization to illustrate how an adjusted mind-set leads to different economic policy proposals. The proximization policy framework is something I proposed a decade ago (McManners, 2008). Back then, it was out of sync with conventional economic thinking; a decade later such a framework is looking ever more relevant and ever more necessary. The framework derives from the mind-set explained in this essay. I realized that proximate economies suit the challenge of the 21st century far better than economic globalization. I use it as an example of the generation of new ideas. I believe it to be a sound frame for economic policy but I accept that it has not yet passed the scrutiny of others, so I present it here as a potential component of the new economics to be considered, debated, and tested.

> "Proximization is selfish determination to build sustainable societies, aimed at social provision and driven by economic policy, whilst minimising adverse impacts on the environment" (McManners, 2008, p. 31).

Beneath this definition lies a set of four supporting principles (McManners, 2008, p. 32; 2010, pp. 12–13):

- decision making on the basis of sustainability – balancing the economic, social and environmental consequences;
- subsidiarity – control left at the lowest possible level;
- the primacy of the state – where power and responsibility reside;
- use of market economics – constrained to fit local circumstances.

My book *Green Outcomes in the Real World* (McManners, 2010) expanded the concept and examined how it played out across the economy:

> "The proximization framework aligns economic levers with the needs of sustainable society. People do not see systematic global problems as their personal business. The key to the success of the proximization framework is that it brings the challenges facing human society within the sphere of national control, with decisions based on local circumstances. However, proximization is not synonymous with isolation and is not a policy of localizing everything. There will still be global trade, but at lower volumes, based on real needs and sustainable ways of satisfying these needs" (McManners, 2014, p. 197).

The 20th-century economic mind-set does not take kindly to the concept of proximization. Using a different mind-set, in which social and environmental issues are uppermost, it looks like an entirely sensible and rational framework to employ. Would it be economically inefficient? Absolutely it would. Economics is allowed to play out to facilitate efficiency at every level but not to overrule the more important high-level objectives and policy choices.

In my view, the set of principles which comprise Proximization should be a fundamental component of 21st century economics. I accept that it needs close scrutiny and examination by a variety of experts across many disciplines. I expect many economists to be deeply sceptical. I accept that it conflicts with current mainstream economic thinking; this does not mean it is wrong. This is a different track, which takes the world in a different direction. Whether I have identified the best alternative track, is for others to judge. Over the last decade, I have reflected and reconsidered proximization. It seems to me that this framework is the best, and perhaps only, sensible pragmatic policy which, if implemented in a timely manner, could prevent civilisation entering a death spiral of over-consumption and environmental overload.

My advocacy for proximization as the framework for future economic policy, shows how altered mind-set leads to different thinking and new proposals. I await whether my proposal gathers support and survives examination. In the context of this essay, it shows how the formulation of economic policy can be enriched by reframing economics as a flexible supporting discipline able to adapt to changing circumstances and evolve to address new challenges.

Switching track

Economics has evolved over the last two centuries, from Adam Smith's ideas to expand into a colossus which is now the dominant driver of policy. This has set the world on a track of extraordinary success, as measured by GDP and wealth. Continuing down this track, defined by economic objectives, leads to levels of consumption beyond our dreams and beyond the capacity of the planet. Success today comes at the price of downfall in the future, as "economic civilisation" hits the buffers of resource limits and unacceptable levels of environmental degradation undermining human health and agricultural capacity. Collapse of civilization has been on the cards for some time, because it is a logical certainty that a civilisation which uses economic glue to hold it together, requiring ever more expansion and ever-increasing consumption, will come to a grinding halt sooner or later. We take solace in the thought that this will not happen in the near future, and perhaps not in our lifetimes. This is no longer the case; collapse could come within two or three decades and certainly within our children's lifetimes – if civilisation continues on the same economic track.

Switching track without delay is vital to a vibrant and successful future for humanity. This cannot be done by good intentions leading to minor tweaks in policy and perhaps global environmental agreements without enforcement mechanisms. The world has to confront the cause of our predicament. That means pushing back against conventional economics. We need to find the courage and determination to insist that economics is demoted from "policy driver" to "policy enabler". This shift of language may seem trivial until you pause and reflect. This shift of mind-set takes economics down a quite different track. This reframed focus of economic policy places social and environmental objectives at a high level in the policy hierarchy, above economic efficiency. This is an alien concept to the current pervading economic mind-set which so dominated the 20th century. In the 21st century, we need to develop a new and better economics which supports a sustainable society living on a finite planet into the long future. Instead of economics driving policy, it should enable policy; instead of being master, economics needs to accept its true place as servant.

References

Ehrlich P.R., and Ehrlich A.H. (2013) "Can a collapse of global civilization be avoided?" *Proceedings of the Royal Society, B* 280, 20122845. [available at: http://dx.doi.org/10.1098/rspb.2012.2845].

McManners, P.J. (2008) *Adapt and Thrive: The Sustainable Revolution*. UK: Susta Press.

McManners, P.J. (2009) *Victim of Success: Civilisation at Risk*. UK: Susta Press.

McManners, P. J. (2010) *Green Outcomes in the Real World: Global Forces, Local Circumstances and Sustainable Solutions*. UK: Gower.

McManners, P.J. (2014) "Reframing economic policy towards sustainability." *International Journal of Green Economics*, Vol. 8 (3/4), pp. 288-305.

Rockström, J., et al. (2009) "Planetary boundaries: exploring the safe operating space for humanity." *Ecology and Society*, 14(2), p. 32.

Sekeris, P. G. (2014) "The tragedy of the commons in a violent world." *The RAND Journal of Economics*, 45, pp. 521–532.

Smith, A. (1759) *The Theory of the Moral Sentiments*. London: Millar.

Smith, A. (1776) *An Inquiry into the Nature and Causes of the Wealth of Nations*. London: Strahan and Cadell.

WWF (2018) *Living Planet Report - 2018: Aiming Higher*. Grooten, M. and Almond, R.E.A. (Eds) Gland, Switzerland: WWF.

Chapter 10

Economism and the Econocene: a coevolutionary interpretation

Richard B. Norgaard

We live in the era of Economism. Human consciousness is deeply etched by economistic beliefs in individualism, materialism, property, markets, economic growth, and freedom as consumer choice. These beliefs are necessary to sustain the system that supports us. But the economy we have is unlikely to support our grandchildren. Natural scientists argue that we are in a new geologic era, the Anthropocene, where people have become the major force in changing the geosphere: the atmosphere, oceans, and land. But it is the economistic beliefs that describe the cosmos of most people, bind people together, support their particular behavior, and sustain the economic system. Economism is altering the physical processes of the geosphere and collapsing the diversity of the biosphere. Econocene is a more appropriate term for the new geologic era. Fossil fuels and their technologies have transformed agricultural and industrial processes, the mobility of goods and people, and the geographies of cities and rural areas. People's values, ways of understanding, and social organization have coevolved with fossil fuels and their technologies, but it is economism[1] that binds people together and girds the economic system we have. We need a new "ism", a new human consciousness, to support a new relationship with Earth and its other inhabitants.

Economistic beliefs are not detrimental because they are mere beliefs. People need a belief system to live together. Yuval Harari develops this argument around the following statement.

> "Any large-scale human cooperation – whether a modern state, a medieval church, an ancient city, or an archaic tribe – is rooted in common myths that exist only in people's collective imagination" (Yuval Harari, 2014, p. 30)

Many critiques of the recent neoliberal economy make the same point that neoliberalism survives on a set of necessary public beliefs, but most critics imply that those who profit from the system orchestrate the beliefs. While not denying that those who most benefit from particular beliefs have helped push them on the masses, the process by which beliefs come to be held and sustained is more complex than this. People need beliefs to explain the system in which they live, and they need beliefs to rationalize their decisions and those of others.

---

[1] Bottomore (1991) describes the diverse ways Lenin, Gramsci, and other Marxists have used the term. Kwak (2017) uses the term to emphasize the unrealistic nature of the ideology of neoliberal economists. Cobb (1999) used the term to designate an era during which economic beliefs organize humanity. My use incorporates Cobb while building on Knight (1932) with respect to how economistic beliefs are necessary, indeed need to be religious in nature, to satisfy the needs of people as well as to keep the economy running.

Furthermore, people are able to choose between alternative beliefs and rationalizations being pushed by religious organizations, interest groups, and social commentators. The dominant choice of Europeans and North Americans switched during the 20th century from Judeo-Christian explanations to neoliberal economism. And the rest of the world also made this shift on their own time scales starting from their own religious bases.

The early Chicago economist, Frank Knight, argued in the 1930s that economics must be included among the beliefs in people's collective imagination. Except Knight used the term "principles", a term that plays an important role in science, but then immediately argues that the "principles" must be essentially religious.

> "The point is that the 'principles' by which a society or a group lives in tolerable harmony are essentially religious. The essential nature of a religious principle is that not merely is it immoral to oppose it, but to ask what it is, is morally identical with denial and attack.
>
> There must be ultimates, and they must be religious, in economics as anywhere else, if one has anything to say touching conduct or social policy in a practical way. Man is a believing animal and to few, if any, is it given to criticize the foundations of belief 'intelligently'.
>
> To inquire into the ultimates behind accepted group values is obscene and

> sacrilegious: objective inquiry is an attempt to uncover the nakedness of man, his soul as well as his body, his deeds, his culture, and his very gods" (Knight, 1932, p. 448–9).

> "Certainly the large general [economics] courses should be prevented from raising any question about objectivity, but should assume the objectivity of the slogans they inculcate, as a sacred feature of the system" (Knight, 1932, p. 455).

Note that Frank Knight argued that economists, mostly unbeknownst even to themselves, should be the surreptitious purveyors of economistic beliefs as religion. Or, to paraphrase and mix Marx with Knight, economists need to be pushers of the opiate to the masses where now religion is economistic beliefs. And yet economists are portrayed to be and tthink of themselves as objective scientists dedicated to reason and reason alone.[2]

Let me be more specific. Economism consists of the shared beliefs that support the market order and capitalist growth upon which most of humanity is currently absolutely dependent. Laborers, white collar "technocrats", entrepreneurs, capitalists, financiers, and specialized scientists including economists work together

[2] While acknowledging Milton Friedman's theoretical and empirical accomplishments, the most important role he played was as a public spokesman for market religion through popular books, a television show, and numerous public appearances. Friedman pushed economistic beliefs as religion in accordance with the argument above of Frank Knight, one of his mentors.

in amazing synchrony through shared economic beliefs that:

a) Explain and rationalize one's place in the economic system,
b) Rationalize the dominant way in which people interact with each other as a process of free choice,
c) Rationalize how "greed is good" in opposition to earlier religious/secularly-based moral teachings with respect to care for others,
d) Divide nature into property that can be owned and traded,
e) Rationalize growth of GDP as progress,
f) Explain the nature, including the emergence, of the economic system,
g) Rationalize transcendence through consumption, the meaning of life is to consume more and more, the mandate of nations is to grow.

Note that as listed here, the belief system is "complete" in that it includes everything that a religion would include: an explanation of the cosmos, of one's place in it, and how to behave. While most people hold other beliefs as well as economistic beliefs, increasingly since mid 20th century, economism has displaced earlier religious beliefs or become syncretic with religious beliefs as in Christian prosperity gospel (Bowler, 2013).

The belief systems that have organized people have changed over time. The beliefs that supported hunter-gathers were different from those that supported agricultural societies that were different from those that have supported industrial societies. This gives us hope for another change that will support people and planet. Yet, paraphrasing Albert Einstein, we cannot get out of the crisis we have created through economic thinking by using economic thinking. A coevolutionary framework for thinking about history and possible futures is an alternative that provides insights.

A coevolutionary framework

Over nearly four decades, I have argued for a coevolutionary framing of people's historical and current relations to nature (Norgaard, 1981; 1994). Others have also found this perspective insightful.[3] Coevolution in biology is a process where two species select on each other (Ehrlich and Raven 1964). Evolution is typically explained in terms of a single species being selected upon by physical conditions of the environment. Tortoises, for example, evolved to be better and better adapted to dry environments through competition for resources and the natural selection of those tortoises more fit for dryness. The Western idea of progress (Bury, 1920; Nisbet, 1980; Lacsh, 1991) easily aligns with the idea of the tortoise becoming more and more fit. Social Darwinists starting in the late 19th century falsely adapted the idea of the survival of the fittest to justify, under a banner of progress, how superior people were outcompeting inferior in the newly emerging corporate industrial capitalist economy (Hofstadter, 1944).

---

[3] The "economic" literature on coevolution is surveyed in Kallis and Norgaard, 2010.

While physical environments are important in the selection process, so are how each species interacts with other species leading to species selecting on the characteristics of each other. More broadly, coevolution is the sum of evolutionary changes of interrelated entities selecting on the characteristics of each other. Each entity in a coevolutionary relationship exerts selective pressures on the others, whereby each affects each other's evolution. Note that with coevolution, there is no equivalent to the concept of progress. The characteristics of species simply change in response to each other's changes.

The concept of coevolution has been extended to the interactions of systems and how they select on the characteristics of each other. [4] A process of social and natural system coevolution is portrayed in Figure 1. The blue arrows portray the direct cause and effect feedbacks between the two systems illustrating how people typically think of how nature affects us and we affect nature. The red arrows in Figure 1, however, also suggests how the two systems can be understood as *coevolving* together with features of the social system favoring the more effective reproduction and survival of particular features in the natural system and *vice versa*.

[4] Lumsden and Wilson, 1981, respond to the cultural critique of Wilson's sociobiology by including cultural systems in the coevolutionary process. Peter Corning, 1983 also provides an independent systems response that is constructive.

Figure 1 The coevolution between nature and society

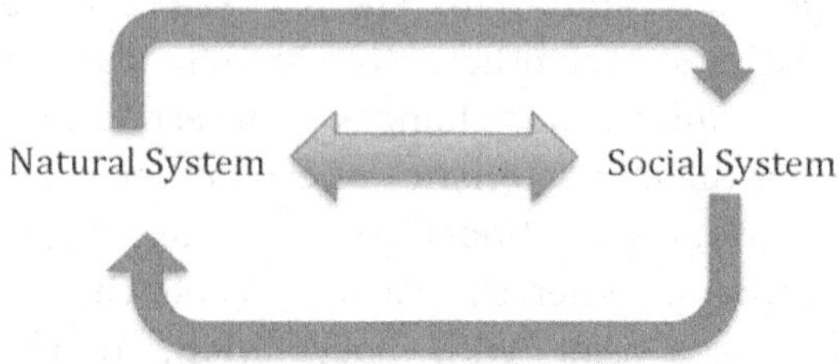

Figure 2 The coevolution of social subsystems with the environmental system

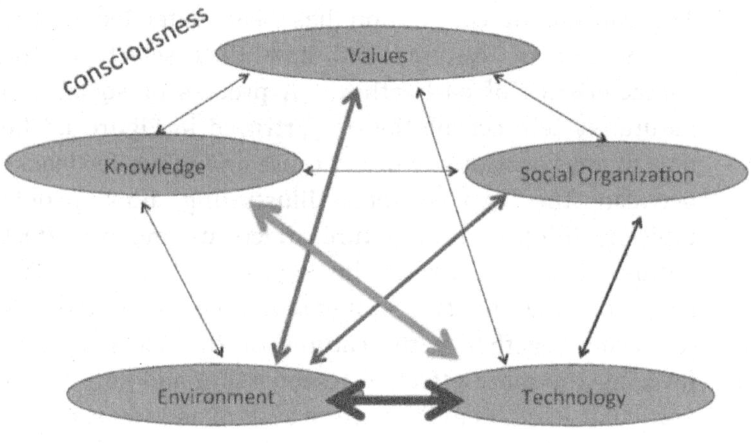

In *Development Betrayed* (Norgaard, 1994), I break the social system into four subsystems: values, knowledge, organization, and technology shown in Figure 2. I envisioned a process wherein each subsystem interacts with the others in direct (mechanical) ways while they also coevolve together through selecting on the characteristics of each other while also interacting and coevolving with the natural system. The distribution of characteristics in each subsystem also changes by innovations and introductions from other areas.[5]

Note that I have put the word "consciousness" between and above the knowledge and value subsystems to indicate that when I use this word, I am thinking of it as a combination of the two. The coevolutionary framing of human interactions, mechanical and evolutionary, with the environment has some special features that are critical to the overall argument of this paper.

First, as in the coevolution between species, things just change in response to each other. There is no presumption of progress. Indeed any criteria for progress are within the value subsystem that is itself coevolving in response to the changes in the other subsystems. And this provides a direct insight into how the nature of progress changed from moral progress during the 17th century to include material progress beginning in the latter 18th century, to become economic progress during the 20th century, and then since 1980 or so to become simply "growing the economy" or GDP growth. Values coevolved with increasingly dominant economic understandings within the knowledge subsystem as well as with the increasingly dominant market organization of

---

[5] This framing is different from the dominant framing within ecological economics of the economy being within, and a subsystem of, the environment (Daly, 1973; Daly and Farley, 2011). Readers may find my portrayal of the environmental system as no larger than any subsystem of the social system rather off-putting. This is a different framework emphasizing a parallel framing of processes rather than of magnitudes of stocks and flows. Note that the social system selects on characteristics of nature, not on the characteristics of natural laws like gravity or thermodynamics.

the social system. As values became more economistic, the criteria of what constitutes progress changed accordingly.

Second, as the previous paragraph clearly suggests, the coevolutionary framework explains path dependence or "lock in" very easily. This characteristic of the framework does not offer much hope for humanity getting out of the current crisis. And yet, coevolution also explains how wholly new features can arise, giving us hope. The environment is changing because of climate change, forcing new direct interactions as well as selecting on the characteristics of the social subsystems. While the lock in was sustained for a decade and a half, especially strongly in the United States, there is now clear evidence that climate change is influencing the consciousness of people around the world.

Third, the mechanical processes also illustrated in this coevolutionary framework suggest how human consciousness, the sum of understandings in the knowledge subsystem and beliefs in the value subsystem, sustains the social organization and technological subsystems that exist. Many authors have noted that particular values and understandings among the people are necessary to sustain a particular economic system.[6] The coevolutionary framing, when the arrows are viewed as cause and effect relationships, illustrates this.

Fourth, the coevolutionary framework illustrates how environmental, organizational, and technological

---

[6] For a recent example with respect to neoliberalism that also reviews the prior literature, see Streeck, 2017.

realities coevolve with people's consciousness, how people understand and value things within the reality they are simultaneously changing. This framework is constructionist, and explicitly so. In this framing, understanding, for example, is recursive, incorporating how prior understanding effected actions taken and the selection processes that changed society and nature. For example, historically we understood soils mostly as physical and then later as chemical systems. While we now understand soils more as biological systems, or biogeochemical systems, our understanding of the agricultural soils that exist today is more complete, and thus better, when we incorporate how we had historically transformed these biogeochemistry systems through plowing and the application of fertilizers based on our earlier, dominantly physical and chemical, understanding of soils.

Understanding how past thinking has created the world "out there" is important for understanding agricultural soils, but it is even more important for understanding our economy. The economy and the problems we have today reflect our past understandings that have been dominated by neoliberal beliefs about markets as self-regulating, about the superiority of markets to government, and about how economic growth supposedly advances well-being and even brings about environmental protection too. People, with the help of the economics profession, have come to worship markets and condemn the supposed inefficiency of governmental "command and control". Yet we ignore the phenomenal rise of the large corporations that employ us and provide us with our daily goods and services. Corporations large, many larger than nation-states, as well as small are organized and supposedly run efficiently by command and control. Somehow, the economics profession fails to teach this, nor do people choose to notice the anomaly either. It is easier to ignore realities that question values, at least for a while. Indeed, as I will try to show, a false consciousness is partly necessary.

Within this framing, let me explain how we reached the crisis we are in.

A coevolutionary history

There have already been 3 substantial transformations in human consciences that have accompanied major organizational changes in societies: 1) from hunter-gatherer societies to agricultural societies, 2) from agricultural societies to nation-building societies, and 3) from building nations to economism (Harari, 2015; Cobb, 1999). A fourth change in consciousness driving and coevolving with other changes, perhaps an Earthism or ecologism, is needed to assure environmental sustainability, social justice, and meaningful lives.

From hunter-gather to agricultural societies. Being smart, especially since the emergence of Homo sapiens a quarter of a million years ago or so, people learned that they could hunt more successfully by hunting together. It also made sense to share what they caught, for some hunting parties were more successful one day, others the next. And young children and elders, best left in camp, needed food too. Sharing was good for the success of all.

Working together and sharing made productive and reproductive sense. Cooperation works best when there are expectations that people can be trusted to meet such expectations, and trust tended to formalize into moral rules. Hence, from the earliest of times, the processes of production and distribution and the human qualities of being trustful and moral, or what we now think of as the separate realms of economics and religion, have been tightly fused.

Religions provide more than simply moral guidance. Hunting, as well as the gathering of nuts, fruits, and vegetables, entailed working with the intricacies of nature. People had practical questions about the timing of events in nature, many of which were important to their material success. For these, people slowly contrived through experience and passed between generations through survival of the fittest increasingly good enough arguments that they composed into stories to document how to work with nature. Some of these stories improved hunting and gathering techniques, partly by cause and effect, partly by selecting on each other. These earthly queries intermixed with larger questions about the heavens and earth, the cosmos, for which existential myths evolved. The ethics of accessing nature and sharing became intertwined in these earthly and existential stories as well.

For the vast majority of human history, people lived in tribes of 50-200 people. The small size of tribes facilitated, though did not guarantee, an organizational structure with information sharing and something close to collective decision-making. People's environmental impacts were largely local and temporary, though people did drive some species to extinction. Most importantly, when a tribe's environment deteriorated, whether by their own doing or an act of nature, there were possibilities of moving to new territory, for population levels overall were low.

From agricultural societies to nation building societies. After many millennia, grazing and farming started gradually within hunting and gathering communities. Dominantly agricultural societies arouse as the effectiveness of agriculture increased and perhaps also as population levels demanded. Agriculture vastly increased people's ability to capture the sun's energy and transform it into food. There were modest increases in well-being, especially for those at the top of the hierachical societies made possible by an agricultural surplus. But most of the productivity gains were absorbed by population growth. Farming facilitated an estimated 225-fold increase in human population during the 12 centuries prior to the rise of industry in 1800, as shown in Table 1.

Cultures largely based on hunting and gathering coexisted with agricultural societies, but they were pushed into mountainous, desert, and other less desirable landscapes. Agricultural societies began having new and larger direct impacts on the environment and put new selective pressures on other species. People in different regions transferred a few seeds, plants, and animals, exchanged ideas about the origins of the universe and the meaning of life, and even traded a few practical items, such as salt and spices, over considerable distances. Yet overall, interconnections between societies were relatively few compared to later times, and thus cultural diversity between the patches was considerable. Such a world might be sketched as in Figure 3, with the coevolutionary processes shown in Figure 2 taking place in each patch. With low interconnectivity, the failure of one culture did not reverberate through and take down societies around the globe.

Table 1 Population and global gross product through history

Date	Population billions	Global Market Activity in trillions 1990 world dollars
2000	6.3	41
1975	4.1	15
1950	2.5	04
1900	1.6	01.1
1850	1.2	00.036
1800	0.90	00.018
1600	0.55	00.0077
0	0.17	0.0018
-10000	0.004	0.000037

Estimates by J. Bradford DeLong 2008

Figure 3 A cultural patchwork quilt with an occasional transfer of a plant or technology

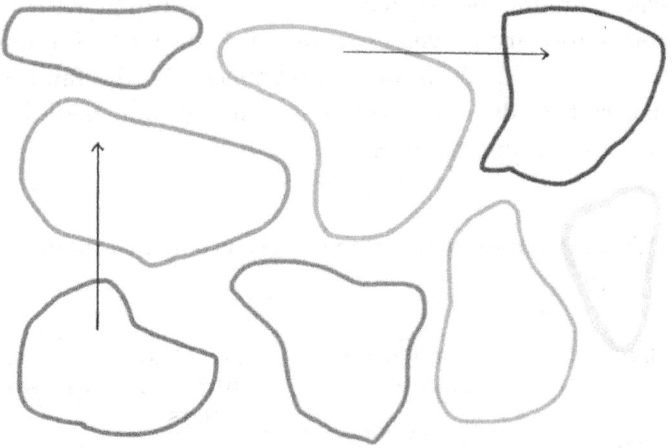

Agriculture, however, was not simply a magnificent human advance through new technology and social organization as conventionally portrayed. There were dramatic transformations, and clearly not all were favorable, in people's consciousness of nature, of their values and knowledge systems, in the process of becoming agricultural societies (Harari, 2014). The tedium of working the soil and harvesting but a few crops rather than dynamically interacting over a wide landscape with a diversity of plants and animals selected against the larger consciousness of nature possessed by hunters and gatherers. Consciousness coevolved toward new, simpler ideas fit for farm laborers. People's social consciousness also needed to be civilized. New rationalizations evolved to support living in larger groups and working as field laborers rather than at a higher level, and vice versa, in the newly formed social hierarchies. Formal religions arose as specialists took on the task of developing, maintaining, and conveying moral principles and origin narratives. Knowledge, values and social organization changed through coevolutionary processes in ways that complemented the changes in farming technology. All of this constitutes the very nature of agri*culture*, the outcome of the agricultural coevolutionary process.

For 10 to 15 thousand years, most peoples lived in multiple, fairly distinct, predominantly agricultural, societies. A millennium ago, the people who were to eventually think of themselves as Europeans were organized around the Catholic Church. Christian beliefs rationalized and supported the feudal social order for centuries. It was an age of Christianism even as Protestantism challenged Catholicism (Cobb, 1999). It was traditional religions that also organized people in agricultural societies pretty much around the world. In the last centuries of agricultural societies, however, knowledge, values, technology, and social organization began to coevolve in new ways.

European intellectuals' sense of the world and their place in it began to change with the Renaissance beginning in 1300. The emergence of modern science proved critical to how people interpreted nature. In the Abrahamic tradition, a single designer created the heavens, sun, planets, and Earth and creatures, plants and people as a whole with Earth at the center of the universe. People, formed in the creator's image and being most favored, had dominion over, yet responsibility for the care of, nature. Modern science succeeded by studying the components of nature separately, it reordered the sun, planets, and Earth, and it ever so slowly set people free from Christian and other religious dogma about nature, though that process is still ongoing. People's sense of dominion began to coevolve with science and technology into the hubris of control of a spiritless world.[7] The new ideas of modern science in Europe coevolved with social organization, specifically the authority of the Catholic Church, over centuries and spread slowly through the population.

Again, taking a European perspective, there were also more and new interactions with other parts of the world, with people of other cultures. Beginning about 500 years ago, Europeans carried plants, animals, and diseases to and from the New World. Soon after the movement of people and goods over the great oceans began to more tightly connect what were separately coevolving patches of social and environmental systems. This created a smaller number of larger patches, beginning the process of reducing the diversity between cultures as well as natures (Crosby, 1973; 1986; Mann, 2011).

Changes in European perceptions of themselves, both with respect to nature and social organization, also coevolved around very important new ideas about individualism that coevolved with the rise in atomism in natural philosophy. Martin Luther's call for reform of the Catholic Church stressed that individuals were

[7] Lynn White (1967) set off an extensive debate about the role of Judeo-Christian teaching and responsibility for the environmental crisis in a famous article in Science. I provided an overview of the responses to White in (Norgaard, 2002), but of course that literature has continued since.

responsible for their own salvation through their own reading of the Bible, the only true source for coming to know Christ and God. Luther's call awakened individualism, expanded education to the masses so people could read, unintentionally further separated church and state, and ignited multiple intellectual Enlightenments: English, Scottish, French and eventually in the Catholic Church and feeding back on Protestantism too (Ryrie, 2017). The natural theology that evolved into natural history and then into natural science was increasingly built on atomism and the assumption that the parts of nature could be understood apart from each other. As a result, modern science split into disciplines with each discipline learning about particular parts of nature. No one needed to understand the whole because it was thought that the parts would naturally unify into the whole. Millgram (2015) characterizes the coevolution of knowledge with technology and social organization since the Enlightenment as the Great Endarkenment. People today, scientists included, are far less conscious of the environmental system in which they live than were hunter-gatherers. The Enlightenments' strong move toward individualism in social thinking and atomism in natural thinking became traits of modern beliefs that are at the core of today's crisis.

Figure 4 A coevolving patchwork quilt of cultures with more connections

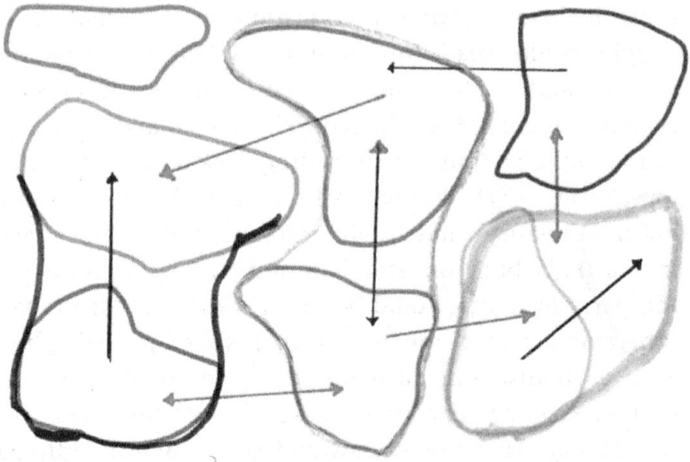

From agricultural societies to nationalism. A period of nation building arose after the Treaty of Westphalia in 1648 that ended the Thirty Years War between the shifting allegiances of royalty to Protestantism and Catholicism. Social philosophers sustained by a new wealth apart from the Catholic Church introduced disruptive ideas about the legitimacy of the power of those in authority to rule over other people, arguing instead that the people should only be ruled by their own consent. The authority bestowed by the Catholic Church on the authority of rulers was already breaking down, and these new ideas further selected against religious authority. Later, there were arguments not only for democratic election of rulers but also for democratic involvement in decision making generally. As Europeans coevolved into nation-states, nationalism became the dominant belief system. It was nationalism that organized the new nations of the new world in the 18th and 19th centuries as well as the breakdown of European colonialism and the rise of nationhood in Africa and Asia during the mid 20th century. Wars in the age of Nationalism were common because nationalist beliefs stressed imaginary ethnic identities, boundaries, and loyalty foremost though modes of governance were also important. Religious influences were still important too, but no longer ruling.

The liberal social thinkers of the Enlightenments who favored independent, free individuals realized that one cannot be free without property. Without one's own land or capital, one can only be someone else's laborer. And if employer's freedom includes letting laborers go, a freedom soon to be derived from market thinking, than freedom certainly requires all individuals to possess property. Around the same time, increased trade changed relative prices of farm products selected for an acceleration of the enclosure movement that started in England and spread to France as well. The transition from feudal societies to market societies separated large numbers of people from the land, causing great misery and great losses of freedom for the masses. The development and spread of liberal philosophy coevolved with the rise in the institution of private property for the few, selecting against the institution of common property with shared responsibilities. The individualism of liberal philosophy selected for individualism over cooperation and care in social organization and rationalized the demise of common property and responsibility under feudalism (Polanyi, 1957).

The demise of land stewardship and the rise of the idea of private property coevolved with notions of atomism in science, the idea that nature could be separated into parts. A new understanding of nature as complex interconnectedness, the science of ecology, would not evolve for another century. In the meantime, the myth that nature could be divided up into parts, without connections remaining, and owned by separate people became not only a part of human consciousness but a key condition of liberal society. The economic concept of environmental externalities has the story backwards. Environmental connections are denied in the concept of private land ownership and were made external in economic thinking from the start.

From nationalism to economism. The changes in how people perceived nature and organized themselves became clearly noticeable in practice around 1800. Europeans, at least those with sufficient property, began to equate freedom with individual choice, sensed a control over nature through technology, the idea of progress began losing its moral base and switched toward the possibility of material abundance for all. These changes coevolved with a dramatic increase in access to energy through the mining and combustion of coal followed by petroleum in the next century. Rather than coevolving with the environment, our social organization, technologies, and even the balance of the ways people understood began to coevolve around fossil fuels.

The economy began to coevolve around fossil hydrocarbons and their associated industrial and transportation technologies beginning with coal in the late 18th century and then petroleum beginning in the latter 19th century. In 1901, Svante Arrhenius documented that carbon in the atmosphere from the combustion of fossil fuels would increase the natural greenhouse effect that keeps the planet reasonably comfortable and warm it further. His calculations of when the warming would become dangerous was grossly in error because he had no way to foresee how rapidly fossil fuel technologies would dominate others and find new niches as well. While this error proved critical, it is important to realize that Arrhenius was making a serious effort to understand the impact of people on the geosphere. The vast majority of theoretical scientists were busily digging deeper, narrower strands of knowledge that occasionally other more applied but still specialized scientists and engineers were turning into technologies that were profitably introduced into human and natural environments with little if any concern for their larger consequences. How could they be concerned given their fragmented training and lives in specialized organizations of specialists who also were oblivious of larger systems? The fragmentation of knowledge and how that coevolved with social organization is a central part of this coevolutionary history of humanity's predicament.

Figure 5 Social system coevolving with fossil fuels

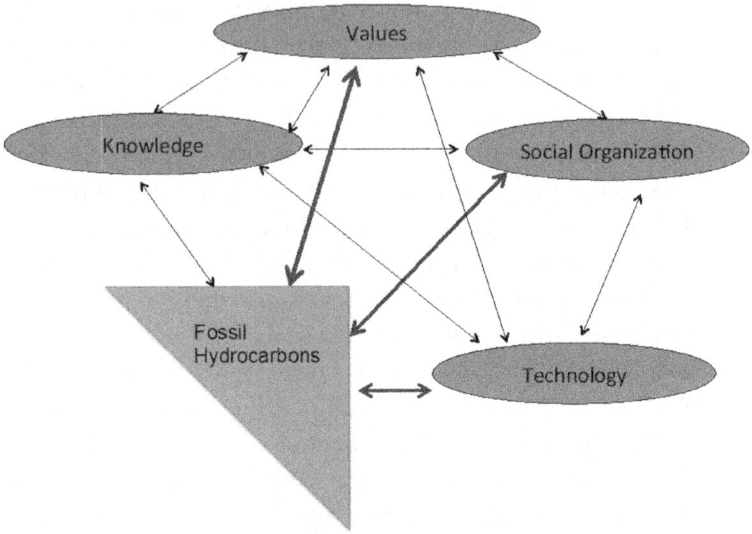

The uniformity across geographies of fossil hydrocarbons and their technologies and the economies of scale of fossil hydrocarbon technologies selected for the corporate industrial order we know today. These direct changes, along with the coevolutionary processes of selection, freed people from coevolving with the complexities of the natural environment. This in turn gave rise to modern economism that pays no heed to nature. With our cosmos being the modern industrial order, economism emerged as the dominant secular religion, an eclectic package of beliefs that explain our place in the economic system, our relation to other people and nature, and how we should live what has been deemed a meaningful life.

Belief in markets spread, indeed was carried around the world, even forcefully so, to counter the rise of the Soviet Union in the Cold War, through efforts to "free" trade globally, and through the implementation of the idea of development. By the second half of the 20th century, much of the world was beginning to look like the market world assumed in economic models. In the late 20th century, the globalization of capital began and the interconnections between the patches of Figure 3 began to look more like Figure 6.

Figure 6 A globalized former patchwork of cultures

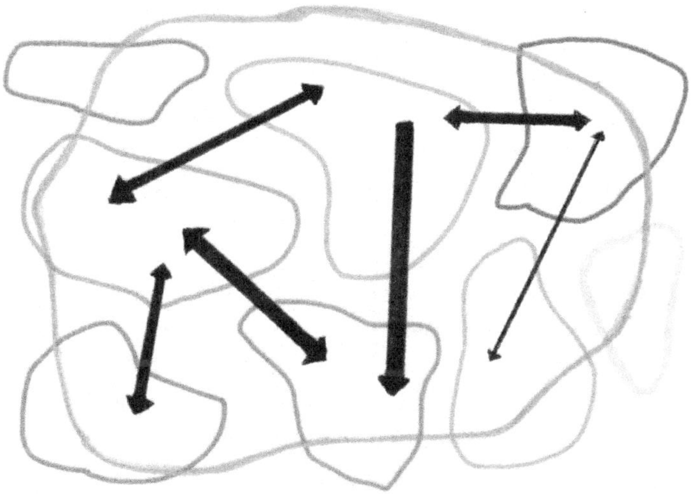

People performing specialized tasks are now so interdependent through markets that if people do not believe in markets and their larger purpose, all markets would collapse, as financial markets nearly have periodically, most recently in 2008. If markets collapse most of our population of 7.7 billion people would very quickly starve. Economism is necessary to sustain the economic cosmos in which people live.

Economism, however, has also become the dominant form of reasoning and the source of metaphors and utopias used in public communication. With the shrinkage of other ways of thinking about systems, economistic terminology has even become critical to how conservation biologists explain nature to the public. Nature, like other forms of wealth, can be thought of as capital that pays dividends in the form of ecosystem services. Saving nature has become a process of designing economic incentives for individual actors to invest in nature in order to reap her ecosystem services. In turn, conservation biologists now frame their research around market terminology to back up the ecosystem market programs they have helped facilitate. Biology is becoming economism.

The industrial order sustained by economism is not sustainable itself. We are in the Econocene maintained and coevolving with economism. Any new social organizational system that is sustainable, socially just, and provides meaningful lives will also need its "ism" to keep it going. This raises a key question. How can we have new system of beliefs / values, ways of thinking, and social organization emerge, a new ism, without crashing the current economic system, with economism maintaining it, on which we depend during the transition?

Figure 7 The coevolution to economism and the industrial order

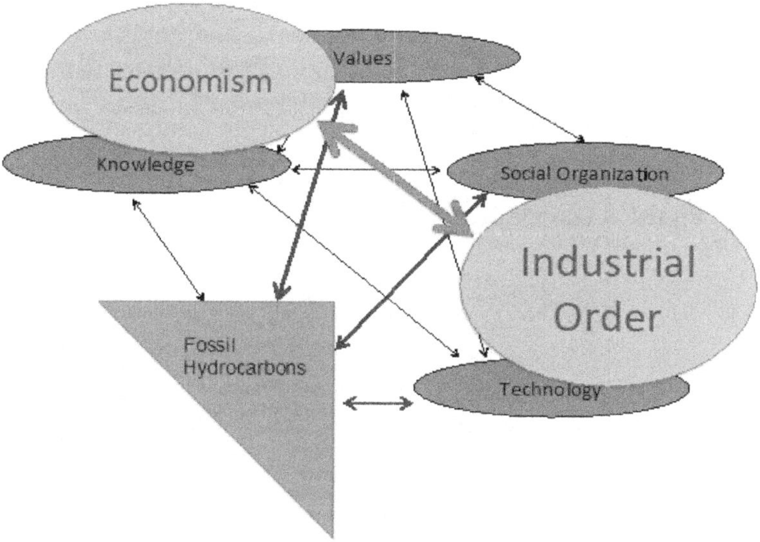

During the 20th century economistic beliefs have supported diverse and coevolving capitalisms as we know them and resulted in spectacular changes. Human population roughly quadrupled from about 1.6 billion people to 6.3 billion people. Global market economic activity during this period increased by nearly a factor of 40, or about 10-fold per capita. This rise of market activity entailed a parallel rise in specialization in work and associated knowledge. We went from a 19th century world in which the vast majority of people on the globe were pretty closely tied to the land and performing a similar mix of comparable agricultural and domestic activities to a 21st century world in which most people are performing specialized tasks using task specific knowledge. People are tied to bureaucratic structures, both public and private, while being globally interconnected by markets.[8] This new system has proved extremely effective at producing material goods while also presenting unprecedented social and environmental challenges. It is this transformation into what I will call the Econocene that must be understood in order to find our way out.

Figure 8 Economism coevolving with the Econocene

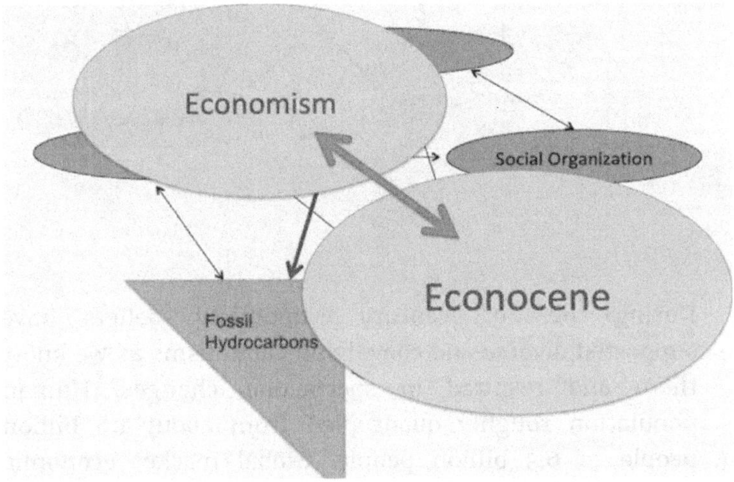

[8] I have skipped over the deliberate role of economists in supporting the most important and global economism of all, neoliberalism. The role of the Mount Pelerin Society and the Chicago School is very well documented. I am also skipping over the role if international institutions established after WWII and their role, in the midst of the Cold War, in establishing a neoliberal economic order that led to economism coevolving with the Econocene.

While social organization, knowledge, and values were coevolving around fossil hydrocarbons and their technologies, however, the geosphere and biosphere systems were operating on a different time scale, accumulating the CO_2 and other greenhouse gases that are now resulting in climate change, sea level rise, and a further quickening of the extinction of species.

The Econocene is a period of rapid transition of the geosphere and collapse of the biosphere. The transition to sustainability, social justice, and meaningful lives will not occur simply through the use of market mechanism to reduce carbon in the atmosphere. The economy has become our cosmos. We awake to stock market reports from financial capitals several time zones to our East, work in command and control hierarchical corporate structures while praising free markets, and are absolutely dependent on others in distant places working for the global economic machine. City lights and polluted air curtain us from the starry heavens, few are even aware of the phase of the moon. Reality is on the screens at our desks and on our cell phones in our hands, we share hearts through social media rather than in person. To face the reality we are in, our consciousness needs to become much more closely aligned with how nature and people function in a rapidly changing interaction. The economism that drives and coevolves with the Econocene must be replaced with a new "ism" that is environmentally sustainable, socially just, and supports meaningful lives.

Figure 9 Economism and the Econocene challenged by the reality of climate change

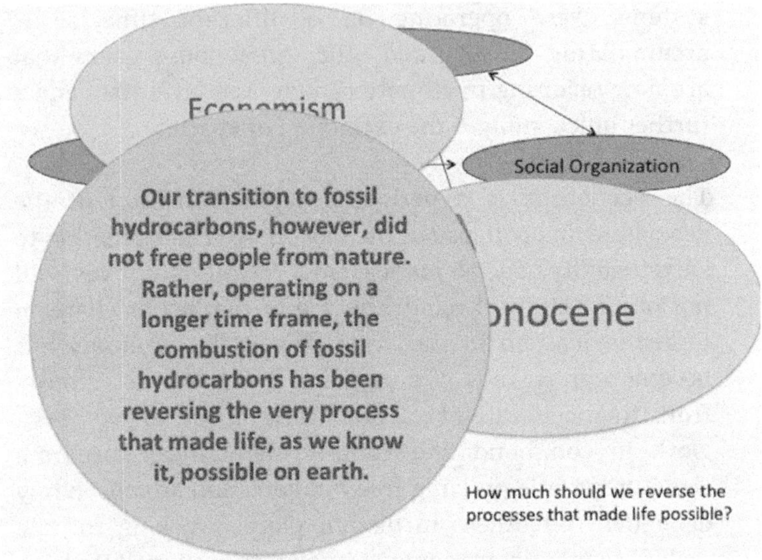

Humanity, fortunately, has been through multiple major transitions before. But now all of humanity is absolutely dependent on a tightly coevolved system of beliefs and social order. If people did not believe in markets, if economism were not equivalent to a religion that frames each person's very existence and *modus operandi*, all markets would collapse, as financial markets have, and 7.7 billion people would starve. How can we change to a new consciousness, to new systems of values, of knowledge, of social organization, and of technology that will coevolve without crashing during the transition and be sustainable thereafter?

Fortunately, capitalist economic order has proven pretty malleable, indeed significantly reconfiguring every quarter century or so. Evolutionary and coevolutionary processes also can occur rapidly. Counter to our mechanical intuition, coevolution explains change, including the evolution/emergence of wholly new properties, even while it explains "interlockedness". This is the good news. The bad news is that the story of progress through conquering nature through better science and technology has been strong for several centuries. While capitalism has indeed changed, it has continually increased specialization and material and energy consumption while also increasing the separation of people, and their knowledge, from each other and nature.

Conscious consciousness changes for survival

The coevolutionary history provided in this article suggests at least the following four ways in which humanity's consciousness needs to shift.

From material progress to holistic survival and morality. The coevolution of economism with the Econocene has led humanity to the brink of disaster. Faith in progress has long been a part of the problem. Actions to stave off climate change have been trimmed and delayed on the presumption that countering environmental destruction has the opportunity cost of foregone human wellbeing through further investments in technology that further increase the production or provide novel forms of material goods. And yet studies show that wellbeing increases little, if at all, with further material assets after basic needs are met. Shifting from faith in progress toward a consciousness of holistic survival would be more appropriate given the challenges of climate change. I include the word holistic to remind us that we need to be more fully conscious of all peoples and other species too.

Most of the questions we face today are moral questions. We have neither fully faced our moral responsibilities to future generations raised by past environmental destruction nor faced climate change over the past three decades. Economists have avoided addressing moral issues in order to meet legislators' and the public's expectations and need for so-called "objective" answers. Hence economists talk of economic efficiency when moral issues are at stake. This shriveling of economists' ability to think and discuss moral issues is the essence of economism.[9] Economics, in theory, cannot say what is moral, but if political processes determine what is moral, economics can talk about alternative efficient economies that meet moral obligations and paths to them. It is past time for economics to work with moral reasoning and political decision-making rather than falsely standing in for them.

[9] Richard Howarth and I used an overlapping generations model to show that if we care about future generations by assuring them environmental rights, for example climate rights, then the efficient solution for future resource allocations changes, the rate of interest goes down, and environmental values go up (Howarth and Norgaard, 1992). In response, Resources for the Future organized a workshop, without inviting us to participate, where participants questioned the need for switching to a model that actually provides the option of addressing intergenerational equity and published the resulting papers in a book (Portney and Weyant, 2000).

From knowledge hubris to knowledge humility. We need to become much more humble with respect to how smart we are. If we were so smart, we would not be in this dire predicament. Science and the scientific community can become part of the solution, but we also need to acknowledge how science has been a part of the problem. Western hubris allowed technologies based on new findings in particular fields of science to be implemented in and spread through whole natural and social systems. Because we had scant knowledge of the whole, specialized innovations transformed the geosphere and biosphere as well as the sociosphere in unexpected ways. Those in denial of climate change are partly caught in the hubris of Western knowledge past. The environmental sciences still evoke a nature that is "out there" and slowly changing at most rather than a nature undergoing rapid change driven by our economy sustained by our beliefs. Science education, research, and participation in management and policy need to shift from the hubris of scientists as agents of material progress through specialization to scientists as humble seekers of understanding of whole and rapidly changing systems.

Given the limited nature of current knowledge, more experimentation in how we interact with nature, with quick corrective steps taken when experiments go wrong, would provide opportunities to learn through experience. Introductions of innovations need to be limited in general until our understanding is sufficient to develop criteria. There will be advantages to de-globalizing. Differentiation in our future economies will allow lessons to be drawn with respect to what might work better. The idea that we can design one best way to transition and sustain a better world is an extension of Western hubris.

From individualism to cooperation and care. Adam Smith wrote two books. We have neglected his first, *The Theory of Moral Sentiments.* Economists found the logic of markets in *The Wealth of* Nations compelling while wealthy converts with political traction spread selected messages. We need a significant shift towards the messages of Smith's first book. It still provides important insights into how empathy can build trust, responsibility, and care that are key to rethinking meaningful lives and social organization. And while Smith did not emphasize care across generations, we now need to care ahead.

From private property to global commons. The belief that land could be owned by a private individual and used however its land-owner saw fit gained traction in the west only centuries ago, an extremely short time in human history. Throughout history, what an individual could do with land has been restrained, but in America in particular, the idea of land ownership as sacred and any restraint considered a deep imposition on liberty and freedom. The interconnectivity of natural systems assured that private land ownership, especially when connected to markets ever more distant, would result in environmental disaster, and it has. The common threads between land need to be managed as a commons, and with today's technologies and markets, those threads have become global. Shifting consciousness in this direction will be difficult but necessary.

Just as a coevolutionary framework helps explain how humanity has come to the brink of social and planetary disaster, it can help us see how we might back off and set out anew. The framing is systemic and evolutionary, it incorporates ecological interactions and the selective processes of evolution, showing how things tightly fit together while also changing. It incorporates the best of postmodernist understanding. Social organization, technology, values, and even science, are "socially constructed", indeed even nature is increasingly being socially constructed, but none are *only* "socially" constructed. The "economy" is important, but to understand how to escape the coevolution of economism and the Econocene, it will be important to concentrate on how other aspects of life besides the material contribute to individual and collective wellbeing and can guide us into the future. We need to both concentrate on survival and consciously expand our consciousness.

References

Bottomore, Tom (1991) "Economism." In Tom Bottomore, Laurence Harris, V. G. Kierman, and Ralph Miliband (Eds). *A Dictionary of Marxist Thought.* Oxford: Blackwell, pp. 168-9

Bowler, K. (2013) *Blessed: A History of the American Prosperity Gospel.* Oxford: Oxford University Press.

Cobb, John Jr (1998) *The Earthist Challenge to Economism: A Theological Critique of the World Bank.* New York: Palgrave Macmillan.

Corning, Peter A. (1983) *The Synergism Hypothesis: A Theory of Progressive Evolution.* New York: McGraw-Hill.

Crosby, Alfred W. (1973) *The Columbian Exchange: Biological and Cultural Consequences of 1492.* Greenwood Press.

Crosby, Alfred W. (1986) *Ecological Imperialism: The Biological Expansion of Europe, 900-1900.* Cambridge, UK: Cambridge University Press.

Daly, Herman E. (1973) *Toward a Steady-State Economy.* San Francisco: W. H. Freeman.

Daly, Herman E. and Joshua Farley (2011) *Ecological Economics: Principles and Applications.* 2nd Edition. Washington, D. C.: Island Press.

Ehrlich, Paul R. and Peter H. Raven (1964) "Butterflies and Plants: A Study in Coevolution." *Evolution,* 18(4), pp. 586-608.

Gowdy, John M. (1994) *Coevolutionary Economics: The Economy, Society, and The Environment.* New York: Springer.

Harari, Yuval Noah (2014) *Sapiens: A Brief History of Humankind.* London: Vintage.

Hofstadter, Richard (1944) *Social Darwinism in American Thought, 1860–1915.* Philadelphia: University of Pennsylvania Press.

Howarth, Richard B. and Richard B. Norgaard (1992) "Environmental Valuation Under Sustainable Development." *American Economic Review,* 82(2), pp. 473-477.

Kallis, Giorgos and Richard B. Norgaard (2010) "Coevolutionary Economics." For a special issue edited by Gowdy, Kallis and Norgaard. *Ecological Economics,* 69(4), pp. 690-699

Knight, Frank H. (1932) "The Newer Economics and the Control of Economic Activity." *Journal of Political Economy,* 40(4), pp. 433–76.

Kwak, James. (2017) *Economism: Bad Economics and the Rise of Inequality.* New York: Pantheon.

Lasch, Christopher (1991) *The True and Only Heaven: Progress and Its Critics.* New York: W.W. Norton.

Lumsden, Charles J. and Edward O. Wilson (1981) *Genes, Mind, and Culture: the coevolutionary process.* Cambridge, MA: Harvard University Press.

Mann, Charles C. (2011) *1492: Uncovering the New World that Columbus Created.* New York: Vintage.

Millgram, Elijah (2015) *The Great Endarkenment: Philosophy in an Age of Hyperspecialization.* Oxford: Oxford University Press.

Nisbet, Robert (1980) *History of the Idea of Progress.* New York: Basic Books.

Norgaard, Richard B. (1981) "Sociosystem and Ecosystem Coevolution in the Amazon." *Journal of Environmental Economics and Management,* **8**, pp. 238-254.

Norgaard, Richard B. (1994) *Development Betrayed: The End of Progress and a Coevolutionary Revisioning of the Future.* London: Routledge.

Norgaard, Richard B. (2002) "Can Science and Religion Better Save Nature Together?" *BioScience,* 52(9), pp. 842-846.

Polanyi, Karl (1957 [1944]) *The Great Transformation: The Political and Economic Origins of Our Time.* Boston: Beacon Press.

Portney, Paul R. and John F. Weyant (eds) (2013) *Discounting and Intergenerational Equity.* New York: Taylor and Francis.

Ryrie, Alec (2017) *Protestants: The Faith that Made the Modern World.* Viking. (No location listed.)

Streeck, Wolfgang (2017) *How Will Capitalism End: Essays on a Failing System.* London: Verso.

White, Lynn Jr (1968) "The Historical Roots of Our Ecological Crisis." *Science,* 3767, pp. 1203-1207.

Chapter 11

End game: the economy as eco-catastrophe and what needs to change

William E. Rees

Prologue

In summer 2018-2019, Australia sizzled in record heat; every state simultaneously suffered temperatures of 40°C to 45°C for days at a time. A particularly searing week in late November killed thousands of spectacled flying-foxes as the mercury soared beyond 42°C, an unprecedented bat carnage that continued on and off through January. When temperatures finally eased in north Queensland at the end of that month, record rains and flooding drowned much of the region; 200,000 people were displaced (several died), hundreds of thousands of livestock drowned, and damage costs soared into the multi-millions.

Hardly anyone on the other side of the world noticed; people were distracted by their own problems. In North America, a weak and wobbly jet-stream allowed the Polar Vortex to bulge far south, engulfing much of Canada and the northern US east of the Rockies in a great amoeba-like lobe of frigid Arctic air. Record low temperatures were recorded at many sites. Late January temperatures in Winnipeg reached as low as -40°C (-40°F), -52°C (-62°F) with the wind-chill; Cotton, Minnesota was the coldest location in the US on 30 January with a low of -49°C (-56°F). At least 22 people across the continent died from the extreme cold.

Australia and North America may have been separated by 90 degrees, but extreme weather united their citizens in common concern over warming-induced global climate change. Indeed, all peoples now face a truly unprecedented communal challenge. We may perceive global warming, biodiversity loss, tropical deforestation, spreading marine dead zones, chronic air/water pollution, land/soil degradation, plummeting sperm counts, etc., as separate problems, but it is more realistic and potentially more productive to recognize that all are symptoms of a singular phenomenon, *gross human ecological dysfunction*. This is a genuine global meta-problem; it is potentially fatal to civilization and, paradoxically, entirely self-induced.

Which begs the question: how is it that the allegedly most intelligent and self-aware species on Earth is systematically destroying its own habitat, the only human-habitable planet in our solar system and the only planet most humans will ever know? The answer is, of course, multifaceted with roots in everything from what was once perfectly adaptive human behaviour, through Newtonian physics, to culturally inscribed (mis)representations of reality.

We cannot in a single chapter explore every dimension of the problem. However, we can show how several of the most important causal mechanism have come together to produce a global economic system whose conceptual framing, operating assumptions and *de facto* practices are pathologically incompatible with the very ecosystems that sustain it. In the circumstances, eco-destruction is inevitable. To understand this remarkable example of maladaptive behaviour we must begin with epistemology – how we know what we know – and a particularly quirky feature of human cognition.

We make it up as we go

> "You may say if you wish, that all 'reality' is a social construction, but you cannot deny that some constructions are 'truer' than others" (Postman, 1999, p. 76).

Many people are startled to learn that most of what they believe to be true, most of what they think they know, is literally *made up*. Foundational cultural narratives and social norms may masquerade as reality but they are nevertheless products of the human mind, massaged or polished by social discourse and elevated to the status of received wisdom by custom or formal agreement. All cultural narratives, worldviews, religious doctrines, political ideologies, and academic paradigms are actually "social constructs".

Indeed, it is not much of a stretch to assert that all formal knowledge is socially constructed. One passively acquires the convictions, values, assumptions, and behavioral norms of his/her tribe or society simply by growing up in that particular milieu. By the time most people have reached maturity they will have adopted their culture's overall "narrative" and will subscribe, consciously or not, to any number of subsidiary religious, political, social, scientific or other disciplinary paradigms.

Some well-known constructs are *entirely* made up – "capitalism", "communism", "civil rights", and "democracy" for example, have no true analogues in the non-human world. These and similar concepts were birthed in words and given legs entirely through socio-political discourse. Other social constructs are created explicitly to describe corresponding real-world phenomena. For example, the domain of science encompasses anything measurable in time and space – gravity, light, energy, matter, etc. Science is unique among formal ways-of-knowing in that scientists explicitly test the validity of tentative constructs (hypotheses) about the real world through observation and experiment and adjust their understanding accordingly. When experimental results are reliably replicable, a hypothesis may be elevated to the level of accepted theory – a social construct that can be used for explanation and prediction concerning a particular entity or phenomenon.

Still other social constructs occupy a middle ground. For example, we can agree that "the economy" is that set of human activities involving the production, distribution and consumption of goods and services. That said, there are many ways of conceiving how an economy should be structured, each reflecting its followers' particular set of socially-constructed values and their beliefs/assumptions about the economy and its relationship to society, governance systems, ecosystems, etc. One approach may give prominence to concepts or activities that are marginalized or omitted altogether from another (e.g., private vs. state ownership, free vs. controlled markets). Things can get complicated – any economic paradigm is an elaborate socially-constructed model that may contain (or omit) other models that are themselves socially constructed.

By now it should be clear that much of what humans take to be "real" may or may not bear any relationship to anything "out there". More remarkably still, most people generally remain unconscious that their collective beliefs may be shared illusions – a cognitive enigma that may well determine the fate of humankind. No matter how well- or ill-founded, entrenched social constructs are perceptual filters through which people interpret new data and information; and, because our constructs constitute perceived reality, they determine how we "act out" in the real world. Millions of lives may be jeopardized if those in positions of authority cherry-pick data guided by some dangerously faulty but comfortable social construct (climate-change denial, anyone?).

Critical thinkers will recognize that, until proved otherwise: a) all constructs belong in the domain of "conjectural knowledge" and that; b) not all conjectures are created equal. Some conjectures will necessarily be "better" than others, particularly in terms of how well they represent biophysical reality. To reiterate, in science

"Conjectures [hypotheses] are our trial balloons, and we test them by criticizing them and by trying to replace them, by trying to show that there can be better or worse conjectures, and that they can be improved upon" ... "So long as a theory stands up to the severest tests we can design, it is accepted; if it does not, it is rejected" (Popper, 1972).

What does all this imply about economic thinking and "the economy"? Let's acknowledge that all economic theories/paradigms are elaborate conjectures and that none can contain more than a partial representation of biophysical, or even social, reality. If this is an important general limitation, we should be particularly concerned about today's dominant neoliberal economic paradigm (the economics of capitalism). Neo-liberal models incorporate a stinted caricature of human behavior, virtually ignore socio-cultural dynamics and make no significant reference to the biophysical systems with which the economy interacts. Moreover, as an economist colleague recently explained, mainstream economists generally do not operate in scientific mode. While natural scientists experiment and subsequently adapt their models better to represent reality, economists, particularly those enamoured with the idea of a self-regulating (free) market, would have the real economy adapt to fit their models.

Economics – a branch of human ecology (not!)

For present purposes, ecology might be defined as the scientific study of the cooperative and competitive relationships that have evolved among organisms in ecosystems and how these relationships serve to allocate energy and material resources among constituent species. Similarly, economics might be defined as the study of economic behaviour and the efficient allocation of scarce resources among competing users in human society. The parallels are obvious; moreover, since humans are of the ecosphere, and the economy extracts resources (energy and materials) from the ecosphere, economics should arguably be a branch of human ecology. Regrettably, the conceptual foundations of the two disciplines have diverged since their beginnings.

Neoliberal economics – not of this world

> "We cannot regulate our interaction with any aspect of reality that our model of reality does not include" (Beer, 1981).

The modern or "scientific" world-view that prevailed from the 18th through much of the 20th Century was framed by French philosopher René Descartes' mechanistic view of the universe as "a vast machine, wound up by God to tick forever" (Berman, 1984, p. 21). Descartes extended his model to include living organisms and even saw human thought as an iterative mechanical process by which the mind observes the world "out there" as separate object (hence the notion of "objective knowledge"). However, it was the genius of Sir Isaac Newton that effectively validated the Cartesian paradigm. Newton's *Principia* gave us apparently universal laws of mass and motion which describe the universe as a mechanical machine of unlimited dimensions behaving according to strict mathematical rules. At last, humans were potentially freed from religious superstition and other forms of unreason. For the first time, European society had a body of science that satisfied Descartes's mechanistic vision, including deterministic predictability, and promised humans the ability to manipulate nature indefinitely toward their own ends. This "scientific materialism" provided the technical foundation of industrial society, and helped entrench a new myth of human dominance over the natural world.

It also supplied a conceptual framing for modern economics. Contemporary neoclassical / neoliberal economics – which has enjoyed a remarkably uncritical sweep through the modern world over the past half century – finds its deepest roots in the concepts and methods of Newtonian analytic mechanics. Inspired by the unprecedented success of Newtonian physics, neoclassical economics was conceived in the late 19th Century as a sister social science in which market behavior could be modeled as "the mechanics of utility and self-interest" (Jevons, 1879). Its founders abstracted the economic process from nature, viewing it as an independent and "self-sustaining circular flow between production and consumption," in which "complete reversibility is the general rule, just as in mechanics" (Georgescu-Roegan, 1975, p. 348).[1]

[1] In the circular flows model, firms pay wages/salaries/dividends to households in exchange for labour and investment; households spend the money on goods and services purchased from firms, enabling the cycle to repeat itself.

From a human ecological perspective, this is pure aberration. It ignores extraction/ (over)harvesting, various material transformations, and the eventual discharge of the entire material flow back into the ecosphere as degraded waste (pollution). To interpret the economy as a self-generating circular flow without considering the unidirectional throughput of energy/material is akin to studying human physiology as a circulatory system with no reference to the digestive tract. One might as well ask biology students to accept that "an organism can metabolize its own excreta" (Daly, 1991, p.197). More generally, neoliberal theory lacks any realistic representation of the energy and resource constraints, functional dynamics, social relationships, interspecies dependencies and time-dependent processes at the heart of ecosystems thinking (see Christenson, 1991). There is no ecology in economics.

A related equally problematic construct is neoliberal confidence that resources are more products of human ingenuity than they are of nature. Theoretically, as scarcity forces up prices it will in turn both encourage resource conservation and stimulate the search for technological substitutes. "If [built] capital and natural resources are substitutes in production then neither can be limiting – if one is in short supply you just substitute the other and continue producing" (Daly, 2012). There are many historical examples of allegedly "near-perfect substitution" such as the partial displacement of dirty coal by solar photovoltaics in electricity generation. It has therefore become part of conventional wisdom that market factors are more than sufficient to overcome emerging resource scarcities (Victor, 1991). Indeed, neoclassical texts have long accepted that "exhaustible resources do not pose a fundamental problem" (Dasgupta and Heal, 1979, p. 205). Neoliberal economists have even dropped land/resources as a separate factor in their production functions, conflating it with finance capital (Wolf, 2010) and again implying that the contribution of nature *per se* to the economy is negligible.

Neither is pollution a serious problem in the neoliberal paradigm. Economists define pollution damage costs not reflected in market prices as "externalities" – market imperfections – that society can "internalize" if it *chooses* to get the prices right, through investment in improved technology, better regulation or pollution charges (e.g., carbon taxes). (This assumes – wholly unrealistically – that we can assign an accurate dollar value to ecological degradation.)[2] In general, society accepts ecosystems degradation as a necessary trade-off against economic growth. We pollute in exchange for jobs or income and see the point at which to stop largely as a matter of negotiated public choice, i.e., there are no unanticipated tipping-points-of-no-return or other serious risks. Significantly, any social construct that conceives of the economy as a self-sustaining system and rationalizes pollution neatly frees the human enterprise for perpetual growth.

H. sapiens as ecological entity

From the ecological perspective, human-induced global change – climate disruption, plunging biodiversity, etc.–

---

[2] Note that if full-cost pricing were possible, many people today could not afford at least some of what they perceive to be basic necessities.

is unambiguous evidence that our prevailing growth-based cultural narrative is seriously flawed. If we wish to re-construct the economy for sustainability our models must include a realistic representation of human behavioral ecology.

We can start by acknowledging that *H. sapiens* is a product of evolution and shares various adaptive genetic traits with other species. Consider bacteria dropped into nutrient broth or deer introduced to a food-rich predator-free island. Both species populations will expand exponentially, spread over the entire "habitat", deplete their "new" resource base and finally collapse. *H. sapiens* is little different. Indeed, *unless or until constrained by negative feedback (e.g., disease, starvation, self-pollution), human populations, like those of all other species, will expand into any accessible habitat and use all available resources – even at the risk of collapse* (though in the case of humans collapse may be delayed since "available" is determined by the state of technology).[3] Evolutionary point: Individuals that employ strategies to secure the most suitable habitat and acquire the most essential resources, on average, survive longer and leave more viable offspring.

Natural selection also generally favors individuals who are most adept at satisfying short-term selfish needs

---

[3] The exploitation of "tight" shale oil/gas, tar-sands, and deep sea drilling for petroleum are examples of advanced technology in pursuit of the last deposits of "available" energy resources. Consider, too, how constantly-evolving fishing technology (e.g., factory-freezer trawlers, sonar tracking) pursues various fish stocks to near-depletion. Credit cards are *virtual* resources that enable people to keep consuming after they have depleted real income.

whether by strictly competitive or in-group cooperative means (see Pratarelli, 2008). (If we don't claim some perishable resource now, some competitor might take it.) Humanity's well-known tendency to favor the here-and-now (i.e., to discount future benefits and costs) has almost certainly evolved through natural selection (and is one ecologically-significant behavioral trait that *has* been incorporated into economic methods and models such as cost-benefit analysis).[4]

Competition is clearly a major evolutionary driver. While not evident to modern urban-dwellers, humans compete, not only with other people, but also with other species for food and habitat. And we usually win – high intelligence and technology have ensured that *H. sapiens*' capacity for habitat and resource domination vastly outstrip those of all other species (Waring, 2010). Our species has the greatest geographic range of any ecologically comparable organisms – we have occupied all suitable, and sometimes even hostile, habitats; in terms of energy use, biomass consumption, and various other ecologically significant indicators, human demands on their ecosystems dwarf those of competing species by ten to a hundredfold. Human consumption of biomass, for example, exceeds the upper 95% confidence limits for biomass ingestion by 95 other nonhuman mammals by two orders of magnitude (Fowler and Hobbs, 2003). *H. sapiens* has become, directly or indirectly, the dominant macro-consumer in all major accessible terrestrial and

[4] A genetic predisposition is not an inevitability. Rather, it is a propensity that is likely to play out in the absence of countervailing circumstances such as moral codes, cultural taboos, legal prohibitions, or other social inhibitors.

marine ecosystems on the planet. Indeed, our species may well be the most voraciously successful predatory *and* herbivorous vertebrate ever to walk the Earth (Rees, 2010). (A remarkable degree of engagement for a species that sees its economy as floating free from the ecosphere.)

H. sapiens: "maximum power" exemplar

Austrian physicist Ludwig Boltzmann famously recognized that the Darwinian struggle for existence is effectively a competition for available energy to do useful work (Boltzmann, 1886/1974). Subsequently, ecologists Alfred Lotka (1922) and later Howard Odum formulated what is now known as the "maximum power principle": Successful systems are those that evolve to maximize their use of available energy per unit time in the performance of useful work (growth, self-maintenance and reproduction) (see Hall, 1995). "Maximum power" is arguably a fundamental organizing force in natural ecosystems.

Unsurprisingly, *H. sapiens* success in "maximum power" terms is unequalled. All species need energy to function and reproduce but most animals are restricted to the chemical energy they ingest in their food – endosomatic ("within body") energy. Humans have an evolutionary leg up in their near-unique capacity to employ "exosomatic" (outside the body) energy to do additional work from harvesting food to engineering the international space station. Thus, the technological history of *H. sapiens* is crudely marked by increasing exosomatic energy use per capita, from 20 giga-joules per year by hunter-gatherers, through 60 GJ/yr by early farmers to 200-300 GJ/yr in typical industrial societies (Fischer-Kowalski and Haberl 2007). At the upper end, this is equivalent to approximately 80 times human endosomatic energy use per person per year.

Human "maximum power" ascendance was kick-started about 10,000 years ago when agriculture generated the first major food energy surpluses. This enabled "civilization" – the emergence of social structure, governments and ruling classes, division of labor, specialization, etc., – and accelerated technological innovation in everything from agriculture through metal-work, and boat-building. Ponting (2011) documents how, for subsequent millennia, humans were able to explore and plunder terrestrial and marine ecosystems over virtually the entire planet, using animal (including human slavery), wood-fire, water, and wind energy alone. These are exosomatic energy sources, but are all derived from contemporary solar energy. It wasn't until the wide-spread use of fossil-fuels (vast stocks of stored ancient solar energy) that the human enterprise was able fully to exercise its "maximum power" muscle.

The effect was spectacular, unprecedented – and likely to be short-lived. It took 99.9% of modern humans' 200,000 year history for our population to reach one billion in the early 19th century. In just the next 200 years (1/1000th as much time) it ballooned to 7.7 billion by 2018. This was an extra-somatic energy revolution.[5]

[5] Abundant cheap energy is the means by which humans acquire food and all the other resources needed to grow and maintain the human enterprise.

From 1800 to 2016, globally fossil energy use increased over 1300 fold![6] By 1997 (when annual consumption was 40% less than in 2018) humanity was already burning fossil fuel containing about 422 times the net amount of carbon fixed by photosynthesis globally each year, or 73 times the global standing stock of carbon in vegetation (Dukes, 2003). (This won't end well.)

Meanwhile, between 1800 and the present, real global GDP increased over 100-fold and average per capita incomes by a factor of 13 (rising to 25-fold in the richest countries) (Roser, 2018). Inevitably, material consumption and attendant pollution have more than kept pace (see graphs in Steffen, et al., 2015), driving an all-too-evident parallel degradation of air, land and water all over the planet. It is particularly worth noting that, with exponential growth, half the fossil energy ever used (and half of the fossil CO_2 ever produced), has been burned / emitted in just the past 25-30 years! (Climate change can only accelerate.) During the 20th Century – economists' "separatist" fantasies aside – H. sapiens' maximum power "success" made our species not only the dominant ecological entity but also the major geological force changing the earth.[7] (It is testament to short cultural memory that people today take the recent spurt of growth to be the norm when it actually defines

[6] The contribution from coal grew from just 97 terawatt hours in 1800 to 43,403 TWh in 2016; petroleum and natural gas were not used at all until later in the 19th century, but by 2016 they were contributing 50,485 and 36,597 TWh respectively to global energy supplies (Richie and Roser, 2019).

[7] It is sadly ironic that destructive global change is actually evidence of humanity's evolutionary success.

the most anomalous few decades in 200 millennia of human history.)

The "second law"

The global degradation accompanying the "great acceleration" underscores why the ecologically relevant flows through the economy are not economist's circular money flows but rather the unidirectional transformations of energy and matter. These transformations are governed not by static mechanics but rather by thermodynamics, in particular the second law of thermodynamics (the entropy law). In simplest terms: every spontaneous change in an isolated system (one that cannot exchange energy or matter with its "environment") increases the entropy (randomness or disorder) of the system; more generally, *every* material transformation irreversibly degrades useable (high-grade) energy/matter to a more disordered, less available, entropic state.

The second law regulates all energy and material transformations in all subsystems of the ecosphere, including the human economy. As Georgescu-Roegen (1975; 1977) tried unsuccessfully to impress on fellow economists, this means that an expanding economic process is ultimately self-destructive: it feeds on useful energy/matter first produced by nature and returns it to the ecosphere as useless waste. A should-be-obvious corollary of second law is that all economic "production" is mostly consumption. Because of second law inefficiencies, the bulk of the energy/matter that enters the production process is emitted almost immediately as (often toxic) land air or water pollution; only a small fraction is embodied in marketable goods and services (and even this eventually joins the waste stream). Again, without reference to this one-way entropic throughput, it is virtually impossible to relate the economy to the environment, yet the concept is virtually absent from economics today (Daly, 1991).

The human enterprise as dissipative structure

Consistent with the second law, an *isolated* system becomes increasingly randomized and disordered with each successive internal transformation: energy dissipates, concentrations disperse and gradients disappear. Eventually, the system reaches at least local thermodynamic equilibrium, a state of maximum entropy in which nothing further can happen.

That said, we are all familiar with real-world systems that are evidently not sliding toward equilibrium. Living organisms and other complex systems "self-organize" in ways that resist the inexorable drag of the second law; they maintain themselves in high-functioning, low entropy, "far-from-equilibrium" states because they are open systems able to exchange energy/matter with their "environments". Consider the ecosphere, a self-organizing, highly-ordered, multi-layered system of mind-boggling structural complexity represented by millions of distinct species, differentiated matter, and accumulated biomass. Over geological time, its biodiversity, systemic intricacy, and energy/material flows have been increasing – i.e., the ecosphere has been moving ever further from equilibrium. This may well be the measure of life. As Prigogine (1997) asserts, "distance from equilibrium becomes an essential parameter in describing nature, much like temperature [is] in [standard] equilibrium thermodynamics".

But there is a wrinkle. Systems biologists recognize that living systems exist in overlapping nested hierarchies in which each component system is contained by the next level up and itself comprises a chain of linked sub-systems at lower levels (see Kay and Regier, 2002). Each sub-system in the hierarchy grows, develops and maintains itself by extracting usable energy and material (negentropy) from its "environment", i.e., its host system one level up. It processes this energy/matter internally to produce and maintain its own structure/function and exports the resultant degraded energy and material wastes (entropy) back into its host. In short, living organisms maintain their local level of organization as far-from-equilibrium-systems at the expense of increasing global entropy, particularly the entropy of their immediate host system (Schneider and Kay, 1994; 1995). All such self-organizing systems are called "dissipative structures" because they self-produce and thrive by continuously extracting, degrading and dissipating available energy/matter (Prigogine, 1997).

Modern interpretations of the Second Law powerfully inform thinking about sustainability. Both the economy and the ecosphere are self-organizing far-from-equilibrium dissipative structures – but with an important difference. Green plants (the *producer* components of the ecosphere) self-produce using photosynthesis to "feed" on an extra-planetary source of high-grade energy, the sun. They use this energy to reassemble carbon dioxide, water, a few mineral nutrients into energy-rich plant biomass upon which most other life depends. Photosynthesis is thus the thermodynamic engine of life, the most important productive process on Earth and the ultimate source of all bio-resources used by terrestrial and marine life, including the human economy.[8] The animal (macro-consumers) and bacterial / fungal (micro-consumer / decomposer) components of ecosystems self-produce by feeding on plant biomass or on each other. Intra-systems negative feedbacks – e.g., predator-prey relationships, disease, temporary scarcities – keep populations of both producer and consumer organisms in check, so the whole system functions in a dynamic far-from-equilibrium "steady-state". Significantly, after bacterial decomposition of dead organic matter, the material elements of life – oxygen, carbon, hydrogen and trace nutrients – recycle completely, perpetuating the system, while degraded energy dissipates off the planet. In short, the ecosphere, an extraordinary assembly of self-perpetuating local order, exists at the expense of increased entropy *elsewhere in the universe*.

By contrast, the human enterprise (human metabolism plus industrial metabolism) functions as a rogue super-consumer. As a fully-contained, growing, sub-system of the non-growing ecosphere, industrial society self-produces by over-exploiting that same ecosphere, super-

[8] This should be old news to economists. As Daly (2012) points out, "Nobel Laureate in chemistry and underground economist, Frederick Soddy... argued [almost 90 years ago] that mankind ultimately lives on current sunshine, captured with the aid of plants, soil, and water."

charged by fossil fuels and the maximum power imperative. Moreover, technology has – at least temporarily – eliminated negative feedback, so growth of the human sub-system is unconstrained. Global society thus elevates itself to a highly-ordered, intricately-structured far-from-equilibrium *non*-steady state by consuming energy/matter extracted from its supportive ecosystems at an ever-increasing pace, and dissipating a growing torrent of entropic waste energy/matter back into the ecosphere. Much of the waste stream is non-recyclable previously unknown synthetics, often toxic or otherwise hostile to life. In short, the admittedly spectacular local order represented by the human enterprise is purchased at the expense of the *entropic disordering of the ecosphere.*

Net primary production by producer species (mostly green plants) has always been more than adequate to sustain the world's entire complement of consumer organisms, including pre-industrial humans. However, *H. sapiens*' growing populations and ever-increasing material demands have become dangerously destabilizing. Any society that dissipates/pollutes its host ecosystems faster than they can regenerate/assimilate is inherently self-destructive. In fact, as a sub-system of the ecosphere, the human enterprise has become parasitic on Earth. This predicament is not merely a matter of bad management that can be addressed through simple reform. There are no exemptions from the 2nd law – beyond certain thresholds, a materially growing economy *necessarily* consumes and disorders the ecosphere; it cannot *not* increase net global entropy.

Complexity, chaos and catastrophe

Consider a final relevant aspect of complex systems dynamics. Both theoretical and empirical studies reinforce the idea that the interaction of the simple laws of physics, chemistry and biology can produce extraordinarily complex systems behaviour. Indeed, "complexity theory" projects a view of nature that, while basically deterministic, is relentlessly non-linear and full of surprises.

Dynamical systems such as the climate and ecosystems are governed by strict rules such that the state of the system at any point in time unambiguously determines the future state of the system. Theoretically, then, if we know the rules and the precise state of a system "right now", we should be able to predict what it will look like at any point in the future. In a model system, this simply requires performing an iterative sequence of calculations; the outcome of each iteration determines the subsequent state (which is the starting point for the subsequent iteration).

In modeling many real-world phenomena, however, analysts find that the state of the model *after just a few iterations* bears no evident relationship to its corresponding external reality. The interplay of even strictly deterministic laws generates patterns of behaviour that are inherently unpredictable even with near-perfect knowledge of the initial state of the system. Complex systems dynamics ensures that the smallest of measurement errors (or seemingly negligible differences in starting conditions) "feed back" and are amplified with each iteration. This means that eventually *any* inaccuracy will confound the model – the tiniest, unavoidable, measurement error can render even a perfect construct useless in predicting real-world systems behavior much in advance.

The general problem is called "sensitive dependence on initial conditions" and the behaviour it produces in both models and real systems – even simple ones – is called "chaos". With vastly increased computing power, analysts have shown that "chaos is everywhere. It is just as common as the nice simple behavior so valued by traditional [simple mechanical] physics" (Cohen and Stewart, 1994, p. 190). Chaos explains why even the best computer models cannot predict the weather next week with complete confidence.

A second related relevant phenomenon is the unexpected, dramatic (i.e., "catastrophic") change that can occur in previously stable systems under stress. Key variables of complex systems, including ecosystems, may range considerably within broad domains or "basins" of stability (e.g., the mean surface temperature on Earth has varied within a few degrees of 15 degrees Celsius for tens of millions of years). We have learned that, within these domains, a variable will normally tend to converge toward a centre of gravity called an "attractor". Initially modellers conceived of attractors as predictable single equilibria (point attractors) or as repeating cycles (periodic attractors). A chaotic systems variable, however, may trace a complex pattern of *individually unpredictable paths* that collectively define a "strange attractor" as internal feedback continually changes the system's internal dynamics. A chaotic system will nevertheless retain its overall structure and behaviour *as long as key variables remain under the influence of their customary attractors.*

And therein lies a potential problem. Although not evident from any previous history, dynamical systems may be characterized by several attractors separated by unstable ridges or "bifurcation points" (picture a terrain of watersheds isolated from each other by irregular hills). "Catastrophe" occurs when a key systems variable, driven by some persistent pressure, is displaced far from its usual attractor. If the variable reaches a bifurcation or tipping-point (the top of a ridge), it may be captured by an adjacent attractor (valley) instead of returning to its familiar domain. There will be large discontinuous breaks in the system's behaviour; its characteristics and quality change dramatically.

Such catastrophic "collapse" may result from incrementally small changes in key variables. As pressure builds, a final marginal change in temperature, pressure, population, or (?), may flip the whole system into a new stable domain, a different attractor. Most significantly, the new domain may be hostile to human interests and there is *no guarantee that the system will ever return to its former state.*

Complex systems ranging from commercial fisheries and disease control, to the Gulf Stream, global climate, and even the economy seem prone to catastrophic behaviour. Indeed, it is precisely the fear of catastrophic tipping points that has stimulated climate scientists to warn:

"....even if the Paris Accord target of a 1.5°C to 2.0°C rise in temperature is met, we cannot exclude the risk that a cascade of feedbacks could push the Earth System irreversibly onto a 'Hothouse Earth' pathway" (Steffen et al., 2018; see also Drijfhout et al., 2015). Similar grave concerns were evident in the Intergovernmental Panel on Climate Change special report (IPCC, 2018) on the consequences of 1.5°C mean global warning (which is now virtually inevitable). We ignore such findings at considerable risk to the human enterprise. As the dominant force in global ecological change, humans may well be driving key biophysical variables toward unknown strange attractors.

End game

> "So, let us not be afraid of the truth. We don't need to avoid our reality... We don't know whether we will survive. We may. But if we do, it will be in an utterly different world, with almost no other living creatures to keep us company, poor food, rampant disease and a wildly uncomfortable environment" (James, 2019).

Our dominant econo-cultural narrative of perpetual growth and ever-progressing technology sees the natural environment as little more than a static aesthetic backdrop to human affairs. It relies on analytic models based on reductionist assumptions about resources, people, firms, and technology that bear little relationship to their counterparts in the real world; in effect, society views the economy as a separate system functioning independently of the ecosphere. Relieved of limiting frictions, mainstream economists and politicians equate "sustainable development" with sustained economic growth abetted by technological progress. We may acknowledge "environmental" problems but as only one of many equivalent sets of competing values and interests. Within a broadly anthropocentric utilitarian framework, society can arrive at a satisfactory (i.e., politically acceptable) trade-off between ecosystem integrity and growth through power-brokering, negotiation, and compromise. Certainly biophysical absolutes have no seat at the bargaining table.

The ecological perspective describes a more dynamic and potentially dangerous world in which the human enterprise is a fully embedded completely dependent sub-system of the ecosphere. Both the economy and the ecosphere are dissipative structures subject to the entropic drag of the Second Law. However, while the ecosphere produces and maintains itself in a far-from-equilibrium more or less steady-state by dissipating solar energy, the human enterprise expands relentlessly by consuming and dissipating the ecosphere. The recent energy-fed/capitalist-led explosion of the human subsystem has made us both the dominant consumer and greatest geological force on Earth. Scientists are concerned that continued incremental growth may push key variables beyond "tipping-points" into unfamiliar domains that might well be catastrophic for civilization. In this framing, the need to maintain ecospheric integrity (climate stability, adequate biodiversity, etc.) is an absolute and, while the boundaries might be fuzzy, there are clearly limits to material growth.

These economic and ecological narratives are virtual polar opposites and it matters greatly which is the "truer" social construct. The empirical evidence is unambiguous and provides a crude test – consider just climate change, biodiversity loss and the direct human demand for biocapacity:

- There is no dispute that the 'greenhouse effect' is real and the main anthropogenic greenhouse gas is carbon dioxide (the greatest industrial metabolic waste by weight of industrial economies). Now increasing by 3 parts per million (ppm) per year, atmospheric CO_2 readings in January 2019 averaged over 410.83 ppm (NOAA, 2019). This is a human-induced increase of 47% above pre-industrial levels and takes atmospheric carbon to its highest levels in 800,000 years; other greenhouse gases (GHGs) such as nitrous oxide and methane are increasing as fast or faster. As a result, mean global temperature has risen by almost 1.0 Celsius degree, mostly since 1980. A statistically improbable 18 of the 19 warmest years in the instrumental record have occurred in this young century: the five warmest years were the last five – 2016 was the warmest; followed by 2015, 2017, 2018 and 2014" (CC, 2019).

- Our best science tells us that the world is currently on track to experience three to five °C warming. There is no dispute that five degrees would be catastrophic, likely fatal to civilized existence. Even a "modest" three degrees implies disaster – enough to destroy economies, destabilize geopolitics and empty cities. As noted, scientists are concerned that even the more difficult Paris target (to limit warming to 1.5°C above pre-industrial levels) may be sufficient to tip the climate down just such a hothouse earth pathway. This should be particularly worrisome since, even if 2019 GHG concentrations are held steady (in fact, they are rising), they are sufficient to commit the world to an additional 0.3 to 0.8 C°C degrees in coming decades (Hansen, 2018).

- Meanwhile, humans are competing with all other vertebrate life for Earth's finite livable space and "surplus" plant biomass – and, thanks to technology, we are "winning". It is, however, a Pyrrhic victory – the human harvest of ten to 100 times more biomass than our ecological competitors is destroying wild nature. With only 0.01% of total earthly biomass, H. sapiens has eliminated 83% of wild animal and 50% of natural plant biomass. From a fraction of one percent ten millennia ago, humans now constitute 36%, and our domestic livestock another 60%, of the planet's mammalian biomass compared to only 4% for all wild species combined. Similarly, domestic poultry now comprises 70% of Earth's remaining avian biomass and the oceans are being depleted by commercial fishing (at the expense of rapidly declining marine mammals) (Bar-On, et al., 2018). "Competitive displacement" of non-human species can be territory-related as when humans appropriate grassland habitats at the expense of wild ungulate ecosystems or more directly energy-related as when over-fishing starves marine mammals and birds: "Fisheries generate severe constraints for seabird populations on a worldwide scale... Indeed, seabirds are the most threatened bird group, with a 70% community-level population decline across 1950–2010" (Grémillet et al., 2018). Overall, the World Wildlife Fund reports an "astonishing" 60% decline in the populations of mammals, birds, fish, reptiles, and amphibians in just over 40 years (WWF, 2018). Other plant and animal groups are also under siege from climate change, habitat loss, insecticide and industrial contamination – even insects are experiencing "Armageddon" (and systemically linked insect-dependent birds, mammals and amphibians are not far behind) (Carrington, 2019; Lister and Garcia, 2018; Hallmann et al., 2017).

- Direct human appropriation of ecosystems (biocapacity) can be estimated using ecological footprint analysis (Rees, 2013). EFA shows that society is in severe "overshoot. The average high-income citizen requires four to 10+ hectares of global average productive ecosystems (gha) to produce the bioresources s/he consumes and assimilate just his/her carbon emissions. Even the average human eco-footprint is 2.8 gha when there are only about 1.7 ha/capita of productive land and water ecosystems on the planet. To sustain just the 2014 world population at North American, European or global average material standards with prevailing technology, we would need four, 2.8 and 0.69 additional earth-like planets respectively (data from GFN, 2019). That is,

> even at the 2014 global average standard of living, the human enterprise exceeded long-term global human carrying capacity by 69%. This is further proof that growth and maintenance of the human system "far-from-equilibrium" is being financed by the liquidation of accumulated stocks of essential natural capital (forests, soils, fish, etc.).

The most important "take-away" from the above is that the data and trends are consistent with and generally predictable from basic ecological constructs. Since the scale of, and demands by, the human enterprise are steadily increasing, and the ecosystems upon which they depend are not, it is inevitable on a finite planet, that material consumption would eventually exceed sustainable rates of production even as the resultant pollution (climate change is a waste management issue) and plunging biodiversity undermine remaining productive capacity.

By contrast, in the absence of ecological content, the true state of the ecosphere remains invisible to neoliberal theory and analyses. Economists' narrow perspective, fixation on growth and rejection of biophysical limits are particularly troublesome. As if to illustrate the point, in a determined show of disciplinary unity, 3333 US economists – including 27 Nobel Laureates – issued a statement on 16 February 2019, asserting that the "most cost-effective lever to reduce carbon emissions" is an escalating carbon tax (WSJ, 2019). This would substitute for "cumbersome regulations", "encourage technological innovation" stimulate the production of "carbon-efficient goods and services", and facilitate the shift to renewables while "promoting economic growth". To "maximize fairness and political viability" (not to mention minimal economic disruption), "all the revenue should be returned directly to US citizens through equal lump-sum rebates" to spend as they will; the majority of families would "receive more in "carbon dividends" than they pay in increased energy prices.

This is simplistic, unreformed, compartmentalised thinking stuck on mechanical market "levers" and technology to solve the climate "problem" and maintain the growth-based *status quo.* There is no hint that climate change is just one symptom of gross human (econo-) ecological dysfunction, no targets, no tipping points and no limits. It rashly assumes the availability of, and a smooth transition to, viable substitutes for fossil fuel. At best the economists' statement is a tiny step forward and a giant (rebate) step back. And why not? The statement displays the same level of ecological (un)sophistication that qualified for the 2018 (pseudo-) Nobel Prize in Economics (see Hickel, 2018).

The ecological framing also raises important questions that would not occur to growth-bound economists: What is the probability that the ecosphere can withstand another doubling of human energy/material demand as is expected before mid-century? Can we devise new social constructs that override rather than reinforce peoples' innate expansionist tendencies and selfish myopia? What might be done globally to avoid resultant tipping points and systemic collapse? Below are some steps that address this last question, all consistent with the evidence. In particular, they recognize that the era of material growth will soon end either through systems implosion or a planned descent. The world community should:

- Acknowledge that collapse is a finite possibility and the usual outcome for societies whose leaders ignore evidential warning signs or are too corrupt or incompetent to act accordingly (Tainter, 1984; also Kemp, 2019);
- Admit the theoretical simplicity and conceptual flaws in neoliberal market economics and capitalist expansionism;
- Embrace the need to "socially construct" a new foundation for economics that is consistent with bio-physical reality, beginning with today's emergent ecological economics (e.g., Daly and Farley, 2010; Victor, 2019);
- Recognize that humans are bio-ecological beings, the most ecologically significant entity in all Earth's ecosystems and subject to ecological and biophysical principles;
- Acknowledge that economic behaviour and processes qualify economics as a branch of human ecology;
- Accept that the human enterprise is a fully-contained dependent sub-system of the non-growing ecosphere;
- Abandon relatively mechanical, linear, single equilibrium models for more dynamic, non-linear multiple equilibrium constructs of the integrated human eco-economic system;
- Forge economic theory that is consistent with the physical, chemical, and bio-ecological concepts governing both economic and ecological material transformations in the ecosphere;

- Accept the limitations of technology – in general, natural capital and manufactured capital are complements, not substitutes; some forms of "natural capital" are essential and non-substitutable;
- Accept that there are "fuzzy" biophysical limits to growth that may not be evident and whose location (tipping points) may be shifting with changes in both natural conditions and exploitation rates. (This implies liberal application of the precautionary principle.);
- Acknowledge that no economy that grows or even maintains itself by depleting essential capital is sustainable;
- Shift the primary emphases of economic planning from quantitative growth and efficiency (getting bigger) toward qualitative development and equity (getting better). At present, the US and many of its OECD allies are growing but un-developing;
- Acknowledge that the vast majority of humans will never leave this planet and that species survival depends on maintaining the functional integrity of the ecosphere. Key controlling variables must remain within the historic "basins of attraction" [stability domains] that enabled human evolution and the emergence of civilized existence;
- Understand that (un)sustainability is a collective problem requiring collective solutions and an unprecedented level of international cooperation, sacrifice and sharing;
- Commit to devising and implementing policies consistent with a "one Earth" civilization;
- Establish as overall goal an ecologically stable, economically secure "steady-state" society (Daly,

1991) whose citizens live more or less equitably within the biophysical means of nature.

- Accept that today's gross and growing income/wealth inequality[9] is a major barrier to sustainability and that one-Earth living requires mechanisms for fair redistribution;
- Recognize that Earth is over-populated even at average material standards. A one-earth life-style for 7.3 billion people requires that humans learn to thrive on the biocapacity represented by 1.7 global average productive hectares per capita (compared to the eight gha/capita require by contemporary North America);
- Begin the public cultural, social and economic discussions and formal planning necessary to reduce fossil energy and material consumption (economic throughput) by up to 69% globally (at least 80% in high-income countries).[10] This is consistent with achieving the IPCC (2018) goal of almost 50% fewer carbon emissions by 2030 and requires six percent per year reductions beginning immediately;
- Conceive and implement a global fertility strategy to reduce the human population to the 2-3 billion people that might be able to live in material comfort on this already much-damaged single planet Earth.

These will seem outrageously radical demands to contemporary economic strategists, invested capitalists, politicians and even most ordinary citizens attuned to the growth imperative and human technological wizardry.

[9] Oxfam estimates that the world's richest 26 billionaires control as much wealth as the poorest half of humanity, or at least 3.6 billion people (Elliott, 2019).

[10] This is technically achievable (von Weizsäcker, et al., 2009).

Nevertheless, it is time for the world to admit that continued adherence to prevailing socially-constructed illusions risks fatal catastrophe. Untransformed, our present system *will* crash. We really have no choice but to act upon what our best science has been telling us for decades. And we have to act urgently: "Effective planetary stewardship must be achieved quickly, as the momentum of the Anthropocene threatens to tip the complex Earth System out of the cyclic glacial-interglacial pattern during which *Homo sapiens* has evolved and developed" (Steffen et al., 2011).

That said, we do have to ask: what is the probability that in the present "post-truth" era, our delusional, leaderless and increasingly fractious world community will be able to agree on the diagnosis advanced here let alone on implementing effective solutions? Some problems may not be solvable. There is certainly no easy solution to humanity's econo-ecological predicament and without an agreed emergency plan for cooperative action there may be no solution at all.

References

Beer, S. (1981) "I said, you are Gods." *Teilhard Review* 15(3), pp. 1–33.

Berman, M. (1984) *The Re-enchantment of the World.* New York: Bantam.

Boltzmann, (1886) "The Second Law of Thermodynamics." In B. McGinness, (ed.), *Ludwig Boltzmann: Theoretical physics and philosophical problems: Select Writings.*Dordrecht, Netherlands: D. Reidel (1974).

Carrington, D. (2019) "Insect collapse: 'We are destroying our life support systems'", The *Guardian*, 15 January.

CC (2019) "The Ten Hottest Global Years on Record", https://www.climatecentral.org/gallery/graphics/the-10-hottest-global-years-on-record

Christenson, P. (1991) "Increasing Returns and Ecological Sustainability." In R. Costanza, (Ed.), *Ecological Economics: The Science and Management of Sustainability*. New York: Columbia University Press.

Cohen, J. and I. Stewart (1994) *The Collapse of Chaos*. New York: Penguin Books.

Daly, H.E. (1991) *Steady-State Economics* (2nd ed). Washington: Island Press.

Daly, H.E. (1992) "Allocation, distribution, and scale: towards an economics that is efficient, just, and sustainable." *Ecological Economics*, 6, pp. 185-193.

Daly, H.E. (2012) "What is the limiting factor?" *The Daly News*, retrieved 6 February 2019, https://steadystate.org/what-is-the-limiting-factor/.

Daly, H.E. and J. Farley (2010) *Ecological Economics, Second Edition: Principles and Applications*. Washington: Island Press.

Dasgupta, P. and D. Heal (1979) *Economic Theory and Exhaustible Resources*. London: Cambridge University Press (205 pp).

Drijfhout, S., S. Bathiany, C. Beaulieu, V. Brovkin, M. Claussen, C. Huntingford, M. Scheffer, G. Sgubin, and D. Swingedouw (2015) "Catalogue of abrupt shifts in Intergovernmental Panel on Climate Change climate models." PNAS 112 (43) E5777-E5786; published ahead of print October 12, 2015, https://doi.org/10.1073/pnas.1511451112.

Dukes, J. S. (2003) "Burning buried sunshine: human consumption of ancient solar energy." *Climatic Change* 61, pp. 31-44 (also at https://core.ac.uk/download/pdf/5212176.pdf).

Elliott, L. (2019) "World's 26 richest people own as much as poorest 50%, says Oxfam", The *Guardian*, 21 January.

Fischer-Kolwalski, M. and Haberl, H. (Eds) (2007) *Socio-ecological transitions*. Cheltenham: Edward Elgar.

Fowler, C. & Hobbs, L. (2003) "Is humanity sustainable?" *Proceedings of the Royal Society of London B* 270 (1533), pp. 2579-2583.

Georgescu-Roegen, N. (1975) "Energy and Economic Myths." *Southern Economic Journal* 41 (3), pp. 347-381.

GFN (2019) "Ecological footprint explorer." *Global Footprint Network*, retrieved 20 February 2019,
http://data.footprintnetwork.org/#/.

Grémillet, D., A. Ponchon, M. Paleczny, M-L. D. Palomares, V. Karpouzi and D. Pauly (2018) "Persisting worldwide seabird-fishery competition despite seabird community decline", *Current Biology* 28 (24), pp. 4009-4013,
https://doi.org/10.1016/j.cub.2018.10.051.

Hall, C.A.S. (1995) *The Ideas and Applications of H.T. Odum*. Boulder: The University Press of Colorado.

Hallmann, C.A., M. Sorg, E. Jongejans, H. Siepel, N.Hofland, H. Schwan, W. Stenmans, A. Müller, H. Sumser, T. Hörren, D. Goulson, H. de Kroon (2017) "More than 75 percent decline over 27 years in total flying insect biomass in protected areas." PLOS ONE; 12 (10): e0185809, retrieved 14 February 2019,
https://doi.org/10.1371/journal.pone.0185809.

Hansen, J. (2018) "climate change in a nutshell: the gathering storm", retrieved 7 February 2019.
http://www.columbia.edu/~jeh1/mailings/2018/20181206_Nutshell.pdf

Hickel, J (2018), "The Nobel prize for climate catastrophe." *Foreign Policy*, retrieved 8 December 2018,
https://foreignpolicy.com/2018/12/06/the-nobel-prize-for-climate-catastrophe/

IPCC (2018) *Global Warming of 1.5°C – Summary for Policymakers*, Intergovernmental Panel on Climate Change, retrieved 11 February 2019,
https://www.ipcc.ch/sr15/chapter/summary-for-policy-makers/.

James, J. (2019) "Why are we afraid of the truth that we are tracking to 4+ °C?", retrieved 15 February, https://countercurrents.org/2019/02/10/why-are-we-afraid-of-the-truth-that-we-are-tracking-to-40-c/.

Jevons, W. (1879) *The Theory of Political Economy*, 2nd ed. London: Macmillan.

Kay J. and H. Regier (2001) "Uncertainty, complexity, and ecological integrity." In: P. Crabbé, A. Holland, et al. (eds), *Implementing ecological integrity: restoring regional and global environment and human health* (NATO Science Series IV: Earth and Environmental Sciences 1, pp.121-156), Dortrecht: Kluwer Academic Publishers.

Kemp, L. (2019) "Are we on the road to civilization collapse?" *BBC Future*, retrieved 24 February 2019, http://www.bbc.com/future/story/20190218-are-we-on-the-road-to-civilisation-collapse

Lister, B.C. and A. Garcia (2018) "Climate-driven declines in arthropod abundance restructure a rainforest food web." PNAS 115 (44) E10397-E10406; retrieved 14 February 2019, https://www.pnas.org/content/115/44/E10397.

Lotka, A.J. (1922) "Contribution to the energetics of evolution." *Proceedings of the National Academy of Science* 8, pp. 147-155.

NASA (2019) "Global Temperature." Retrieved 20 February 2019, https://climate.nasa.gov/vital-signs/global-temperature/.

NOAA (2019) "Trends in atmospheric carbon dioxide", retrieved 20 February 2019, https://www.esrl.noaa.gov/gmd/ccgg/trends/

Ponting, C. (2011) *A New Green History of the World – the Environment and the Collapse of Great Civilizations*. London: Vintage (464 pp).

Popper, K. (1972) "Conjectural knowledge: my solution of the problem of induction." In Karl Popper, *Objective Knowledge: An Evolutionary Approach*. Oxford: Oxford University Press.

Postman, N. (1999) *Building a Bridge to the 18th Century*. New York: Alfred Knopf.

Pratarelli, M. (2008) *Myopic Man: On the Nature and Universality of Self-Deception and its Long-Term Effects on Our Environment.* Beulah, CO: Medici Publishing.

Prigogine I. (1997) *The end of certainty: time, chaos and the new laws of nature.* New York: The Free Press.

Rees, W.E. (2010) "What's blocking sustainability? Human nature, cognition and denial." *Sustainability: Science, Practice & Policy* 6 (2), pp.13-25.

Rees, W.E. (2013) "Ecological Footprint, Concept of", in S.A. Levin, (ed.), *Encyclopedia of Biodiversity*, second edition, Volume 2, pp. 701-713, Waltham, MA: Academic Press.

Ritchie, H. and M. Roser (2019) "Fossil Fuels", retrieved 15 February 2019, https://ourworldindata.org/fossil-fuels.

Roser, M. (2018) *Economic Growth.* Retrieved 15 February 2019, https://ourworldindata.org/economic-growth.

Schneider, E.D, and J.J. Kay (1994) "Complexity and thermodynamics: toward a new ecology." *Futures* 26, pp. 626-647.

Schneider, E.D, and J.J. Kay (1995) "Order from disorder: the thermodynamics of complexity in biology." In M.P. Murphy and L.A.J. O'Neill, (eds), *What is life: the next fifty years*, Cambridge, UK: Cambridge University Press.

Steffen, W., Å. Persson, L Deutsch, J. Zalasiewicz, M. Williams, K. Richardson, C. Crumley, P. Crutzen. C. Folke, L. Gordon, M Molina, V. Ramanathan, J. Rockström, M. Scheffer, H,J. Schellnhuber and U. Svedin (2011) "The Anthropocene: From Global Change to Planetary Stewardship," *Ambio*, 40 (7), pp. 739–761, doi: 10.1007/s13280-011-0185-x.

Steffen, W., W. Broadgate, L. Deutsch, O. Gaffney, and C. Ludwig (2015) "The trajectory of the Anthropocene: The Great Acceleration." *The Anthropocene Review*, 2 (1), pp. 81-98.

Steffen, W., J. Rockström, K. Richardson, T. M. Lenton, C.Folke, D. Liverman, C.P. Summerhayes, A.D. Barnosky, S. E. Cornell, M. Crucifix, J.F. Donges, I. Fetzer, S. J. Lade, M. Scheffer, R. Winkelmann, and H.J. Schellnhuber (2018)

"Trajectories of the Earth System in the Anthropocene." *PNAS* 115 (33) 8252–8259, retrieved 10 February 2019, https://doi.org/10.1073/pnas.1810141115.

Tainter, J. (1988) *The Collapse of Complex Societies*, Cambridge. UK and New York: Cambridge University Press.

Victor, P.A. (1991) "Indicators of sustainable development: some lessons from capital theory", *Ecological Economics,* 4 (3), pp.191-213.

Victor. P. (2019), *Managing without Growth: Slower by Design, not Disaster* (2nd ed), Cheltenham, UK: Edward Elgar.

von Weizsäcker, E., Hargroves, K., Smith, M., Desha, C. and Stasinopoulos, P. (2009) *Factor 5: Transforming the Global Economy through 80% Increase in Resource Productivity.* London, UK and Droemer, Germany: Earthscan.

Waring, T. (2010) "New evolutionary foundations: theoretical requirements for a science of sustainability." *Ecological Economics,* 69 (4), pp. 718–730.

Wolf, M. (2010) "Why were resources expunged from economics?" *Martin Wolf's Exchange*, retrieved Feb 5, 2019, https://archive.li/uF7LN.

WSJ (2019) "Economists' statement on carbon dividends – The Largest Public Statement of Economists in History", originally *Wall Street Journal*, retrieved 18 February 2019 at https://www.clcouncil.org/economists-statement/.

WWF (2018) *Living Planet Report.* Gland, Switzerland: Worldwide Fund for Nature.

Chapter 12

An ecosocialist path to limiting global temperature rise to 1.5°C

Richard Smith[1]

The wildfire that wiped out Paradise, California, viewed from space. The fire took 85 lives, 11 remain missing, and it destroyed 13,972 homes, 528 commercial buildings and 4,293 other structures. [NASA]

---

[1] Thanks to David Klein, Nancy Holmstrom, Bill Tabb, Ted Franklin, Marie Venner, and William Neil for their helpful suggestions and criticism. I remain responsible for the remaining errors and inadequacies of this paper.

Introduction

After decades of empty promises, backsliding, and failed cap and trade and carbon tax schemes, we now find ourselves facing a climate emergency: either we drastically suppress global CO_2 emissions, and soon, or global warming will soar beyond any human power to restrain it, ecologies, farming and fishing will collapse and then civilization itself, perhaps before the century is out. To date, all efforts to suppress emissions have foundered on the contradiction between the need to prioritize economic growth over saving the environment. Given capitalism, climate change will kill us in the long run but reversing economic growth will kill us in the short run – and so we keep kicking the can down the road until now we find ourselves at the precipice. This paper proposes a four-point Emergency Plan to suppress both emissions and economic growth in the United States without precipitating economic collapse – by nationalizing the fossil fuel industrial complex with buyouts to phase out fossil fuels, creating a WPA-style jobs program to guarantee re-employment for redundant fossil fuel workers, launching an emergency system-wide transition to renewables, and downsizing the U.S. industrial economy, reducing our "carbon footprint" on planet earth by closing and retrenching superfluous and harmful industries. While this process must of course begin within the framework of capitalism, I envision the from-below struggle to "decarbonize" the economy leading to increasingly radical demands for popular control over economic and political decision-making across the economy tending in the direction of ecosocialism. Indeed, that is, in my view, the essence of the message of the amazing Sunrise children's crusade against Democrats and Republicans in the U.S. Congress who refuse to back Alexandria Ocasio-Cortez's Green New Deal.[2] The specifics of this model are geared to the U.S. economy and wouldn't be applicable to industrialized countries where fossil fuels are either absent or already nationalized. But the essential idea of asserting popular power to prioritize people and planet over growth and profit and of de-growing much of our over-industrialized economies is widely applicable.

I. The IPCC Report "Global Warming of 1.5°C" and the imperative to immediately suppress fossil fuel production

The much-awaited report from the UN's top climate science panel describes the enormous gap between where we are and where we need to be to prevent dangerous levels of global warming. The 2015 Paris climate accord committed industrial nations to reduce their emissions sufficient to keep global temperatures within a 2°C rise over pre-industrial levels. In the final accord, highly vulnerable island nations and faith communities represented at the UN pressed the authors to include the 1.5°C limit as an aspirational target in the final draft of the accord with 2°C as the backup target.

[2] Paul Blest, "Dianne Feinstein is not happy with kids asking her to vote for the Green New Deal," Splinter News, February 22, 2019: https://splinternews.com/dianne-feinstein-is-not-happy-with-kids-asking-her-to-v-1832833216. And here are Sunrise kids occupying Senator McConnell's office on February 25, 2019:

Soaring GHG emissions over the past five years, rising atmospheric CO_2 concentrations, ice-cap retreats, intensified storms, forest fires above the Arctic circle, and die-offs of the world's coral reefs have all raised concerns about what even a little bit more warming would bring. Parts of the planet including the Arctic and many inland areas, have already warmed beyond 1.5°C. California is on fire most of the year. The worst hurricanes are twice as severe (more precipitation, slower passage, greater wind speeds) than they used to be. And this is just a start. Rapid climate breakdown is one reason why climate scientists now think that the goal just five years ago of limiting warming to 2°C "increasingly seems disastrous in this context."[3] The Paris pledges were never sufficient even to keep warming below 3°C let alone 2°C. Few of the signatories have even managed to meet the low bars they set for themselves. The world's largest countries including China, the U.S., and Canada have us on track to a 4-5°C warming. As CO_2 concentrations continue growing, preventing runaway warming is going to require ever deeper, truly draconian cuts in emissions, which will mean great economic disruption. IPCC estimates already show us needing to achieve a near vertical drop in emissions in the early 2020s. Every day we delay getting off of fossil fuels increases the probability that we won't be able to save ourselves.

The 2018 IPCC special report painted a stark portrait of how quickly the planet is heating up and called on

---

https://www.facebook.com/sunrisemvmt/videos/994371184095235/

[3] Chris Mooney and Brady Dennis, "Climate scientists are struggling to find the right words for very bad news," *Washington Post*, October 3, 2018.

governments to take *immediate* steps to suppress emissions:

> "If emissions continue at current rate, atmosphere will warm by as much as 2.7° Fahrenheit, or 1.5° Celsius, above preindustrial levels by between 2030 and 2052. Further, warming is more extreme further inland of large water bodies. [To keep temperatures from rising beyond 1.5° degrees] anthropogenic CO_2 emissions [must] decline by about 45% *worldwide* from 2010 levels by 2030 ... [This] would require rapid and far-reaching transitions in energy, land, urban and infrastructure (including transport and buildings), and industrial systems ... These systems transitions are unprecedented in terms of scale ... and imply deep emissions reductions in all sectors, a wide portfolio of mitigation options and a significant upswing in those options."[4]

Preventing ecological collapse requires transforming the world economy at a speed and scale that "has no documented historic precedent."[5] What would this take?

---

[4] IPCC, "Global Warming of 1.5°"(October 7, 2018): http://report.ipcc.ch/sr15/pdf/sr15_headline_statements.pdf.
Note, the subsequent 9th UNEP Emissions Gap Report 2018 states up to a 55% cut over the time line:
"In contrast, global GHG emissions in 2030 need to be approximately 25 percent and 55 percent lower than in 2017 to put the world on a least-cost pathway to limiting global warming to 2°C and 1.5°C respectively" (UNEP, 2018: p. xv): https://www.unenvironment.org/resources/emissions-gap-report-2018.

[5] Ibid. Also Coral Davenport, "Major report describes a strong risk of crisis as early as 2040," *New York Times*, October 7, 2018.

Myles Allen, Oxford University climate scientist and an author of the report said "It's telling us we need to reverse emissions trends and turn the world economy on a dime." To prevent 2.7 degrees of warming "greenhouse emission must be reduced by 45 percent from 2010 levels by 2030, and by 100 percent by 2050. Use of coal as electricity source would have to drop from 40 percent today to 1-7 percent by 2050."[6] Drew Shindell of Duke University, another author of the report said: "It would be an enormous challenge to keep warming below a threshold of 1.5 degrees... What might that look like? In part, it would include things such as no more gas-powered vehicles, a phaseout of coal-fired power plants and airplanes running on biofuels," he said. "It's a drastic change," he said. "These are huge, huge shifts... This would really be an unprecedented rate and magnitude of change."[7]

In response to the report, UN Secretary General António Guterres warned world leaders to "**Do what the science demands before it's too late.**"[8]

II. Capitalist priority to growth and profits over people and planet

Given this unprecedented existential crisis one might expect governments would responsibly meet this climate emergency with emergency plans to prevent ecological

---

[6] Davenport, "Major report" op cit.
[7] Mooney and Dennis, "Climate scientists struggling," op cit.
[8] Somini Sengupta, "Projection on climate is ominous. Now what?" *New York Times*, October 10, 2018.

collapse bold proposals for "deep emissions reductions in all sectors," for "far-reaching transitions in energy, land, infrastructure, and manufacturing" and so on. After all, the 2018 IPCC 1.5°C report makes clear that on present trends we could be facing the collapse of agriculture in California, the Great Plains, India, China much of Africa, mass famine, submerging cities, destruction of the world's last forests and worse, possibly as soon as 2040, well within the lifetimes of many leaders today and certainly their children's and ours. On February 5th U.K. Met Office meteorologists warned that the climate is warming so fast that global temperatures could exceed 1.5°C, the lowest of the Paris targets set for the end of the century, in *five years*.[9]

What's more, the solution to our climate crisis is astonishingly simple and doesn't require any new tech breakthroughs. The first step is to stop doing what we're doing: immediately begin shutting down fossil fuel production, stop new drilling, stop producing and registering fossil fuel-powered vehicles, drastically curb air travel, ration fossil fuels, curtail manufacturing and construction. The second step is to force through an immediate transition to renewable energy across the economy (and do what we can to enable this transition around the world).

A. Where are the bold proposals?

Yet we hear no bold proposals to meet the challenge from any governments – not from European socialist parties,

[9] Jonathan Watts, "Met Office: global warming could exceed 1.5C within five years," *Guardian*, February 6, 2019.

not from Canadians or Australians (the leading exporters of the world's dirtiest fuels), certainly not from the Chinese (the world's largest polluters who, moreover, are now abandoning the limits on coal-burning they just imposed last year in order to restore growth in the face of the trade war),[10] let alone from the Trump administration. Trump's response to his own government's prediction of a 4°C warming by 2100 is "the planet's fate is sealed" so we may as well abandon Obama's federal fuel-economy standards for cars and light trucks and "burn baby burn".[11] To the extent we hear any proposals at all, it's just renewed calls for more of the same carbon taxes, the same fantasy tech fixes like carbon capture and storage that have manifestly failed to staunch emissions to date.[12] Why is that?

The reason why no government dares take the obvious steps to save the humans is because no one has come up with a magic fix to suppress emissions without suppressing economic growth and profits. Given capitalism, economic growth and profit maximization must be systematically prioritized over all other considerations including emissions reduction or companies will fail, the economy will collapse, and mass unemployment will be the result: global warming may kill us in the long run but economic collapse will kill us

[10] Jane Cai, "Northern China's clear air get lost in the smog of the trade war," *South China Morning Post*, November 15, 2018: https://www.scmp.com/news/china/politics/article/2172956/northern-chinas-clean-air-targets-get-lost-smog-united-states.

[11] Juliet Eioperin et al., "Trump administration sees a 7-degree rise in global temperatures by 2100," *Washington Post*, September 28, 2018.

[12] For example, Jim Snyder and Mark Drajem, "FutureGen's demise means carbon capture for coal faces woes," *Bloomberg*, February 5, 2015: https://www.bloomberg.com/news/articles/2015-02-05/futuregen-s-demise-shows-carbon-capture-for-coal-faces-long-road.

in the short run. This is the ultimate contradiction of capitalism: We have to destroy our children's tomorrows to hang onto our jobs today.

That's why from the very first climate negotiations around the Kyoto Protocol in the 1990s, all efforts to contain emissions have been subordinated to maintaining economic growth. Year after year, decade after decade, for 21 straight years to COP21 at Paris in 2015, UNFCC annual summit negotiations invariably collapsed in failure and acrimony. Despite the pleas of climate scientists, desperate submerging Pacific islanders, Africans, Indians and others who contribute few emissions but suffer disproportionately from global warming-induced drought and crop failures, no industrial nation has been willing to accept binding emissions limits because they all understand that caps would suppress economic growth. As George Bush Sr. infamously told the 1992 Climate summit, "The American way of life is not negotiable." And if America will not accept binding emissions caps, why should China? Facing growing protests over their do-nothing annual summits, the only thing negotiators at Paris could agree on was to stop holding their embarrassing annual farces (henceforth they agreed to meet every five years instead) and contrive another "agreement" in which industrial countries pledged to reduce their emissions somewhat some day but are under no legal obligation to do so – prompting James Hansen, the world's foremost climate scientist, to complain that

> "It's a fraud really, a fake. It's just bullshit for them to say: 'We'll have a 2°C warming

> target and then try to do a little better every five years'. It's just worthless words. There is no action, just promises. As long as fossil fuels appear to be the cheapest fuels out there, they will continue to be burned."[13]

B. Halting global warming requires degrowth, substantial de-industrialization

If there's no magic tech fix then phasing out fossil fuel consumption has to mean shutting down or at minimum, drastically retrenching companies, beginning with the fossil fuel producers like Peabody Energy (coal), Exxon Mobil, Chevron, but continuing down the petrochemical food chain through the fossil fuel dependent industries. After all, it's easy to blame the fossil fuel companies for global CO_2 emissions.[14] Environmental groups have focused too narrowly on fossil fuel producers, their pipelines and such, while ignoring the downstream industrial and personal consumers. Not to put too fine a point on it but the oil producers don't burn the oil. We burn the oil producing, processing, transporting and refrigerating food, driving our cars, building our homes, heating and cooling our homes, manufacturing this and that, jetting off on vacations, and so on. Fossil fuels are pervasive. As this table shows, fossil fuel emissions are produced across the entire economy.

---

[13] Oliver Milman, "James Hansen, father of climate change awareness, calls Paris talks 'a fraud'", *Guardian*, December 12, 2015. And it was not just Hansen. The "Paris pledge" drew blistering condemnation from scientists around the world. See Tom Bawden, "COP21: Paris deal far too weak to prevent devastating climate change, academics warn," *Independent*, January 8, 2016.

[14] As does, for example, Tess Riley, "Just 100 companies responsible for 71% of global emissions, study says," *Guardian*, July 10, 2017.

Total U.S. Greenhouse Gas Emissions by Economic Sector in 2016

Transportation	28.5%
Electricity	28.4%
Industry	22%
Commercial & Residential	11%
Agriculture	9%

Source: EPA: https://www.epa.gov/ghgemissions/sources-greenhouse-gas-emissions.

If governments in the industrialized economies had listened to climate scientists in the 1980s and taken steps then to radically suppress emissions, perhaps we wouldn't be in the desperate fix we find ourselves in today. But they didn't. They dithered, stalled, insisted on market "solutions" that were really designed to fail with the result now, after decades of "green capitalism" we find ourselves facing an existential crisis that admits of only one proximate solution: state intervention to slam the brakes on emissions by shutting down the emitters. If we're serious about suppressing fossil fuel emissions, then we have to drastically retrench and in some cases completely shut down thousands of downstream fossil fuel-dependent companies in transportation, petrochemicals and plastics, manufacturing, construction,[15] agribusiness, tourism and more in the U.S.

[15] I was surprised to discover that the construction industry where I worked for many years is the largest single consumer of (petrochemical-derived) plastics and plastic packaging, accounting for about a third of all plastics production in the U.S. Though not always visible in buildings, they're used in a wide and growing range of applications including weather insulation, piping, window frames, protection and finishes. The majority of construction waste is also plastic. Roma Saini, "Plastic waste! Why all the fuss?," Willmot Dixon Interiors, February 14, 2018: https://www.willmottdixoninteriors.co.uk/plastic-waste-fuss/.

alone. In cases like plastics, disposable products and others, we would have to virtually abolish entire industries because there's just no other way to suppress their emissions and make them sustainable.

1. *Transportation, now the leading source of CO_2 emissions in the U.S.*

Transportation emissions are surging because cars, trucks, planes, trains, and ships are burning more fossil fuels as we commute, travel, and ship more goods.

Airliners line up for take-off at Los Angeles International airport, June 2001. [Photo: PBS]

Aviation emissions are the fastest-growing source of transport sector CO_2 emissions. Air travel is booming around the world, growing by 10% in recent years and projected to double over the next twenty years, driven by cheap flights, Airbnb, and a growing global middle class. Surging economies are also lifting air-cargo demand, up

9% in 2017 (Amazon and FedEx overnight deliveries, strawberries delivered from Chile to New York in January, etc.). Though aircraft fuel economy has increased, every gain in efficiency engineers wring out of engines, airframes and so on, is quickly overwhelmed by ever more planes. In result, since 1990, global aviation emissions have doubled and absent some tech miracle, could double again in the next 20 years as air traffic grows. As of 1999, aviation emissions accounted for just 3.5% of global CO_2 emissions, but on present trends they could account for 15% by 2050.[16] That would significantly dent improvements won in other sectors like power generation. Furthermore, it is estimated that 4.9% of global radiative forcing, including cirrus cloud effects, was attributable to aviation.[17] And in the case of aircraft, unlike power plants, there is no practical tech fix for this problem in the foreseeable future. Airplanes lack the alternatives available for road and sea transport like renewable electricity, hydrogen or lower-emitting natural gas, which aren't dense enough to provide the huge amounts of energy needed for take-off. Replacing jet kerosene with biofuels would just compete with food production and accelerate global deforestation, replacing one problem with others.[18] Electric batteries are still too heavy for large aircraft so there are no electric airliners on anyone's drawing boards. Both solar-power or

---

[16] IPCC, *Aviation and the Global Atmosphere: A Special Report of the International Panel on Climate Change* (Cambridge: Cambridge University Press,1999), chapter 9: http://www.grida.no/climate/ipcc/aviation/index.htm.

[17] Bethan Owen et al., "Flying into the future: aviation emissions scenarios to 2050," *Environmental Science Technology*, 44 (2010), pp. 2255-2260.

[18] Abrahm Lustgarten, "Palm oil was supposed to help save the planet. Instead it unleashed a catastrophe," *New York Times*, November 20, 2018: https://www.nytimes.com/2018/11/20/magazine/palm-oil-borneo-climate-catastrophe.html.

hydrogen-power are still in the realm of sci-fi, or at best far-off dreams, for commercial passenger jets.[19]

For the present, the only way to radically suppress emissions is to stop flying so much, ground planes, sharply cut back and ration air travel, and sharply reduce aircraft manufacturing. That means drastically retrenching Boeing, retrenching or closing down United Airlines, Delta, etc. with all that implies. That's a big problem. But global warming is a bigger problem. Look at California: Swaths of the state are on fire blanketing northern and southern California with toxic smoke-filled air that's as bad as Beijing or Bombay with serious long-term health implications.[20] And this is with just a 1°C warming. Imagine what California will look like with a 2°C or 3°C let alone a 4°C warming.

2. Auto-related emissions are the largest share

Auto emissions account for 80 percent, of transportation-related GHG emissions in the U.S, 28 percent of all anthropogenic emissions in California.[21] Auto emissions are surging despite the advent of all-electric and hybrid cars because electric-car sales comprise such a tiny share, less than 2% of current sales, gasoline-powered auto sales are surging, and especially because of the industry and consumer preference for

---

[19] Sara Stefanini, "Biofuel options for aviation running low," *POLITICO*, January 16, 2018:
https://www.politico.eu/article/biofuel-options-for-aviation-running-low-sustainable-aviation/.
[20] Julie Turkewitz and Matt Richtel, "Air quality in California: devastating fires lead to new danger," *New York Times*, November 16, 2018.
[21] Oliver Milman, "Vehicles are now America's biggest CO_2 source but the EPA is tearing up regulations," *Guardian*, January 1, 2018.

ginormous gas-hog SUVs and 6-7,000 pound "light" trucks which have become the biggest selling "cars" in the U.S. today – so much so that Detroit has all but abandoned producing ordinary automobiles, especially gas-sippers.[22]

The San Diego Freeway ((Highway 405), Los Angeles

Environmentalists like Bill McKibben think the solution is simply to replace America's half-billion fossil fuel-powered cars with a half-billion solar and wind-powered electric cars.[23] There's a place for electric cars in a sustainable society but that's not *the* solution. Why? First, because the bulk of all the pollution produced over the

[22] Jaime Condliffe, "Cheap gas and big cars are killing Obama's fuel economy push," *MIT Technology Review*, July 19, 2016: https://www.technologyreview.com/s/601929/cheap-gas-and-big-cars-are-killing-obamas-fuel-economy-push/. Rober Ferris, "Ford is basically giving up on US car business, and GM is not far behind," CNBC, April 26, 2018: https://www.cnbc.com/2018/04/26/ford-is-basically-giving-up-on-us-car-business-and-gm-is-not-far-behind.html.

[23] Bill McKibben, "World at War," *The New Republic*, August 15, 2016: https://newrepublic.com/article/135684/declare-war-climate-change-mobilize-wwii.

life cycle of any vehicle, gasoline or electric-powered, is generated *before it leaves the showroom*, in the production of the car (the extraction and transport of raw materials, the steel, aluminum, rubber, plastics, fabrics, leather, adhesives, electronic components and so on that go into the car, the manufacturing process itself) and in the final disposal of the car. A German cradle-to-grave study found that gasoline-powered cars emit 56 percent of their pollution before they ever hit the road, and 4 percent after they are junked.[24]

Producing electric cars is even more polluting than producing gasoline-powered cars, a lot more. Take Elon Musk's Tesla S which he touts as a "zero emission vehicle". That's false advertising. The Tesla could emit zero CO_2 emissions during its "driving" life phase if it were always charged with 100% renewable energy, which it is not because the national grid is still, on average, 65% powered by fossil fuels and likely to remain so for decades to come. But even if it were, when emissions from its production and disposal phases are taken into account, it would be more accurately described as a "50% lower emissions vehicle" – at best. In a life cycle comparison of GHG emissions from electric vs. gasoline cars, the Union of Concerned Scientists found that BEVs [battery electric vehicle] release significantly more emissions (15 percent more for a mid-size Nissan Leaf, 68 percent more for the full-size Tesla S) during their

[24] See John Whitelegg, "Dirty From Cradle to Grave," (1993) a translated summary of the German study: http://www.worldcarfree.net/resources/free.php. Union of Concerned Scientists, *Cleaner Cars from Cradle to Grave* (November 2015): http://www.ucsusa.org/sites/default/files/attach/2015/11/Cleaner-Cars-from-Cradle-to-Grave-full-report.pdf.

production phase than comparable gasoline vehicles, mostly due to materials and fabrication of the lithium-ion batteries.[25] The UCS study still concludes that

> "From cradle to grave, BEVs are cleaner. On average, BEVs representative of those sold today produce less than half [47–49%] the global warming emissions of comparable gasoline-powered vehicles, even when the higher emissions associated with BEV manufacturing are taken into account" (p. 1).

So BEVs are cleaner by half, but nowhere near zero emissions. What's more, electric cars typically weigh 25% more than comparable gasoline-powered cars because of those batteries. With its 1,200 pound 85 kWh battery, the full-size luxury Tesla Model S weighs 4,960 pounds, nearly 1,300 pounds heavier than a comparable gasoline-powered Cadillac CT6 or Mercedes-Benz C 300. To compensate, Musk specifies aluminum for the body and chassis of the Tesla. But aluminum smelting is extremely energy intensive – releasing 10 times more CO_2 emissions than steel production.[26] So if producing the aluminum body and chassis generates 10 times more CO_2 emissions than producing steel Cadillacs or Mercedes-

---

[25] Union of Concerned Scientists, *Cleaner Cars from Cradle to Grave* (November 2015), pp. 3, 5:
http://www.ucsusa.org/sites/default/files/attach/2015/11/Cleaner-Cars-from-Cradle-to-Grave-full-report.pdf

[26] Geoffry Brooks, "The trouble with aluminum," *The Conversation*, May 24, 2012:
https://theconversation.com/the-trouble-with-aluminium-7245.
Also: C.A. Ungureanu et al., "Life-cycle cost analysis: aluminum versus steel in passenger cars" (2007):
http://www.phinix.net/services/Carbon_Management/Life-cycle_Cost_Analysis.pdf.

Benz, and then you consume 25% more energy (from our still mainly fossil fuel-powered grid) to drive your 5,000 pound Tesla S a few miles to work or pick up some bags of groceries at the mall than if you drove a lighter-weight gasoline-powered car, how "green" is that? What quantity of emissions, if any, are you really saving? And that's just the Model S sedan. The new Tesla Model X SUV is so obese with its 6,768 pound Gross Vehicle Rating that it's actually *illegal to drive over the Brooklyn Bridge* because the bridge has a posted vehicle weight limit of 3 tons. Who knew that Americans would one day be climbing into 3-ton passenger cars to go pick up a quart of milk?[27]

Given the foregoing it's difficult to see how replacing a half-billion fossil-fueled-powered cars with half-a-billion 1,200 pound heavier, somewhat less-polluting electric cars, is going to save the world.

---

[27] Lloyd Alter, "New Tesla Model X is too heavy to legally drive over the Brooklyn Bridge," *Treehugger*, February 25, 2016: https://www.treehugger.com/cars/new-tesla-model-x-too-heavy-legally-drive-over-brooklyn-bridge.html. Elon Musk is not alone. There are at least 8 other cars currently sold in the U.S. that are also illegal to drive across the Brooklyn Bridge, and probably many other bridges across the country which are similarly rated for cars and light trucks: the Chevy Silverado 2500HD: 7,552 lbs., the Ram 2500 Power Wagon Laramie: 7324 lbs., the Nissan Titan Crew Cab S: 6,709 lbs., Ford -250 Super Duty Super Cab: 6,230 lbs., Lincoln Navigator L: 6,609 lbs., Cadillac Escalade ESV: 6,040, Lexus LX570: 6,000 lbs., Toyota Sequoia: 6,000 lbs. But not to worry, you won't risk a summons if you're driving your Rolls-Royce Phanton Drophead Coupe: 5,995 lbs or your Land Rover LR4: 5941 lbs. or your Bentley Mulsanne: 5,519 lbs. or your Porsche Cayenne S E-Hybrid: 5,809 lbs. Who knew that "cars" would one day weigh more than 3 tons? Even the Cadillac Coup de Ville land yachts of the mid-1950s weighed only 4,400 pounds. Will Sabel Courtney, "8 cars that legally can't cross the Brooklyn Bridge," *The Drive*, January 18, 2016: http://www.thedrive.com/new-cars/1720/8-cars-that-legally-cant-cross-the-brooklyn-bridge.

Lincoln Navigator: 6,609 lbs (also illegal to drive over the Brooklyn Bridge) [Photo: The Drive].

For what it's worth, here's my vote for THE WORLD'S MOST ECOLOGICAL CAR:

1953 Chevrolet Bel Air (3,600 lbs)[Photo: author].

This is 1953, or thereabouts, Chevrolet Bel Air (in Havana). Nineteen fifty-three Chevrolets got 14.4 miles per gallon, a bit better than the current 2018 Chevy Impala, rated at 14.0 mpg city driving (and note that this is after *more than half a century* of Detroit fuel economy improvements).[28] But the mileage isn't so important. What's important is that this car has a massively smaller carbon footprint than any Tesla or Prius or Volt because *it was only produced once and Cubans have been (rather stylishly) driving it for 66 years.* Since the U.S. imposed its blockade on Cuba in 1962, we Norte Americanos have gone through at least 6 or 8 cars with all the pollution resulting from their production and disposal, while the Cubans have thriftily and ecologically repaired those old cars and kept them on the road (not their own choice to be sure, but nonetheless).

If the bulk of CO_2 emissions from cars are produced before the car leaves the showroom then, obviously, *the best way to suppress vehicle emissions is to produce as few cars as we need and make them last as long as possible.* But of course that runs directly counter to the needs of the capitalist auto industry which must seek to maximize sales by driving repetitive consumption, the faster the cycle the higher the profits. Ever since the 1920s, the auto industry has been based on designed-in and advertising-driven obsolescence as the industry ritually pushed "new" but trivially different models each year. Detroit's

---

[28] https://www.fueleconomy.gov/feg/noframes/38824.shtml. 1953 Chevy convertible: https://www.automobile-catalog.com/make/chevrolet_usa/full-size_chevrolet_3gen/full-size_bel_air_1gen_convertible_coupe/1953.html. Note also that the '53 Chevy weighed just 3,600 pounds vs. 5,000 for the Tesla 6.

Holy Grail was to get you to trade in your "old" car every year. They used to focus on style – grills and tailfins – and of course they've always pushed the biggest "fully loaded" models like the ponderous Cadillac Escalades and Lincoln Navigator barges one sees all over my home town New York City – "big car big profit, small car, small profit". Today they've ramped up the technology larding their cars with high-tech features and gadgets: hybrid or all-electric motors, driver assist, AppleCarPlay, rear cameras, even interior cameras, dangerously distracting infotainment systems (Unsafe at any speed!), radar, automatic braking, computerized suspensions, heated (and massaging) seats, heated steering wheels, mood lighting, and more. Much of this high-tech is rapidly obsolesced, can't be upgraded, or is prohibitively expensive or impossible to repair.[29] Rapid tech obsolescence and the high cost of high-tech repairs is driving consumers to lease cars short-term instead of buying them, and short-term use is accelerating disposal of perfectly good but "obsolete" vehicles.[30] That brand new 2018 Fiat 500e electric car with its 84 mile range will be obsolete in a year or two as new models boast figures well into the triple digits. Like your perfectly functional

---

[29] I recently had to replace a headlight on my 2004 Volvo station wagon. That lightbulb, which of course today comes with built-in electronics, cost me $1450. The car has a Bluebook value of around $4,000 so replacing both headlights would cost almost as much as the car is worth – for lightbulbs! And for all the electronics, it's not apparent to me that those headlights illuminate the road any better than the standard $35 sealed-beam bulbs on my old VW Beetle.

[30] Jack Flemming, "Cars full of tech that get outdated fast – so people are leasing, not buying," *Los Angeles Times*, July 7, 2017. Charlie Sorrel, "New tech-filled cars break down more often than older cars, according to AAA," *Fast Company*, July 22, 2016: https://www.fastcompany.com/3062006/new-tech-filled-cars-break-down-more-often-than-older-cars-according-to-aaa.

iPhone 5 that Apple refuses to upgrade because they'd rather you to buy the latest model, chances are it will become e-car waste, junked long before it's worn out. The replacement battery for a Tesla S costs $44,000, more than half the base cost of the car ($75,000). How many people are going to buy a second-hand Tesla that needs a $44,000 battery? Especially when the 2022 model will have double or triple the range? If the shift to electric cars ends up shortening the life cycle of cars, that could drive up resource consumption and CO_2 emissions from automobile production, just as it has done with phones, computers, and appliances, instead of reducing it. But industry consultant McKinsey & Co. cheers on disposable cars, predicting that "technology-driven trends", "shorter lifecycles" and "faster replacement rate" will drive up profits.[31]

In short, the entire auto industry – electric or gas-powered – is completely unsustainable. We don't need an auto industry that produces tens of millions of cars a year. The solution to minimizing pollution is to redesign the entire *transportation system* on the basis of rational social needs, not individual vehicles, not individual corporate profit, to minimize resource consumption instead of maximizing it. The only way to suppress emissions from the auto industry is to drastically reduce vehicle production, ban the production of needlessly obese SUVs and giant pickup trucks (except for those with a demonstrated need for such work-vehicles), vastly

[31] McKinsey & Company, "Automotive revolution – perspective towards 2030," January 2016: https://www.mckinsey.com/~/media/mckinsey/industries/high%20tech/our%20insights/disruptive%20trends%20that%20will%20transform%20the%20auto%20industry/auto%202030%20report%20jan%202016.ashx.

expand many modes of public transit and biking, discourage private ownership of cars and encourage the use of shared vehicles. And to the extent that we need cars, if we're going to conserve resources and minimize pollution we would have to make them small, simple, durable energy-sippers, endlessly re-buildable, easily upgradeable.[32] Such radical but necessary changes would demolish Detroit's business plan. But either we save GM and Ford and Tesla for a few decades till we collapse, or we save the humans, which is it to be?

3. Death by plastic

Suppressing emissions from petrochemicals, plastics, pharmaceuticals, and industrial farming: According to a recent International Energy Agency (IEA) report, rapidly growing greenhouse gas emissions from the petrochemical industry – which includes plastic, fertilizer, and pharmaceutical companies – are eroding climate gains from reductions in other sectors such as electricity generation. Fatih Birol, executive director of the IEA told the *New York Times* "When we look at the years to come, the petrochemical sector is by far the largest driver of global oil demand growth, much higher than cars, much higher than trucks, aviation, and shipping."[33]

The exponential production of plastics since WWII, especially single-use disposable plastic packaging, bags,

---

[32] Like, say, VW Beetles: Christopher Ross, "The vintage Volkswagen beetle goes electric," *Wall Street Journal*, April 23, 2015.
[33] Kendra Pierre-Louis, "Greenhouse gas emissions from plastics are predicted to rise," *New York Times*, October 4, 2018.

synthetic fabric microfibers, microplastics used paints, cosmetics and other products, is turning the world's oceans into seas of plastics, wiping out marine creatures, seabirds, mammals. According to one study, plastic trash will soon outweigh all the fish in the oceans.[34] We can't strain the oceans. Here again, there's no market solution for any of this. There's no tech fix either. Plastic recycling is no solution.[35] The only solution is to just stop making most plastics: abolish production of disposable bottles, plastic packaging, disposable plastic shoes, clothes, microfibers and micro plastics if they can't be rendered sustainable. The EU has just approved a ban on single-use plastics, though this is subject to member approval so can't be assumed. Plus, inexcusably, this would not take effect till 2021 (sorry, whales and seabirds). British MPs have called for a ban on microplastics used in cosmetics.[36] These are important steps but far from what we need to be doing right now. After all, we lived without all this disposable plastic junk in the 1940s and '50s and we weren't living in caves. We can have the "convenient" throwaway economy or we can have a habitable planet. We can't have both.

---

[34] Rex Weyler/Greenpeace, "The ocean plastic crisis," October 17, 2017: https://www.greenpeace.org/usa/the-ocean-plastic-crisis/. Lauren McCarthy, "Report warns that plastics will soon outweigh fish in world's oceans," *Common Dreams*, January 19, 2016: https://www.commondreams.org/news/2016/01/19/report-warns-plastics-will-soon-outweigh-fish-worlds-oceans. Leah Messinger, "How your clothes are poisoning or oceans and food supply," *Guardian*, June 20, 2016.

[35] Matt Wilkins, "More recycling won't solve plastic pollution," *Scientific American*, July 6, 2018:

[36] Ceylan Yeginsu, "European parliament approves ban on single-use plastics," *New York Times*, October 25, 2018. John Vidal, "Microplastics should be banned in cosmetics to save oceans, MP says," *Guardian*, August 23, 2016.

Sperm whale dies off Indonesia, November 21, 2018 from ingesting 13 pounds of plastic garbage including 115 plastic cups, a pair of flip-flops, 4 plastic bottles, 25 plastic bags, yards of plastic string etc. Another died off Thailand on June 3rd after swallowing 80 plastic bags. Still another died off Spain in April after ingesting 60 pounds of plastic trash, fish netting and garbage bags.[37] And those are just the ones that have been reported in one year. No one knows how many others died, or how many smaller creatures, dolphins, fish, sharks, sea birds, turtles and such perish every day from ingesting our trash, not to mention our toxic chemicals [Photo: Inertia].

4. Death by toxics

Same with toxic chemicals. We face an out-of-control global toxics crisis, the result of largely unregulated

---

[37] Matthew Haag, "64 pounds of trash killed a sperm whale in Spain, scientists say," *New York Times*, April 12, 2018; Dylan Heyden, "Dead sperm whale that washed up in Indonesia ingested 13 pounds of plastic," *Inertia*, November 19, 2018: https://www.theinertia.com/environment/dead-sperm-whale-that-washed-up-in-indonesia-ingested-13-pounds-of-plastic/. Rosie Perper, "A whale died after swallowing 80 plastic bags in Thailand," *Business Insider*, June 3, 2018: https://www.businessinsider.com/thailand-whale-died-swallowed-plastic-bags-2018-6.

production and irresponsible management of toxic chemicals: wanton spraying of toxic pesticides of large swaths of the planet, irresponsible dumping of chemicals, irresponsible production and marketing of untested, unsafe chemicals including household "miracle" products from Teflon to flame retardants, falsely alleged to be safe but in fact are carcinogenic, poisoning our waters, farms, food, and other life forms.[38] No doubt in any modern industrial society we are going to need some toxic chemicals. But we need to rationally and democratically plan what we need and what we can do without. This can't be left up to private corporations with a vested interest in their own products.[39] Chemical production should follow the Precautionary Principle such that chemicals are produced only after they've been proven safe, and safe handling procedures have been installed so that the public and nature are protected – as for example, the protocols set up by the NGO Safer

[38] Tiffany Kary and Christopher, "Cancer-linked chemicals manufactured by 3M are turning up in drinking water," *Bloomberg*, November 2, 2018: https://www.bloomberg.com/graphics/2018-3M-groundwater-pollution-problem/?srnd=premium. Elizabeth Grossman, "Flame retardant exposure poses a significant health risk to women," *Truthout*, May 28, 2016: https://truthout.org/articles/flame-retardant-exposure-poses-a-significant-health-risk-to-women/. Abraham Lustgarten, "How the EPA and the Pentagon downplayed a growing toxic threat [Teflon]," *Truthout*, July 14, 2018: https://truthout.org/articles/how-the-epa-and-the-pentagon-downplayed-a-growing-toxic-threat/. "Evaggelos Vallianatos, "Spraying poisons, chasing ghosts," Counterpunch, November 15, 2018: https://www.counterpunch.org/2018/11/15/spraying-poisons-chasing-ghosts/. Samuel S. Epstein, *Toxic Beauty* (Dallas: Benbella Books, 2009) (toxins in cosmetics and personal care products). Philip and Alice Shabecoff, *Poisoned Profits: The Toxic Assault on Our Children* (New York: Random House 2008), Phil Brown and Edwin J. Mikkelsen, *No Safe Place* (Berkeley: University of California Press, 1990), etc.

[39] Marie-Monique Robin, *The World According to Monsanto: Pollution, Corruption, and the Control of the World's Food Supply* (New York: New Press, 2009); Robert Kenner, *Food, Inc.* (Magnolia Pictures, 2008), etc.

Chemicals Healthy Families[40] – instead of the system we have now in which chemical companies flood the market with thousands of new chemicals each year with little or no regulation and we and other species become their test guinea pigs.[41] If chemical companies like Monsanto, Dow Chemical, and 3M, or factory farming operations like ConAgra, Tyson, and Perdue can't be redesigned and restructured to produce safe, ecologically sustainable products, safe and sustainable foods, then what choice is there but to close them down and secure their chemists and production workers alternative employment elsewhere?

Last year, China stopped accepting our plastic waste, paper waste, and electronic e-waste. This put the West, with our throwaway economies, on notice, forcing us to reconsider all the superfluous packaging, disposable electronics, and waste produced by capitalism. The trash crisis calls out for a radical solution and the radical solution is, again, all too obvious. As the Los Angeles Times recently editorialized, "*California has a recycling crisis. The only way to solve it is to stop making so much*

---

[40] Safer Chemicals Healthy Families at https://saferchemicals.org. See further discussion of this point and other sources and organizations in Smith, *Green Capitalism: The God That Failed* (London: World Economic Association Press, 2016, pp. 45-46.

[41] Vicky Xiuzhong Xu, "Palau bans many kinds of sunscreen, citing threat to coral," *New York Times*, November 2, 2018.

trash."[42] If jobs will be lost, profit will be lost, so be it. We ecosocialists have a plan for that (Part IV below).[43]

5. Ships and their "supersized pollution problem"

Tourism is yet another growing contributor to CO_2 emissions, accounting for nearly a tenth of emissions and growing fast. Cheap airfares, Airbnb, and the growing global middle class spur the growth.[44] And not just planes. Cruise ships are by far the biggest per-capita polluter in the history of travel. One large cruise liner puts out more, and far filthier, emissions in a day than 5 *million cars* (the equivalent of all the cars in Beijing). The coastal provinces of China, the Los Angeles basin, and Mediterranean Europe are smothered under layers of bunker fuel fumes.[45] Here again, with no solar-powered cruise ships on the horizon, the only way to suppress emissions from this staggeringly wasteful and polluting industry is to shut it down.

Suppressing shipping emissions: Cruise ships are bad enough. But the bulk of shipping emissions today come from the thousands of container ships that course

---

[42] May 26, 2018: http://www.latimes.com/opinion/editorials/la-ed-recycling-crisis-20180526-story.html. Yet instead, so far at least, instead of suppression production, cities are burning the plastic trash. Brilliant. Oliver Milman, "'Moment of reckoning'" US cities burn recyclables after China bans imports," *Guardian*, February 21, 2019.

[43] On green manufacturing and its limits, see Smith, Green capitalism, pp. 85-90.

[44] Farhad Manjoo, "'Overtourism' and the role of technology," *New York Times*, August 29, 2018. Zach Wichter, "Paris beckons as a fare war turns Europe into a bargain," *New York Times*, July 3, 2018.

[45] John Vidal, "The world's largest cruise ship and its supersized pollution problem," *Guardian*, May 21, 2016. Janet Wilson, "California fights shipping pollution as international shippers push back," *Inside Climate News*, October 2009: https://insideclimatenews.org/news/20091012/california-fights-shipping-pollution-international-shippers-push-back.

between China and the U.S. and Europe. Since China joined the WTO in 1991 the volume of world shipping has quadrupled[46] and the bulk of that is container ships full of mostly disposable products produced by police-state shackled, semi slave-labor in China's Special Exploitation Zones – all manner of plastic junk, disposable shoes, clothes, tools, and household goods, quickly obsolesced electronics and so on – products that in a rational sustainable economy would never be produced in the first place.[47]

Who really needs all this?

As of 2015, shipping emissions accounted for 3% of global CO_2 emissions but are projected to grow by 250% by 2050 when they will account for 17% of global

[46] Linda Qiu, "China's growth fuels boom in world shipping traffic," *National Geographic*, November 27, 2014.
[47] Note, China became a full member of the WTO in 2001. On production of disposable products by disposable workers in China, see my "China's communist-capitalist ecological apocalypse, *Real-World Economics Review* no. 71 (2015): http://www.paecon.net/PAEReview/issue71/Smith71.pdf.

emissions.[48] The shipping industry has stalled on suppressing emissions because, as with aviation, there's no currently available replacement for fossil fuel-powered propulsion for large vessels.[49] Electric boats may be possible someday, but not today, and not soon enough to replace diesel in the brief timeframe of the decade or so we have to radically suppress emissions before it's too late. Shipping industry analysts report that

> "While improved battery technology has helped the new generation of electric European ventures get afloat, long-haul, ocean-going vessels currently do not have the option of docking regularly to plug in small batteries – meaning that they are unlikely to become completely electrified in the near future... With current limitations on technology, it seems that electrification will be limited to small craft undertaking short, ferry-type voyages."[50]

Thus for the present, the only way to suppress shipping emissions is to stop shipping so much stuff around the world. No major container companies are registered in the United States but the U.S. is the biggest consumer of container-shipped goods from China. Thus the best way for the U.S. to suppress shipping emissions is to stop

[48] Editor, "Shipping emissions 17% of global CO_2, making it the elephant in the climate negotiations room," *Transport and Environment*, November 23, 2015.

[49] Phil McKenna, "An ambitious global effort to cut shipping emissions stalls," *Inside Climate News*, October 26, 2018.

[50] Jess Smee, "Are electric vessels the wave of the future in shipping?" *Deutsche Welle*, March 20, 2018: https://www.dw.com/en/are-electric-vessels-the-wave-of-the-future-in-shipping/a-43046309.

importing China's disposable products whether manufactured by Chinese or Western companies.[51] Radical? Extreme? No doubt. But what's the alternative? How else can we suppress maritime emissions in the brief time we have left?

Yet, to introduce note of caution here, even if electric container ships were available today, given capitalism, this would most likely just accelerate the destruction of our world because the problem is not shipping. The problem is the capitalist driver of ever-growing shipping. For example, China's domestic and export overproduction is based on relentless looting of global natural resources to feed the gaping maw of its industrial engines. From New Guinea and Indonesia to Congo to the Amazon, China is strip-mining the planet, mostly to produce products neither the Chinese nor we really need.[52] Electrifying China's export shipping could speed

---

[51] As I've argued elsewhere (Smith, *RWER, issue* 71; and *Green Capitalism: The God That Failed,* World Economics Association Books), we can never have sustainable economies until we banish the production of designed-to-be-obsolesced, disposable products from plastic shoes to iPhones to Chevrolets and Teslas. The bulk of what we manufacture from China to the U.S. is not produced to meet the needs of people but to meet the needs of industries to sell and sell again to people. When such products are, instead, produced to be durable, repairable, rebuildable, upgradeable and shareable, we will not need anything like the size of manufacturing industries we have today and virtually all the manufactured goods the U.S. economy truly needs can easily be produced in the U.S. and at prevailing U.S. wages.

[52] See the many papers by Professor William Laurance and his associates at TESS Centre for Tropical Environmental & Sustainability Science, James Cook University: Queensland and their publication ALERT: http://alert-conservation.org. For example, "Is the global era of massive infrastructure projects coming to an end? YaleEnvironment360, July 10, 2018: https://e360.yale.edu/features/is-the-global-era-of-massive-infrastructure-projects-coming-to-an-end. And again, Smith, "China's communist-capitalist ecological-apocalypse, op cit.

this destruction. The only way to suppress it is to shut down China's engines of overproduction.[53]

In sum, "de-growing" so much of the economy, wrenching down and phasing out polluting, harmful, wasteful and useless industries would certainly be "huge," "unprecedented shifts." But isn't this exactly the sort of "systems transitions ... unprecedented in terms of scale" that we need to be making?

C. Why carbon taxes fail

Three years on from Paris nothing has changed: Soaring emissions, shocking increases in CO_2 concentrations, arctic fires and floods notwithstanding, governments, economists, even climate scientists continue to proffer the same old market-based fake solutions: steeper carbon taxes, unproven, impossibly expensive "carbon removal" technologies, and fabulist geoengineering schemes.[54] Everything but the obvious – "deep emissions reductions in all sectors" at "unprecedented speed and scale."

The UN IPCC scientists themselves called for steep increases in carbon taxes, pushing governments "to set a price high enough to spur truly deep as reductions in carbon emissions."[55] I contend that this approach is

---

[53] Smith, *China's Engine of Ecological Apocalypse* (Verso 2019).
[54] Brad Plumer, "Scientists push for a crash program to scrub carbon from the air," *New York Times*, October 24, 2018. Cf. Howard J. Herzog, "Why we can't reverse climate change with 'negative emissions' technologies," *The Conversation*, October 9, 2018: http://theconversation.com/why-we-cant-reverse-climate-change-with-negative-emissions-technologies-103504.
[55] Brad Plumer, "U.N. climate report urges putting a price on carbon," *New York Times*, October 9, 2018.

hopelessly untenable. Both cap-and-trade schemes and carbon taxes grew out of the Reagan-era Milton Friedmanite hostility to government "command and control" and the preference of Reagan, Clinton and subsequent governments for market solutions to all problems. The idea was to incentivize businesses and consumers instead of imposing state-mandated quotas or rationing.

1. Theory and practice

The carbon tax idea was straightforward: tax fossil fuels and consumers and companies will seek non-fossil fuel alternatives. As taxed coal-fired power plants and fossil fuel-powered motor vehicles became more costly to operate relative to untaxed (and even subsidized) renewable energy and electric cars, over time coal and gas-fired power plants and petrol-powered motor vehicles would fade from the scene. The theory seemed compelling, even obvious in the abstract world of economic modeling.[56] In the real world, they've

[56] It might seem surprising that leading oil companies like ExxonMobil, Chevron, BP officially *support* carbon taxes. They do so because they understand that at some point governments are going to impose some kind of tax on carbon emissions and they want the best deal they can get if they have to be taxed at all. They've deeply resisted "cap and trade" schemes, effectively dooming them everywhere, because cap and trade would impose a limit to growth. Carbon taxes are the lesser evil. So long as carbon taxes are modest, they pose no existential threat to their business because they impose no cap on production, no limit to growth. They're just another cost of doing business, like other taxes, and one that can be passed on to the consumer. Meanwhile, growth can continue. BP's 2018 Energy Outlook projects that global energy demand will grow by a third between now and 2040. BP and the other oil giants aim to cash in on that growth. That's why the day after the IPCC report was released, ExxonMobil contributed a million dollars to jumpstart a carbon tax program. They're not fools. They're not looking to put themselves out of business. They're looking to maximize returns, to grow their business

completely failed to significantly suppress emissions. The whole idea was doomed from the start.[57]

The fault lay in the economics: If the tax is too light, it fails to suppress fossil fuels enough to help the climate. But if it's heavy enough to really suppress them, then companies and consumers balk and resist the tax – because without any safety net for businesses and consumers, the entire burden falls on them, so they rationally resist – to save profits and jobs. Thus to date, the only carbon taxes that have proven acceptable to governments and the voting public are those which are too light to do the job. More than 40 governments including EU, California and British Columbia have imposed taxes on carbon but none has put more than a trivial dent in emissions.[58] A report from the Organization for Economic Cooperation and Development (OECD) found that the average carbon price across 42 major economies was around $8 per ton in 2018 (that's equal to 7 cents on a gallon of gasoline) – "far below the level that the IPCC scientists say is necessary to address

---

while posing as good corporate citizens, paying their taxes (if they have to), contributing to the "solution." See: Ben Gamen, "Big oil companies want a price on carbon. Here's why," *The Atlantic*, June 1, 2015: https://www.theatlantic.com/politics/archive/2015/06/big-oil-companies-want-a-price-on-carbon-heres-why/446637/. Smith, *Green Capitalism*, pp. 59-64. Ed Crooks, "ExxonMobil gives $1m to campaign for a carbon tax," *Financial Times*, October 9, 2018: https://www.ft.com/content/9665a09a-cba9-11e8-b276-b9069bde0956.

[57] See my discussion of this in *Green Capitalism*, pp. 61-64.

[58] For example, Food and Water Watch, *The British Columbia Carbon Tax: a Failed Experiment in Market-Based Solutions to Climate Change,* October 24, 2016: https://www.foodandwaterwatch.org/insight/british-columbia-carbon-tax-failed-experiment-market-based-solutions-climate-change. Also: Lenore Taylor, "The carbon tax is dead and there is nothing credible to take its place," *Guardian*, July 16, 2014 (on the failed Australian tax).

climate change."[59] No government will set a price high enough to spur truly deep reductions in carbon emissions because they all understand that this would force companies out of business, throw workers out of work, and possibly precipitate recession or worse. What government wants that?

The carbon tax "blueprint for destroying the world economy" (Heritage Foundation)

Given relentlessly growing emissions, and given the failure of modest carbon taxes to suppress fossil fuel production to date, the desperate IPCC climate scientists called on governments to impose truly draconian taxes – on the order of $135 to $5,500 per ton or more – whatever it takes to suppress fossil fuel consumption enough to keep global warming below 1.5°C.[60]

The pro-market Heritage Foundation, not surprisingly, skewered this recommendation arguing that taxing industries by anything like these numbers "would bankrupt families and businesses and trigger a global economic disaster." The IPCC plan, they said is "a blueprint for destroying the world economy."[61] In point of fact, given capitalism, they're right about that. In a world of abstract models, the carbon tax strategy works perfectly. But in the real world, with real investors and real employees – *and without a rationally planned, carefully*

[59] Plumer, "Putting a price on carbon," op cit.
[60] Plumer, "Putting a price on carbon," op cit.
[61] Heritage Foundation, "IPCC climate report merely a blueprint for destroying the world economy," October 17, 2018: https://www.heritage.org/energy-economics/commentary/un-climate-report-merely-blueprint-destroying-the-world-economy.

managed drawdown and phase-out combined with guaranteed state support for the investors and guaranteed "just transitions" for the affected workers, the imposition of draconian carbon taxes would bankrupt some of the largest companies in the world, precipitate a stock market crash, throw millions out of work, and most likely "destroy the world economy". Voters intuitively understand this. That's why they just voted down the carbon tax in Washington State. And if voters won't pass Washington's trivial $15 a ton tax (equal to 13 cents on a gallon of gasoline), why would scientists think they would pass a tax of $135 (or $1.20 on the gallon), let alone $5,500 a ton ($48.95 a gallon!)?[62] To ask the question is to answer it. Imposing the entire cost of driving out fossil fuels onto companies and consumers without providing them alternatives and a safety net is hardly a winnable plan. Just the opposite: it's guaranteed to turn people against the whole project.

Just look at France. Président Macron wanted to promote an "ecological transition" and "liberate households from dependence on petrol" – a legitimate objective in itself – but as a former investment banker he naturally preferred to use market incentives and penalties to do the job

[62] Hat tip to Professor David Klein for the ton-gallon conversions. Climate scientist James Hansen has proposed a variation on this theme, his "carbon tax and dividend" scheme. Under this plan, consumers would pay a $1 tax per gallon on gasoline but they would receive a "dividend" check for the amount they were taxed back from the government at the end of the year. Painless! Great! The folly of this scheme is that if there's no pain there's no gain. The average American driver consumed 656 gallons of gasoline in 2016. If he or she were to get a refund of the $656 carbon tax, what's to prevent them from spending that refund on, say, a new flat-screen TV (the production of which produces lots of CO_2 emissions), or a Jet-Blue round trip flight to Cancun, $268 from NYC (far more emissions than their car produces in a year) either of which they could easily afford with that tax refund? No pain – but no gain either.

instead of *dirigiste* state intervention and planning. So he imposed a carbon tax on petrol, arguing that raising the price of fuels would force consumers to reduce their use of cars and/or buy more fuel-efficient vehicles or electric cars. Those with oil heating would likewise replace boilers that use oil with wood-burning or gas boilers. Instead, he got insurrection. Hundreds of thousands of French workers donned "yellow vests" and have been protesting across the country every weekend since November.

Why? First off, rising housing costs in recent decades, especially in Paris and other big cities, have forced more and more workers to move out to the suburbs or even rural villages. Far from urban public transit networks, they have no choice but to drive to work. Macron's banker buddies can easily afford to go buy a Tesla to drive into the city from their weekend chateaux. But French workers have been squeezed by stagnant salaries and sharply rising social taxes in recent years, so few of them can afford to replace their home heating boiler (5,000 euros and up) let alone go out and buy even the least expensive electric car sold in France, a Renault Zoe at 27,000 euros. Macron's (flat) carbon tax squeezed them even harder making that "ecological transition" even less affordable. It's not that French workers are bone-headed and don't care about the environment. It's that Macron's carbon tax puts the burden on those who can least afford it and doesn't suppress emissions anyway. Further, while Macron abolished the wealth tax on the rich – stoking endemic class hatred (this is France after all)[63] – he

[63] Richard Greeman, "As Macron prepares new repressive measures yellow vests and red unions strike together," Portside, February 15, 2016:

simultaneously halved the energy transition tax credit (CITE) that subsidized households to install more efficient boilers. And while levying carbon taxes on those who can least afford it, many of France's most polluting industries including maritime and air transport, are exempted.[64] Little wonder 77% of French people support the Yellow Vests and despise the arrogant banker-président who despises the "gilets jaunes" as "a hate-filled mob."

2. But if carbon taxes worked, what then?

Yet the foregoing are the least of the problems with the carbon tax strategy. For suppose governments actually adopted those draconian carbon taxes. After all, the whole point of carbon taxes is not just to reduce consumption somewhat. That's not going to stop global warming. The point is to drive fossil fuels out of the market and replace them with renewables. Draconian taxes would certainly drive them out – but would precipitate economic chaos and collapse in the process. ExxonMobil, BP, Shell and so on, among the biggest companies in the world, would be bankrupted. So would the industrial fossil fuel consumers: refineries and distributors, petrochemical companies, auto manufacturers, airplane manufacturers, airlines, tourist hotels, shipping companies, synthetics producers, plastics and disposable industries and more from Europe

https://portside.org/2019-02-15/macron-prepares-new-repressive-measures-yellow-vests-and-red-unions-strike-together.

[64] Maxime Combes, "Yellow vests" Macron's fuel tax was no solution to climate chaos," *rs21*, December 4, 2018: https://www.rs21.org.uk/2018/12/04/the-yellow-vests-why-macrons-fuel-tax-was-no-solution-to-climate-chaos/.

to China. In short, making carbon taxes "work" would result in chaos, mass unemployment and "destroy the world economy".

The patent bankruptcy of the carbon tax strategy to "get us off fossil fuels" points to the need for comprehensive solutions – indeed *dirigiste* government planning and large-scale expenditures to transform the entire economy in a sustainable direction instead of simple one-shot taxes. If Macron really wants to suppress transportation and provincial home heating-generated CO_2 emissions, he's going to have to discourage urban sprawl by consolidating cities, creating more affordable housing in and near the cities, stop building endless express motorways and ring roads, expand rail, express bus and other public transit to existing suburbs, encourage car sharing and other ways to reduce what in California they call "vehicle miles traveled."[65] At minimum, if he wanted to help suburban, rural and small-town families get off fossil fuels, he could have subsidized replacement home heaters by doubling or tripling the energy tax credit instead of halving it. He could have instituted a "cash-for-clunkers" program to take petrol cars off the roads and give workers cheap electric cars and home chargers. He could set up free electric bike stations in inner suburbs. There are lots of ways he could directly suppress emissions. But none of them are market solutions. All of them require direct government planning and subsidies.

---

[65] Alissa Walker, "When electric isn't good enough," Curbed, October 24, 2018: https://www.curbed.com/a/texas-california/electric-cars-climate-change-sacramento-california.

III. FDR's "command-and-control" economy that worked

If the U.S. government really wanted to suppress emissions it could simply ration fossil fuels like FDR did during WWII when he rationed private consumption to divert maximum supplies to the war effort. They could inaugurate a crash program to phase out fossil fuels. They could simply order America's private companies to shut down private vehicle production and change their production output from fossil fuel vehicles to, say, electric cars, public transit, wind turbines, solar power plants, just as FDR ordered GM, Ford and other companies to stop producing private cars and start producing tanks, bombers, fighter planes, artillery, machine guns, ammunition, just as he ordered Packard to build fighter aircraft and PT boat engines, just as he ordered Chrysler to build diffusers to process uranium for atomic bombs. And so on. FDR's virtual takeover of the commanding heights of American industry during WWII was tantamount to a temporary nationalization but his "command and control" was accepted by labor and capital and it succeeded – brilliantly – providing the industrial base to win the war.[66]

We would no doubt require the declaration of an emergency comparable to WWII to win the approval of Congress and the American people. The war against fascism certainly qualified as an emergency. Even so, fascism killed tens of millions but it did not threaten the extinction of life on earth. Global warming threatens

[66] Arthur Herman, *Freedom's Forge* (New York: Random House, 2012); A.J. Baime, The Arsenal of Democracy (Boston: Mariner, 2015).

extinction. Is saving the humans, not to mention the whales, less of an emergency than saving General Motors or Boeing or Apple? If Roosevelt could turn the economy on a dime to meet the emergency of WWII, why can't we do the same today? If we don't organize an emergency industrial shutdown and retrenchment of unnecessary production, superfluous manufacturing, superfluous electricity usage, wasteful over-consumption, nature is going to do it for us in a most unpleasant manner.

Yet today we can't get a fair public debate on the need for state intervention in the economy because since the age of Thatcher and Reagan, Milton Friedman's maxim "capitalism good, government bad" has been the reigning ideology of Republicans and Democrats, Tories and Labour alike. Since the 1980s Friedmanite economists and their Koch-funded think tanks have produced a relentless stream of propaganda "proving" that "command and control" or almost any "government intervention" is "bad" and "can't work".[67] Reagan told us businesses don't want to be ordered about and told what to do like FDR did. They want to be "incentivized." Since the 1980s Republicans and Democrats alike have steadily whittled away and gutted government regulation and control wherever possible,[68] clearing the way for

---

[67] For a critique of this hypocrisy and nonsense see Richard Smith, "Six theses on saving the planet," *Next System Project*, 2017: https://thenextsystem.org/six-theses-on-saving-the-planet, PDF pp. 20ff.

[68] For example, President Obama's much-ballyhooed program to "double vehicle fuel efficiency by 2025" was a con job, custom designed to service Detroit's desire to build humongous cars by including a gaping loophole based on "vehicle footprints" which let the Big Three and others build ginormous gas-hog trucks and SUVs – Sierras and Ticonderogas and Denalis and Armadas and Suburbans that get worse gas mileage that Cadillac land yachts of the 1950s – but still meet their "fleet mileage"

Trump's *coup de grâce* at the EPA this past year. In the process, their "market solutions", like carbon taxes, have completely failed to staunch emissions.

IV. The ecosocialist path to 1.5°C sustainability

We ecosocialists have a practical answer. We accept the science that to prevent runaway global warming "greenhouse emission must be reduced by 45 percent from 2010 levels by 2030, and by 100 percent by 2050." We agree with the IPCC that this will require "deep emissions reductions in all sectors." We agree that it will require "far-reaching transitions in energy, land, infrastructure, and manufacturing," that it will require "systems transitions" (indeed, more than they imagine). And we understand that this must all be done at "unprecedented speed and scale." "We understand that we desperately need to "**do what the science demands before it's too late**".[69]

---

targets (As I noted elsewhere, "Your typical 1955 Cadillac coupe DeVille got 12.9 mpg in city driving whereas your typical 2013 Cadillac Escalade gets 10 mpg in the city... and this is after *six decades* of Detroit's fuel economy "'improvements'": *Green Capitalism: the God That Failed*, p.131). Thus Obama let the industry push gas-hogs at the expense of gas-sippers while his "drill baby drill" program to frack the whole country conveniently supplied so much cheap gas that Americans all but abandoned small cars in favor of giant over-accessorized gas-hogs which have become the biggest sellers. In result, U.S. vehicle fleet mileage actually declined under Obama. Brilliant. Worse, it "incentivized" Detroit to abandon producing actual cars altogether. By spring 2018, Detroit auto companies announced they would stop producing economical small cars to focus on producing their profitable luxury hogs. So much for "win-win" market solutions. Roger Ferris, "The steadily disappearing American car," *CNBC News*, April 6, 2018: https://www.cnbc.com/2018/04/06/the-steadily-disappearing-american-car.html.

[69] See the extensive discussions of this topic at http://systemchangenotclimatechange.org.

But we also understand that imposing drastic cuts in fossil fuel production has to translate into industrial shutdowns and retrenchments across the economy. There is just no way around this. We reject the carbon-tax-to-collapse scenario. Instead, we propose a strategy of rationally planned, democratically managed, wind-down and phase-out of fossil fuels and a coordinated transition to renewable energy that avoids economic collapse and guarantees reemployment for the affected workers. Our strategy for the United States is based on a four-point:

Emergency plan to meet the climate emergency:

1. Declare a State of Emergency to suppress fossil fuel use: ban all new extraction, ration gasoline, ban production of new fossil-fuel vehicles. Nationalize the fossil fuel industry to phase it out. We propose to do this by means of a government buyout at fair value (fair to both owners and society). Nationalize downstream fossil fuel industrial consumers from pipelines, refineries, distribution networks to autos, aviation, airlines, shipping, petrochemicals, some manufacturing, some tourism and others whose business is irreversibly based on fossil fuels and which without a government buyout would be bankrupted.

2. Institute a new federal Public Works Administration-style jobs program (on the model of the Works Progress Administration, Civilian Conservation Core and other programs set up under FDR) to re-employ every worker in the fossil fuel-related industries at equivalent pay and benefits in other useful but low-

emission work. Those workers in coal, oil, plastics, toxic chemicals and so on, deserve jobs, just not the jobs they have now.

3. Launch an emergency state-directed program to phase-in renewable electric generation, replace fossil-fuel powered transportation with electric propulsion, discourage individually-owned vehicles, and encourage public transit, shared vehicles, bicycles and other non-fossil fuel modes of transportation.

4. Develop emergency plans to phase out wasteful, destructive and polluting industries from arms production to needless toxics, chemicals, pharmaceuticals etc, disposables like single-use plastics, designed-to-be obsolesced iPhones, cars etc., and useless inventions.[70] Develop emergency plans to shift from fossil-fuel dependent factory farms to fully-organic agriculture.

A. The argument for nationalization

If fossil fuel companies can't reverse their suicidal growth-to-bust trajectory, then what alternative is there but to nationalize them, socialize them, put them under public ownership to phase them out?[71] We say to

---

[70] "Alexa has 80,000 skills and none you need," *Bloomberg Business*, March 18, 2019, pp. 22-23.

[71] I'm hardly alone here. Others have also argued for nationalization to phase-out fossil fuels. See Carla Skandier, "Nationalize the fossil fuel industry," *In These Times*, November 17, 2017 and "Quantitative easing for the planet," *The Next System Project*, August 30, 2018. Also Peter Gowan, "A plan to nationalize fossil fuel companies," *Jacobin*, March 2018.

Democrats, Republicans, capitalists, and pro-fossil fuel trade unions too, "If you have a better strategy to save the planet, where is it?" As *Juliana vs. United States* (the Our Children's Trust suit which is now working its way through the courts) contends, the federal government has a responsibility to preserve a habitable planet to maintain "life, liberty, property" for the next generations. The plaintiffs argue that preserving a habitable planet requires that the government immediately begin enforcing strict limits on CO_2 emissions, immediately develop plans to phase out of fossil fuel production, and ban all further extraction (Section 7, p. 80ff).[72] Since the companies can't put themselves out of business because they have a legal and fiduciary responsibility to their shareholders, not to society, the only way to effect the phase-out of fossil fuels without precipitating economic collapse is for the government to nationalize the companies so we can dismantle them and redeploy their capital and labor with as little economic pain as possible.

---

whereas Gowan looks to hiring freezes, voluntary redundancies, and welfare state and union-run retraining schemes "to smooth out the transition as much as possible," I argue for iron-clad government-funded job guarantees at comparable pay and benefits for laid off fossil fuel workers. Given the need for immediate and rapid shutdowns and retrenchments, nothing less than this has a chance of winning worker support in those industries. Thirdly, this plan insists on a parallel and coordinated emergency phase-in of renewables to replace the industries we need to phase out. Fourthly, this plan calls for transitioning to broad democratic planning.

[72] Our Children's Trust, Juliana v. United States, Youth Climate Lawsuit: https://www.ourchildrenstrust.org/juliana-v-us.

"Hang the capitalists?"

That's why I propose to nationalize them via federal buyouts. Many on the Left ask with justification, "why should we pay to buy out the fossil fuel capitalists who already have too much money and who after all, are directly responsible for global warming?" Indeed, some call for trying fossil fuel executives for crimes against humanity.[73] I'm all for that – but I'm for buying out the shareholders. For three reasons: First, as noted above, the oligarchs drill the oil but they don't burn it, or much of it. We all bear responsibility for that. Second, roughly a quarter of fossil fuel stocks are held by pension funds, i.e. workers. Ecosocialists are not in the business of expropriating workers. Third, and most importantly, the timeframe: speed is of the essence. The Left does not have the power to expropriate the fossil fuel companies and we can't afford to waste decades waging a class war we can't win. With just a decade or so to suppress emissions by 50%, we need to make the transition as smooth and fast as possible. The path of least resistance is to buy out the corporations and the capitalists – a hostile buyout perhaps, but a buyout nonetheless.

Of course politicians will holler about the cost. The cost is significant but surprisingly affordable, a bargain actually. As of December 2018, the ten largest American oil and gas companies claimed a combined value in 2018 of $968.1 billion (Exxon Mobil is valued at $344.5 billion, Chevron $239 billion, ConocoPhilips $79.3, and the

[73] Kate Aronoff, "It's time to try fossil-fuel executives for crimes against humanity," *Jacobin*, February 5, 2019.

others from $68 to $33 billion).[74] The two major coal companies have trivial net worth (Peabody at $3.6 billion, Arch at $1.5 billion). But the IEA says that in truth, the world's fossil fuel industries are worth a fraction of their claimed value because most of their assets – the oil and gas and coal in the ground – are fast becoming valueless "stranded assets" as electric utilities and vehicle manufacturers shift to renewable power and because of growing political pressure to "leave it in the ground."[75] Given their looming existential profits crisis, the companies might actually welcome a buyout. But if society is to pay a fair price for those companies, their nominal retail value would have to be discounted by the harm their production has already done to people and planet. On any fair assessment, that would leave these companies owing the government, not the other way around. Yet even at their current retail value, just under a trillion dollars, by the standards of wasted U.S. expenditures, this is easily affordable. President Trump just *gave away* $2.3 *trillion* in tax cuts to the rich this year alone.[76] Just rescinding that inexcusable giveaway would cover the cost of nationalizing the entire fossil fuel producing industry and leave enough left over to buy out the bulk of America's fossil fuel-burning industries as well including Boeing, the major airlines, the American auto industry, the worst polluting chemical industries, and leave billions to spare. Boeing's net worth is $95

---

[74] *Statista*: https://www.statista.com/statistics/241625/top-10-us-oil-and-gas-companies-based-on-market-value/.

[75] Julian Ambrose, "IEA warns $1.3 trillion of oil and gas could be left stranded," *Telegraph*, March 20, 2017.

[76] David Rogers, "POLITICO analysis: At $2.3 trillion cost, Trump tax cuts leave big gap," *POLITICO*, February 28, 2018: https://www.politico.com/story/2018/02/28/tax-cuts-trump-gop-analysis-430781.

billion, the seven largest U.S. airlines have a combined market capitalization of $130 billion, the American auto industry, Ford, GM and Tesla (excluding Chrysler which was bought by Italy's Fiat in 2011), has a combined value of $277 billion, the big dirty three of Dow-Dupont, Monsanto, and 3M combined are worth $225 billion. The government could buy all these companies, even without discounts for their social and environmental crimes, for a paltry $727 billion. Add in the bulk of private and shareholder-owned gas and electric utility sector, 20 companies with a combined market value of $557 billion, and the government could buy up all of America's fossil fuel producers and the bulk of its fossil fuel-burners for $2.26 trillion and still have some change left after rescinding Trump's tax giveaway to the rich.[77] In effect, we buy out the fossil fuel capitalists with their own money. What's not to like about that – from the standpoint of the working class?

As to the cost of replacing fossil fuel infrastructure with renewables, Stanford's Professor Mark Jacobson estimates the capital cost of building renewable energy to supply electricity generation, heating, cooling, and power an all-electric ground transportation system at $9.5 trillion for the U.S. or $950 billion per year amortized over, say, the ten year conversion plan envisioned by the Green New Deal Resolution No. 109 now before the House of Representatives. That's around a third more than our imperial war department spends ($686 billion for 2019) bombing and wrecking countries around the world.

---

[77] Sources: company financial statements. Utilities: Statista: https://www.statista.com/statistics/237773/the-largest-electric-utilities-in-the-us-based-on-market-value/.

Furthermore, it's important to note that the Jacobson study also maintains that the "levelized cost of energy" (including upfront capital costs, fuel cost over time, operation and maintenance over time, and decommissioning cost) for a renewable energy powered system is about 90% less over the lifespan of the system than the fossil fuel-powered systems it would replace, meaning the capital costs would be offset for society by large savings in the day-to-day cost of energy (not to mention savings in healthcare, reduction in mortality and so on that would also accrue from a transition to renewables).[78]

Of course there are many ways to pay for this. We could rescind the dozens of other needless tax cuts and handouts to the rich since the days of Reagan. We could impose the fines on the banks that Obama and Trump failed to impose.[79] We could restore the progressive tax structure we use to have before Reagan as Rep. Ocasio-Cortez has called for. We could impose wealth tax as in France as Sen. Warren has proposed, and so forth. We could gut the imperial war budget (around three-quarters of a trillion a year). According to the most recent study, by the end of 2019, the U.S. government will have spent a staggering $5.9 trillion on its illegal criminal wars in

---

78 Jacobson et al., "Matching demand with supply at low cost in 139 countries among 20 world regions with 100% intermittent wind, water, and sunlight (WWS) for all purposes," *Renewable Energy*, 123 (2018), Table 4, Case C (with heat pumps), p. 245. Table 4 does not break out the capital cost for the U.S. from North America. But Jacobson says that "The capital cost for the U.S. + Canada is $11.3 trillion. The U.S. is 84% of this, or $9.5 trillion." (personal communication, 2/25/18). House Resolution 109.

79 Ben Protess et al., "Trump administration spares corporate wrongdoers billions in penalties," *New York Times*, November 3, 2018.

Afghanistan, Iraq and Syria since September 11, 2001[80] – wars that were never fought for democracy and certainly not for defense since none of those countries attacked us but were, instead, imperial wars to protect "our" oil under "their" sand.[81] In short, there are many ways to pay for the transition from fossil fuels to renewables without massively indebting future generations.

B. The argument for a new Public Works Administration-style jobs program

Under capitalism, closing and retrenching companies means throwing workers out on the street. We're socialists. We don't accept that. If capitalists won't provide the jobs then it's the government's responsibility to do so. We face an existential threat to our survival and the only way to prevent collapse is to shut down the industrial emitters. It's not the workers' fault if the industries they work in need to be closed or cut back to save our children and theirs. They deserve jobs, different jobs, better jobs. If society is going to abolish their present jobs then it owes them new jobs with comparable pay and benefits. This is not only morally right as a "just transition" but it's also the only way we can win the support of those workers in the struggle for the common good.

---

[80] Neta C.Crawford, "US Budgetary Costs of the Post-9/11Wars through 2019: $5.9 Trillion Spent and Obligated," Watson Institute: Boston University, November 4, 2018.

[81] Though in truth, money saved here should first be paid out in reparations for the hundreds of thousands of people the U.S. military murdered and the countries we've wrecked.

There's no end of low-carbon work to be done, starting with the buildout of renewable energy. As Carla Skandier writes, "The good news is that the energy transition requires a lot of workers. An investment of US$ 200 billion annually in renewable energy and energy efficiency, as estimated by economist Robert Pollin and others, could create 4.2 million jobs in the US, a net gain of 2.7 million when jobs lost from the fossil fuel sector are counted."[82] Beyond this we will need to create jobs in environmental restoration, infrastructure repair and upgrades, expansion of public transit, commuter biking, development of share-vehicle commuting, heating and environmental retrofitting, upgrading and improving our schools, expanding medical care, expanding services and facilities for our retired workforce, expansion of organic farming, and more. If private corporations can't figure out how to make a profit doing these then society through *our* government must employ them – just as FDR's New Deal public works programs of the 1930s built tens of thousands of projects – dams, electric utilities, airports, highway systems, bridges, tunnels, new railways, gorgeous train stations, glorious city halls and post offices, incredibly beautiful parks, beautiful public schools, modern hospitals, public universities and more.[83] FDR's construction programs built the bulk of this country's national infrastructure that we still rely on today. And if in the midst of the Great Depression, the government could afford to provide full-time government-funded jobs for tens of millions of workers

---

[82] "Quantitative easing," op cit.

[83] Nick Taylor, *American-Made: The Enduring Legacy of the WPA: When FDR Put the Nation to Work* (New York: Bantam, 2008). Joseph Maresca, *WPA Buildings: Architecture and Art of the New Deal* (Atglen, PA: Schiffler Publishing Ltd., 2016).

from 1933 through the 1940s, becoming the nation's largest employer by far, our immensely wealthier society and government can easily afford to re-employ the millions of workers from the fossil fuel-based industries to construct a permanently sustainable economy.

C. The imperative of economic planning and democratic management

After they wrecked the economy and plunged us into the Great Depression in 1929, America's capitalists couldn't pull themselves together to restore economic growth. In 1932 Franklin D. Roosevelt campaigning on a state-interventionist, help-the-poor "NewDeal," won a landslide victory over "do-nothing" Herbert Hoover who was deservedly blamed for letting the Depression get as bad as it did and reviled for his failure to intervene to rescue the economy and the citizenry. Over the next decade President Roosevelt turned the government into the biggest economic engine and the biggest employer in U.S. history, creating a state-directed capitalism, organizing a powerful state planning apparatus, setting up federal employment programs, and superintending construction of the country's first basic social safety net – transformations that handed him three more landslide victories in 1936, 1940, and 1944. Roosevelt and Congress established the National Resources Planning Board (NRPB) in 1933 to organize public works programs to put people back to work, revive the economy, and modernize our infrastructure. This was the first and so far only national planning agency in U.S. history. It evolved from public works planning to broader social and economic planning and then to mobilize and direct resources for the war effort. Regional planning groups were created in New England and the Pacific Northwest. Most states established planning agencies while planning boards emerged in many cities. Capitalist ideological hostility to economic planning per se – they couldn't stand the threat of a good example – eventually forced Congress to abolish the national board in 1943. But NPRB legacies included wartime and postwar planning, the first version of the G.I. Bill of Rights, the Second (Economic) Bill of Rights (a manifesto for postwar liberalism), an institutionalized policy planning process via the Council of Economic Advisors, and the annual federal budget process established by the Employment Act of 1946.[84]

The climate emergency, the existential crisis we face is to say the least, far more dire and urgent than the Great Depression. And here again, capitalism has no solution to the crisis it created because the capitalist solution to everything is the same growth that drove us to this precipice in the first place. That's why we argue that the only way to brake the drive to collapse is to socialize the commanding heights of the economy. We don't need to nationalize the entire economy. Small producers, worker co-ops, family farmers, mom & pop shops, restaurants and so on are not killing the planet. Large corporations are killing the planet. They can't help themselves. To preserve a habitable world we need to take them under public ownership so we can abolish the harmful companies and rationalize, reprioritize and restructure

[84] Marion Clawson, *New Deal Planning: The National Resources Planning Board* (Johns Hopkins University Press, 1981). Jess Gilbert, *Planning Democracy* (New Haven: Yale University Press, 2015).

production to create a permanently sustainable, if somewhat less industrialized, economy.

D. System change not climate change

It's perhaps conceivable, taking FDR's war-emergency industrial reordering as a precedent, that the three-point plan just described for fossil fuels buyout-nationalization, state-directed transition to renewables, and creation of a large federal jobs program, could be enacted within the framework of capitalism, though the result would be a largely state-owned economy. Roosevelt created his state-directed capitalism with only modest, if latent and increasing ruling class resistance because, first, he revived a broken capitalism from the depths of the worst depression in history and saved it from communism, secondly, he never needed to actually nationalize the industries he effectively took over because he was paying their owners handsomely to expand and grow their companies, if by producing different products, and thirdly, this was all just a temporary encumbrance, limited to the duration of World War II.

We face the opposite problem: we face a booming capitalist economy at the top of its form with a powerful entrenched ruling class in full command of their economy and their state. What's more, we need to nationalize industries not to grow them but to phase many of them out, retrench others, and transform some like GM from an automaker into, say, a wind-turbine maker or a train manufacturer, and permanently. Given the "unprecedented scale" of the problems we face, given the "far-reaching transitions in energy, land, urban and infrastructure … and industrial systems" we need to make, and given the speed with which we need to make these changes, it's difficult to imagine how this could be done within the framework of any capitalism. Private ownership of the means of production, profit maximization, and market competition have been the leading drivers of global ecological collapse and pose the main barrier to the rational re-organization, restructuring and reprioritization of the economy we need to make to save the humans. The depth and urgency of our climate crisis cries out for something like an immediate transition to ecosocialism. We don't have many decades left before it's too late to bother trying.

The IPCC climate scientists tell us we have just a dozen years or so to rein in our growth-to-bust economic system and come up with an effective pollution-suppression plan and mechanisms to staunch CO_2 emissions. Ocean scientists tell us we need a "Five-Year Plan" to save the oceans.[85] We need global plans to save the oceans, save the forests, save species from going extinct.[86] This requires generalized large-scale economic planning. These problems can't be solved by private corporations competing in an anarchic market.[87] Saving the world requires the sort of large-scale economic planning that only governments can do. We need to replace market anarchy with rational planning and

[85] Jenna Iacurci, "Report Calls for Five-year Plan to Save World's Over-fished Oceans," *Nature World News*, June 24, 2014.
[86] Damian Carrington, "Humanity has wiped out 60% of animal populations since 1970, report finds," *Guardian*, October 29, 2018.
[87] Richard Smith, "How did the common good become a bad idea? The eco-suicidal economics of Adam Smith," in idem, *Green Capitalism* op. cit.

management of a mostly, though not necessarily entirely, publicly-owned economy.[88]

E. Planning for the common good requires democracy

Rational planning requires democracy. Polls show that 69% of Americans (85% of those ages 18-29), 71% of Chinese, 87% of Europeans, nine-in-ten Ugandans, Tanzanians, South Koreans, Chileans, and Brazilians want binding limits imposed on CO_2 emissions.[89] Corporations don't want binding limits. Well, why don't we get to vote on such questions? We don't need to be experts. Corporate boards aren't composed of experts. They're composed of major investors and politically influential VIPs who solicit experts to advise them, then vote to decide what they want to produce and sell, and hire engineers to figure out how to get it done. Why can't society do the same – but in the interest of the common good? We need to establish democratic institutions to plan and manage our social economy -- planning boards at local, regional, national and international levels. We have plenty of examples from the Paris Commune to Polish Solidarity in 1980. We have the example of FDR's National Resources Planning Board – established by an elected president and congress. And as Greg Palast and co-authors describe in their book on regulation of public utilities, we have a working prototype right now:

---

[88] Smith, "Six theses," op cit.

[89] Yale University, "Americans support limits on CO_2," May 28, 2014: http://climatecommunication.yale.edu/publications/americans-support-limits-on-co2/. Bruce Stokes et al./Pew Global, "Public support for action on climate change," Pew Research Center, November 2015: http://www.pewglobal.org/2015/11/05/2-public-support-for-action-on-climate-change/.

> "Unique in the world ... every aspect of US regulation is wide open to the public. There are no secret meetings, no secret documents, individuals, industrial customers, government agencies, consumer groups, trade unions, the utility itself, even its competitors. Any and all citizens and groups are invited to take part: Everyone affected by the outcome has a right to make their case openly... In public forums, open to all citizens, the principles of social dialogue and transparency come to life. It is an extraordinary exercise in democracy – and it works."[90]

We see no reason why this cannot be scaled up to the whole economy. If the major U.S. political parties offer no solution to the climate crisis, if they abdicate their responsibility to preserve a habitable planet for our children, then it's up to the American left to provide vision and leadership in this struggle – and right now that looks to be the DSA, the Democratic Socialists of America (of which I am a proud member). As Maria Swart wrote to members on October 23rd:

> "Limiting global warming below 2 degrees is going to take a monumental mobilization of people from all walks of life, demanding that our lives matter more than corporate interests. Neither Republicans nor

[90]Greg Palast, Jerold Oppenheim, and Theo MacGregor, *Democracy and Regulation: How the Public can Govern Essential Services* (London: Pluto Press, 2003), pp. 1-4.

> Democrats have risen to the task of leading this effort. In the absence of their political leadership, we must rise to the occasion."

V. From FDR to AOC: the Green New Deal

Congressional Representative-elect (D. New York) Alexandria Ocasio-Cortez (aka AOC), the whip-smart self-confident 28 year-old DSA-er who won office on the Democratic ticket, rose to the occasion, drafting a proposed Plan published in early November for a Green New Deal she put before Congress in January 2019. The Plan called for the achievement within 10 years from start of execution the following:[91]

1) 100% of national power generation from renewable sources,
2) building a national, energy-efficient, "smart" grid,
3) upgrading every residential and industrial building for state-of-the-art energy efficiency, comfort and safety,
4) decarbonizing the manufacturing, agricultural and other industries,
5) decarbonizing, repairing and improving transportation and other infrastructure,
6) funding massive investment in the drawdown and capture of greenhouse gases,
7) making "green" technology, industry, expertise, products and services a major export of the United States, with the aim of becoming the undisputed

[91] Select Committee for a Green New Deal (November 2018): https://ocasio2018.com/green-new-deal.

> international leader in helping other countries transition to a completely carbon neutral economies and bringing about a global Green New Deal.

AOC's Green New Deal does not stop with decarbonization. She writes that

> "The Plan ... shall recognize that a national, industrial, economic mobilization of this scope and scale is a historic opportunity to virtually eliminate poverty in the United States and to make prosperity, wealth and economic security available to everyone participating in the transformation. In furtherance of the foregoing, the Plan (and draft legislation) shall:
>
> 1) provide all members of our society, across all regions and all communities, the opportunity, training and education to be a full and equal participant in the transition, including through a job guarantee program to assure a living wage job to every person who wants one;
> 2) take into account and be responsive to the historical and present-day experiences of low-income communities, communities of color, indigenous communities, rural and urban communities and the front-line communities most affected by climate change, pollution and other environmental harm;

3) mitigate deeply entrenched racial, regional and gender-based inequalities in income and wealth (including, without limitation, ensuring that federal and other investment will be equitably distributed to historically impoverished, low income, deindustrialized or other marginalized communities);
4) include additional measures such as basic income programs, universal health care programs and any others as the select committee may deem appropriate to promote economic security, labor market flexibility and entrepreneurism; and
5) deeply involve national and local labor unions to take a leadership role in the process of job training and worker deployment."

1. What's said and what's not said

To be sure, the process of "decarbonization" outlined above is abstract, lacks specifics, and is far from fully worked out. Most obviously, it's hard to imagine how the government could decarbonize fossil-fuel producers and industrial consumers without taking them into public hands. What's not said is that decarbonization has to translate into shutdowns and retrenchments of actual companies. How does one decarbonize ExxonMobil or Chevron or Peabody Coal? To decarbonize them is to bankrupt them. Further, the same is true for many downstream industrial consumers like Boeing, United Airlines, Duke Energy, Dow Chemical, Dupont, General

Motors and others. These companies live and breathe fossil fuels. Someday there may be electric airliners but not in the foreseeable future. And it's the near term, the next decades that are the most critical. Given the imperative need to radically suppress CO_2 emissions from aviation, the only way to do that in the here and now is to drastically reduce flying and thus aircraft production: ground many if not most planes, and ration flying, reduce or eliminate most air freight. But there is no mention of shutdowns, retrenchments, buyouts or nationalization. That will need to be addressed if the GND Plan is to move from the abstract to the concrete.

Perhaps the biggest weakness of AOC's GND from an ecosocialist perspective is that it's not based on the fundamental understanding that an infinitely growing economy is no longer possible on a finite planet. There's no acknowledgement of the imperative need to "de-grow" the overproducing overconsuming economies of the industrial North including China. There's no acknowledgement of the imperative need to abolish entire unsustainable industries from bottled water to toxic pesticides to disposables to arms manufactures and many others. This too is going to have to be addressed if the GND aspires to create a truly sustainable economy.

2. "Decarbonization": a self-radicalizing transitional demand

And yet, the audacity and breadth of the technical and socio-economic transformations here envisioned call to mind nothing so much as the economic and social revolution that FDR's New Deal brought about (albeit within the framework of capitalism), and goes beyond him in some respects. Occasio-Cortez may be just 28 but she is a bold, feminist, anti-racist, and socialist-inspired successor to FDR. With this Green New Deal she's taking the global warming discussion to a new level, changing the conversation and challenging the political economy. She's not calling for cap and trade or carbon taxes or divestment or any other "market solutions". She's issuing a full-throated call for "decarbonization", in effect throwing the gauntlet down to capitalism and challenging the system because, as we know, there is no way to decarbonize an economy based on endless growth, endless resource consumption, and thus inevitably, endless pollution and CO_2 emissions. Thus the push for decarbonization must inevitably raise the question of nationalization because, how else can government enforce the retrenchments and shutdowns needed to save the planet without precipitating economic collapse? The nationalization of the coal and oil producers, the obvious first targets, would in turn raise the question of what to do about all the industries that are based on fossil fuels – autos, aviation, petrochemicals, plastics, construction, synthetics, manufacturing, shipping, tourism, and so on. And consideration of how to decarbonize those industries would in turn raise larger questions about what society should or shouldn't produce, and who should decide such questions – private companies or society as a whole through democratic processes. Thus as I see it, the push for decarbonization is simultaneously a push for an increasingly radical democratization and that in turn could open the way to a transition to some form of ecosocialism – because at the end of the day, that's the only way to reconcile the jobs vs. environment dilemma.

3. Government good, capitalists not so much

AOC's Plan pushes us in that direction even though it's situated entirely within the framework of capitalism. Firstly, her Plan is a definitive break with the Reagan-Thatcher-Friedman doctrine that "capitalism good, government bad," that the best role for government is to "get out of the way and just incentivize the private sector" that has been the religious dogma of the ruling class and its puppets in congress since the 1980s. Instead, the Plan calls for robust expansive government to drive the needed changes, for two reasons: (1) scale and (2) time:

No. 1: "The level of investment required will be massive. Even if all the billionaires and companies came together and were willing to pour all the resources at their disposal into this investment, the aggregate value of the investments they could make would not be sufficient." Besides, "private companies are wary of making massive investments in unproven research and technologies; the government, however, has the time horizon to be able to patiently make investments in new tech and R&D, without necessarily having a commercial outcome or application in mind at the time of the investment."

No. 2: "The speed of investment required will be massive. Even if all the billionaires and companies could make the investments required, they would not be able to pull together a coordinated response in the narrow window of time require to jump-start major new projects and major new sectors."

AOC explicitly rejects the Reaganite claim of the superiority of the private sector and the efficacy of incentives:

> "We've also seen that merely incentivizing the private sector doesn't work – e.g. the tax incentives and subsidies given to wind and solar projects have been a valuable spur to growth in the US renewables industry but, even with such investment-promotion subsidies, the present level of such projects is simply inadequate to transition to a fully carbon neutral economy as quickly as needed... we're not saying there isn't a role for private sector investments; we're just saying that the level of investment required will need every actor to pitch in and that the government is best placed to be the prime mover."

4. There is no Plan B

Secondly, AOC makes a powerful and explicit case for state planning: She calls for a new Select Committee "with a 360° view" to serve the specific function of "examin[ing] emerging issues that do not fit clearly within existing standing committee jurisdictions or cut across jurisdictional boundaries." The Select Committee requires "a mandate to develop a plan for the transformation of our economy to become carbon neutral." Having its own Select Committee

> "ensures constant focus on climate change as the standing committee deals with that and many other issues of the day – such as wild fires in California, infrastructure, clean water issues, etc. First, they would put together the overall plan for a Green New Deal – they would have a year to get the plan together, with the plan to be completed by January 1, 2020. The plan itself could be in the form of a report or several reports. Second, they would also put together the draft legislation that actually implements the plan."

Her request for a new Select Committee was shot down but this is only the beginning of a long battle.

This is not yet FDR's Planning Board, nor as presently envisioned would it have the power to nationalize industries, nor is it clear that it would even have the power to order companies to stop producing, say, cars, and shift to producing wind turbines *à la* FDR. But as the environmental crisis deepens, as it becomes more and more urgent to effect radical change, fast, it will soon become apparent that nationalization with buyouts is the only way to force the pace of decarbonization and reallocate capital and labor to the sustainable projects we need.

Of course all this sounds wildly utopian at the moment and this Plan is sure to go nowhere under Trump and the Republican Senate. But it won't seem so extreme in a few years as the western states burn up, as Florida sinks under the waves, as food production across the Great Plains and California begins to collapse, as temperatures reach Saharan levels in the Southwest – and as people everywhere look for government to "do what the science demands before it's too late." There is no Planet B. Let's hope with a developing vision and that monumental mobilization around this Green New Deal we can derail that dystopia and build an ecosocialist civilization to save the human race.

Chapter 13

Toward sustainable development: from neoclassical monopoly to democracy-oriented economics

Peter Söderbaum

Introduction

Research and education in universities is subdivided into disciplines. There are departments of economics and departments of political science for example. Specialization and division of labour is thought of as being fruitful; Economics is about resource allocation at the micro and macro levels while political science is about democracy and governance. Something is sometimes gained through specialization but there are losses as well. This opens the door for counter-movements in terms of transdisciplinary research. Should "efficiency", for example, be exclusively a matter for economics and economists and democracy exclusively something for political scientists?

Sustainable development is a challenge in contemporary society. It is a complex, multidimensional issue where contributions from all university disciplines can make a difference. Social sciences such as economics, business management, political science, economic history, sociology, psychology, all have something to offer. And barriers between disciplines become less relevant.

Present development is unsustainable in essential ways. Climate change and biodiversity loss are examples. This process of unsustainable development has been going on for some time and we have every reason to try to identify factors explaining the failures. This is not easy but the difficulties are no reason to refrain from attempts.

For many years there has been a common view among more or less influential actors in society about progress indicators such as GDP-growth and monetary profits in business. Such thinking patterns have largely been made legitimate by mainstream neoclassical economics. It can therefore be argued that neoclassical economists have been successful in propagating their conceptual framework and many actors have benefitted in some respects from referring to the ideas. Theories and methods in economics and business management have become popular in many circles. These days it is however increasingly understood that while some actors have benefited in the short run, the same actors and all other actors and citizens have lost something at a more fundamental level. How can one speak of progress if essential development trends are unsustainable?

Through education and research neoclassical economists have had an impact on development in single nations and globally. What is more of a problem is that those employed at university departments of economics have largely neglected alternative schools of thought. The neoclassical monopoly in introductory economics education for example has been protected. Neoclassical theory may have a role among other schools of thought but the monopoly position can no longer be defended. A more pluralistic attitude in university departments of economics would, as I see it, have been more helpful in reconsidering visions and progress indicators.

Neoclassical economists may still argue that their approach is useful also when attempting to turn development in a sustainable direction. While neoclassical environmental economics may have something to offer I am sceptical regarding statements about the *sufficiency* of neoclassical theory and method. Something more is needed. I argue strongly that it is time to open up university departments of economics for alternative schools of thought such as institutional economics and ecological economics.

The call for a strengthened democracy is relevant for economics in two respects:

1. Standardization of economics to one single paradigm in teaching and research should be abandoned in favour of a pluralistic philosophy where different schools of thought can coexist. This has to be reflected in the organization of university departments of economics, recruitment of PhD-students etc.
2. The ideals of democracy can also be recognized in the very definition of economics as a discipline and in its

conceptual framework, theories and methods. Individuals and organizations can be understood in political terms rather than in terms limited to markets. Sustainability assessment of investment projects in society will differ for example between a technocratic approach (such as neoclassical Cost-Benefit Analysis, CBA) and a democracy-oriented approach (such as Positional Analysis, PA) as will be explained later on in this essay.

I start from the observation that "democracy" is a word that is largely absent from textbooks in economics (Mankiw and Taylor, 2011). My question is: What happens if we bring in democracy seriously into economics?

The illusion of value-neutrality

Neoclassical theory is positivist in terms of theory of science. Individuals and firms interact in markets and are understood in mechanistic terms, the presumption being that some of the ideas about good science from physics are applicable also for economics. This is by critics referred to as the "physics envy" position of neoclassical economists. The purpose is to explain and predict behaviour at the micro level of individuals and firms as well as performance of the economy as a whole. Hypotheses are tested and experiments are carried out when possible.

The neoclassical economist is ideally standing outside watching what goes on in an alleged value- neutral manner. It is assumed and believed that only one paradigm exists and that since there is no alternative perspective, the neoclassical one represents the "truth". Among economists Gunnar Myrdal, known for his studies of development in various parts of the world, has questioned the neoclassical position arguing that "values are always with us in research and education":

> "Valuations are always with us. Disinterested research there has never been and can never be. Prior to answers there must be questions. There can be no view except from a viewpoint. In the questions raised and the viewpoint chosen, valuations are implied.
>
> Our valuations determine our approaches to a problem, the definition of our concepts, the choice of models, the selection of observations, the presentation of conclusions – in fact the whole pursuit of a study from beginning to end" (Myrdal, 1978, p. 778).

Since values are unavoidable according to Myrdal and we live in a democratic society, value issues have to be dealt with openly rather than hidden "behind a veil of neutrality". In her study of different schools of thought in economics, Tanja von Egan-Krieger (2014) similarly argues that there is no value-free economics. In her comparative study she scrutinizes mainstream neoclassical economics, feministic economics, institutional economics and ecological economics. The term "feminist" in feminist economics for example suggests that values and ideology are involved. Ecological economists take environmental and development issues seriously into account and so on.

In my own writings I claim to respect some traditional ideas of good science while adding others. Respecting democracy is one where the plea for many-sidedness in analysis is reducing the possibilities of manipulation. But my person and subjectivity will still influence the kind of problems I am choosing for study and how I frame my analysis.

Neoclassical economists sometimes try to escape from the above criticism by making a distinction between "positive statements" which are "descriptive" and "normative statements", the latter being "prescriptive", as claims about "how the world ought to be" (Mankiw and Taylor, 2011, p. 32). But even descriptive statements are normative and specific in value terms. There is always a choice about how to frame problems and what to describe.

Individuals and organizations as political actors

In neoclassical theory individuals and organizations are related to each other in markets for commodities, financial capital and labour. According to Homo Oeconomicus assumptions the individual as consumer maximizes "utility" within the scope of her monetary budget constraint. Self- interest is emphasized and there is little or no concern for the interest of others. The only kind of organization in neoclassical theory is the "firm" which is assumed to "maximize profits" in the interest of shareholders. Shareholders are assumed to be exclusively concerned about dividends in monetary terms.

The emphasis on self-interest and otherwise narrow interests are sometimes defended as just assumptions that simplify analysis. But a simplified analysis may entail losses in relevance. Some of us economists and other actors are worried about the possibility that neoclassical theory systematically legitimizes self-interest and narrow interests among actors in the economy. Today we need an economics that – without denying the existence of self-interest – rather pushes individuals and organizations in the direction of broadening their interests where ethics, responsibility and even ideology play a role. It is no longer reasonable to believe that markets automatically can solve our sustainability problems for example.

As alternative to Homo Oeconomicus, a Political-Economic Person (PEP) is suggested. This understanding of human beings is based on social psychology with concepts such as role, relationship, trust, network, motive, dissonance, conflict, cognition, learning, etc. Individuals are actors guided by their "ideological orientation" and this ideological orientation may be narrow or broad. Rather than assuming that all individuals as actors are guided by the same motives, the ideological orientation becomes something to be empirically investigated in each case. There may be individuals who are close to the self-interest position assumed in neoclassical theory but also others with broader concerns and interests. And for one individual the ideological orientation varies over time in relation to context and decision situation.

The neoclassical profit-maximizing firm is similarly replaced by a Political-Economic Organization (PEO), i.e. an actor guided by its ideological orientation or "mission". A sub-set of organizations are joint-stock companies ("firms" according to neoclassical vocabulary) but even for them, the possibilities of broader missions need to be investigated empirically. Corporate Social Responsibility (CSR) and "fair trade" are increasingly discussed. As economists we could open the door for such possibilities rather than systematically support actors with narrow interests.

The concepts of ideology and ideological orientation

"Ideology" and "ideological orientation" are contested concepts (Connolly, 1993) which suggests that when used the concepts need to be defined. Among economists Douglass North has proposed the following definition:

> "By ideology, I mean the subjective perceptions (models, theories) all people possess to explain the world around them. Whether at the microlevel of individual relationships or at the macrolevel of organized ideologies providing integrated explanations of the past and the present, such as communism or religions, the *theories* individuals construct are *coloured* by normative views of how the world should be

> organized" (North, 1990, p. 23, emphasis in original).

Ideology can be described as a "means-ends" relationship. It is about where you are (present position), where you want to go (future positions) and how to get there (strategy), bringing desired positions and available means together. Politicians and political parties in a democratic society refer to their ideologies or ideological orientations. These ideological orientations are not static but the subject of repeated reconsideration. When turning to us as citizens, political parties refer to their ideological orientation. As individuals and members of groups we respond in one way or other to the ideological elements and programs of specific politicians (political parties). This suggests that all of us are guided by something that can be referred to as ideological orientation.

Neoclassical theory is largely limited to quantitative analysis. "Only that counts which can be measured" is the motto. But when studying sustainability something more is needed. Sustainable development cannot easily be defined in one-dimensional quantitative terms. In addition to quantification we need to refer to qualitative and visual elements. In her attempt to replace neoclassical quantitative analysis with other thinking patterns, Kate Raworth points to the importance of words and pictures "Our brains are wired for visuals" (2017, p.11). Raworth refers to media theorist John Berger who suggests a dominant role for visualization in human cognition "Seeing comes before words. The child looks and recognizes before it speaks" (Berger, 1972 p. 7). An ideological orientation, however understood, is seldom built on complete information as in neoclassical modelling. It is rather in most cases fragmentary and uncertain.

Other terms can be used with content similar to "ideology". "Worldview" is one, "vision" another, "narrative" a third. I have chosen ideology because it goes well with the political and democracy- oriented perspective emphasized in this essay. But also "narrative" is a term that can be used. Neoclassical theory can be understood as a narrative about consumers, firms, markets and economic growth. This narrative is no longer functioning so well. What can we do to open the door for other perspectives? Eva Kras, former chairperson of Canadian Society for Ecological Economics, suggests that we should "listen to visionaries" (Kras, 2007). Her examples of visionaries include Vandana Shiva with her book *Earth Democracy* (2006), David Korten, author of *When Corporations Rule the World* (2001) and Naomi Klein's *This Changes Everything. Capitalism vs the Climate* (2014). While something can be achieved through modernization and modification of mainstream ideology, we also need to consider alternatives at the level of perspectives. Alternative schools of thought in economics can be helpful in this respect.

Sustainable development as ideological orientation in a democratic society

Neoclassical economists refer to value-neutrality and regard their personal values as a non-issue. As an institutional ecological economist, I am turning things around; Value or ideological issues should be openly discussed and be part of analysis. The present call for sustainable development is a typical example of an ideological issue. SD is a contested concept in the sense that it can be interpreted in more ways than one. Some actors prefer a *business-as-usual* interpretation in the sense of minimizing changes from the established mainstream view and activities. Other actors are ready to modify and *modernize* thinking patterns, activities and institutions while a third category of actors may have internalized a need for *radical change* (Söderbaum, 2008 pp. 13-22).

In all these cases of interpretation, "ideological orientation" appears as a useful and very relevant term to describe differences between thinking and motives among actors. Sustainability issues are ideological and any attempt to avoid value issues or politics will make the analysis less relevant and meaningful. The role of economists is to articulate alternative visions, narratives and ideological orientations that appear relevant to politicians and other actors and discuss their advantages and possible weaknesses. The economist can claim a degree of independence (but not value-neutrality). Many-sidedness in analysis reduces opportunities for manipulation.

As is well known, sustainable development became seriously part of an international dialogue through the Brundtland Commission (World Commission on Environment and Development, 1987). Emphasis was on ethical issues between human beings in one region and those in other regions and between the present population and future generations. A philosophy of cautiousness in decision situations has been emphasized by some (Harremoës et al., eds, 2002). Too often policies have been implemented which today are regretted at least by some actors. More recently 17 sustainable development goals have been sanctioned by the United Nations (2015).

It may be concluded that a power game is going on in society between single actors and groups of actors concerning ideological orientation. How should we as economists or other social scientists relate to such a power game? A first observation is that we as economists are political-economic persons much like other individuals in the economy. This means that we have specific roles as professionals but that we at the same time are citizens and part of a democratic society. As economists we should respect normal ideas of democracy and, when possible, contribute to a strengthened democracy. We should not limit our studies to one ideological orientation but rather "listen to many voices". Competing ideological orientations in relation to a decision situation should be made visible for all actors involved or concerned. Arguments about best alternative will then be conditional upon each ideological orientation considered. In this way conflicts of interest will become more visible and actors behind each ideological orientation have a chance to reconsider their ideas and will be held responsible and accountable for their voting in political assemblies and decision situations more generally.

A political dictatorship at the national level can be understood as a reliance on one main ideological orientation (or a set of ideological orientations being close to each other) while dialogue about this ideological orientation is systematically avoided or impeded. Actors who do not share this main view and present their opinions openly risk harassments of various kinds.

But dominance of one ideological orientation may exist even in nations normally described as democracies. In nations such as Sweden and globally, an economic growth and market ideology is dominant to such an extent that one can refer to this specific market ideology as a kind of dictatorship. Behind this are, as I see it, university departments of economics (with neoclassical theory in a monopoly position) but also international organizations such as the European Union with its specific organizational infrastructure, the World Trade Organization (WTO), the World Bank and the International Monetary Fund (IMF). Transnational corporations with their lobbyists also play a role in defending this market ideology. Actors advocating radical versions of sustainable development are facing this "partial ideological dictatorship". This explains why there is a lot of inertia and that progress has been, and still is, limited. Even political dictatorships such as China have had to adapt to the global market ideology for their relations with other nations. The idea advocated by some that market ideology automatically leads to democracy, appears far-fetched.

Arguments for democracy are helpful by the insistence on a dialogue between advocates of different views. Today debate about radical change in ideological orientation and institutional framework is too often avoided among political actors in powerful positions and media actors who control essential arenas. Since the dominant market ideology does not go well with sustainable development, a part of our role as economists must be one of arguing in favour of pluralism and democracy.

Defining economics in a new way

The discussion so far can be summarized as follows:

- A number of development trends nationally and globally are unsustainable
- The theories and conceptual framework of neoclassical economics has been dominant in governance nationally and internationally while development has become increasingly unsustainable
- In the latest decades university departments of economics are characterized by a close to monopoly position of the neoclassical paradigm
- Other schools of thought in economics exist but have played a peripheral role in university departments of economics in the Western world and even globally
- Economics and other social sciences cannot be value-free or value-neutral. A degree of independence is however possible for economists
- The neoclassical paradigm is specific not only in scientific terms but also in ideological terms. The ideology of neoclassical theory and method is close to market fundamentalism

- In terms of ideological orientation, the neoclassical theory and conceptual framework has contributed to make neoliberalism legitimate
- Neoliberalism as ideological orientation has contributed to make the present political- economic system legitimate
- For economics to constructively contribute to sustainable development a first step is to eliminate the neoclassical monopoly in education and research at university departments of economics and open the door for competing schools of thought
- The monopoly for neoclassical economics at university departments of economics can be described as a "local monopoly". Heterodox economists are often connected with other social science departments such as economic history, political science, sociology or business management
- In a democratic society this kind of dialogue between schools of thought grounded in partly different ideological orientations is very much needed.

From the above it follows that there are good reasons to bring democracy into the definition of economics. I suggest that economics is defined as "multidimensional management of (limited) resources in a democratic society".

Why "multidimensional" management?

There is an emphasis on the monetary dimension in neoclassical theory and method. Non-monetary impacts are reduced to monetary ones to make analysis more tractable using a "trade-off philosophy". In neoclassical Cost-Benefit Analysis (CBA) actual market prices and a kind of hypothetical market prices are used to transform non-monetary impacts of various kinds to the monetary dimension. Even different non-monetary dimensions are "traded" against each other in this way.

Those indoctrinated in the neoclassical paradigm tend to see the mentioned simplification of analysis as smart and elegant. Prices in monetary terms on ecosystem services are estimated and regarded as "correct" as are prices on each unit of CO_2 pollution and such prices are determined in a technocratic manner by the neoclassical economist as expert.
This kind of transformation of all kinds of impacts to their alleged monetary values is here referred to as "monetary reductionism". Instead a more holistic and multidimensional approach is recommended where the judged importance of different kinds of impacts is left to the ideological orientation of each individual as actor or each group of actors.

The distinction between a "technocracy-oriented" philosophy and a more "democracy-oriented" approach is illustrated in Table 1.

Table 1 Roles and relationships in the cases of technocracy-oriented respectively democracy-oriented analysis

Actor categories:	**Technocracy-oriented** (Cost-Benefit Analysis, CBA)	**Democracy-oriented** (Positional Analysis)
Analyst	Expert on correct market prices and CBA-method resulting in "optimal solution"	Facilitator, expert on PA-method and dialogue resulting in conditional conclusions
Stakeholder	Expected to be essentially passive. Possibly asked about "willingness to pay"	Is encouraged to express opinion and participate in dialogue and analysis
Concerned citizen and other non-stakeholder actor	Silence will facilitate analysis and decision process	Is encouraged to express opinion and participate in dialogue and analysis
Politician / decision-maker	Expected to accept the authority of analyst and the result of analysis	Participates in dialogue. Decisions are based on the ideological orientation of each politician / decision-maker who is thereby made a responsible actor

Source: Adapted after Söderbaum, 2000, p. 84.

The technocracy-oriented view (left-hand column in Table 1) positions the analyst at the centre expected to carry out the Cost-Benefit Analysis in a correct way. The ethical imperatives built into the CBA-method should be applied, for example actual market prices of various impacts. Other actor categories such as stakeholders and politicians related to the decision situation are expected to regard the analyst as expert and accept the optimal solution produced through analysis.

The democracy-oriented view regards the ethical and ideological imperatives built into CBA as just one possibility. The analyst gets a different role of identifying competing ideological orientations among politicians, stakeholders and other concerned actors. This can be done in various ways by studying official documents, by listening to different actors and inviting them to participation etc. Ideological orientations can be formulated as narratives for example beliefs in private business initiatives and economic growth in GDP-terms versus ideas that emphasize some interpretation of sustainable development. When related to a specific decision situation such ideological orientations (narratives) can be expressed in more concrete terms. There is no single optimal solution as result of analysis, only conditional conclusions and preference orders connected with each ideological orientation considered.
Within the scope of the alternative, multidimensional view of economics, the expected impacts of implementing specific alternatives can be classified in four categories:

- Monetary flows (referring to periods of time)
- Monetary positions (referring to points in time)
- Non-monetary flows (referring to periods of time)
- Non-monetary positions (referring to points in time)

Neoclassical analysis is generally carried out in monetary terms. Economic growth is measured for single periods such as years and thus exemplifies monetary flows. Business accounting emphasizes monetary flows (e.g. profits) and monetary positions (balance sheet with its different components).

When considering the results of sustainability politics, non-monetary flows and positions play a central role. Pollution of various kinds can be measured as non-monetary flows with impacts on non-monetary positions. Plastics are used for various purposes in society and parts of these flows end up and are accumulated in the oceans. CO_2 pollution in flow terms has impacts on the atmosphere measured as states or positions of carbon dioxide in ppm terms. The existence of inertia and possibly irreversibility in non-monetary terms is a reason to avoid CBA with its trade-off philosophy in monetary terms. Instead non-monetary positional changes should be an essential part of the picture. House construction, road building and other development projects need to be evaluated in such non-monetary terms. It is increasingly understood that this is also true of life-style changes.

The possibility of irreversibility suggests that decision-making often needs to be looked upon in multiple-step terms. Each step opens the door for some future steps (with connected impacts) and at the same time excludes future possibilities, much like positional changes in a game of chess. This kind of options in positional terms can be described in decision-trees like the ones used in game theory but with positions rather than pay-offs as the result. And such positions only represent the beginning of new moves. The kind of thinking in positional terms indicated is an essential part of Positional Analysis (Söderbaum, 2000, Brown et al., 2017) to be discussed further in next section of this essay.

Even the use of terms such as "cost" and "benefit" may need to be reconsidered as part of the distinction between monetary and non-monetary impacts. Once more, a classification in four categories appears relevant:

- Monetary cost
- Monetary benefit
- Non-monetary cost
- Non-monetary benefit

Among institutional economists I have mentioned Gunnar Myrdal and Douglass North. William Kapp (1976) is a third person who at an early stage emphasized environmental and development studies. Neoclassical economists admit that third parties may be negatively (or positively) affected by a market transaction. Reference is made to single cost (benefits) for so called "externalities", again measured in monetary terms. In his early book "The Social Costs of Private Enterprise" (1950), and at other places Kapp referred to "costs" in a much broader sense:

> "as an economist I have long held the view and continue to believe that the institutionalized system of decision-making in a market economy has a built-in tendency of disregarding those negative effects (e.g.

> air and water pollution) which are "external" to the decision-making unit ... Thus, a system of decision-making operating in accordance with the principle of investment for profit cannot be expected to proceed in any other way but to try to reduce its costs whenever possible by shifting them to the shoulders of others or to society at large" (Kapp, 1970, p. 18).

Rather than thinking in terms of single externalities, Kapp pointed to a more systemic tendency of "shifting the costs of environmental disruption to third persons or to society" (ibid. p. 20). To judge the validity of such claims in relation to empirical decision situations, I believe that we once more need the distinction between monetary and non-monetary impacts and consider them as separate.

While a number of largely accepted standardization systems for measurement of environmental and other impacts exist in present society, there may still be conflicting views of the importance of specific impacts and the importance of all expected impacts following the choice of one alternative of choice. This is where the concept of ideological orientation and the reference to democracy in our definition of economics comes in. If one respects the existence of more than one ideological orientation in a society and in relation to a decision situation, then there may be conflicting views about costs and benefits connected with specific alternatives of choice. What is perceived as a cost or negative impact of one actor may be regarded in positive terms, i.e. as a benefit by another actor. Such differences in situation and perspective need to be respected. The idea of one common and standardized idea of "cost" or "benefit" is abandoned.

The purpose of Positional Analysis

As has already been made clear, Cost-Benefit Analysis is the neoclassical approach to investments in infrastructure, such as energy systems, roads, airports. Analysis is carried out in monetary terms where all kinds of impacts are traded against each other. Even impacts in different periods of time are traded against each other using a so called discount rate. The result of aggregation is a "present value" for each alternative considered.

An attractive element in the CBA-method is that one single optimal solution can be produced. When thinking of alternatives to CBA, attempts have been made to construct other methods that can match CBA in attractiveness in this sense and thus point to one alternative as the best or optimal. While all kinds of alternative methods can be discussed, such methods suffer from reductionism to one dimensional calculation and in ideological terms.

Positional Analysis claims to be more compatible with a strengthened democracy. The purpose is to listen to many voices and then illuminate an issue in a many-sided way with respect to:

- Alternatives of choice that appear relevant;

- Ideological orientations that are part of the ongoing dialogue;
- Estimated impacts of alternatives considered.

Conditional (rather than unanimous) conclusions follow based on each ideological orientation considered. Ideological orientation A will suggest one order of preference among alternatives while ideological orientation B may point to a different order of preference. For reasons of tractability, only a limited number of ideological orientations and alternatives of choice are considered, for example 3 or 4 but they should clearly differ from each other according to the principle of many-sidedness. Not all actors will find that their ideological orientation is considered in detail but they will hopefully still be helped by the analysis carried out. And there is always a possibility to complement the existing analysis or demand a new study.

This is certainly not an approach that solves all problems but it represents a considerable step forward when compared with neoclassical CBA. Conflicts of interest among politicians, stakeholder and other concerned actors are illuminated rather than hidden. Dialogue may be initiated on the basis of such competing views and each actor may confirm or reconsider her/his view. It should be made clear that this approach differs from a "**consensus view of democracy**" often attributed to the writings of Jürgen Habermas (Martin, 2005). The ideological orientations of specific actors can be modified or changed as a result of dialogue and the decisions made in a political assembly but it is not realistic to believe that all differences disappear. It may be a wise strategy sometimes to settle on a compromise but again the word compromise tells us that the parties involved differ in terms of agenda or ideological orientation. Consensus in the sense that all actors agree about one ideological orientation and one alternative as the best is still possible but an exception. Instead a "**conflictual view of democracy**" is advocated. Chantal Mouffe is an author who acknowledges the adverse aspect of political relationships and uses the term "agonistics" (2013). A degree of conflict can be regarded as constructive in solving different problems or challenges in society. Is it possible to deal with the problems of climate change without allowing for tensions between actors in society?

In a democracy-oriented study of alternative policies or decisions, analysis should be many-sided and match the existence of competing opinions. But there is always a risk that the ideological orientations, the alternatives of choice or the estimates of impacts are systematically manipulated by the analyst or other actors using their power. To reduce such risks an open dialogue about the issue and the study carried out is needed. A well-functioning democracy can here be seen as a security system.

A model of social and institutional change toward sustainability

"Institution" is another "contested concept" in the sense that it can be defined in more ways than one. "Ideology" and "power" are other contested concepts. As I understand it, neoclassical economists tend to limit attention to concepts that can be quantified and therefore avoid or reduce the role of contested concepts of the mentioned kind. Precision in quantitative terms is preferred while there may be considerable losses in relevance. But the mentioned contested concepts are used in real life and if one wishes to construct a conceptual framework that is relevant in relation to sustainability and other practical problems perceived by actors, it is probably wise to incorporate some of these terms into ones conceptual framework. The important thing then becomes one of clarifying how the terms are used.

"Institution" is here understood as phenomena that contribute to regularities or a degree of repetition in the behaviour of actors (individuals or organizations). Governmental laws and guidelines exemplify institutions as do organizations of a governmental or non-governmental kind. Even habits of thought and habits of behaviour shared by many actors can be referred to as institutions.

Douglass North suggests a slightly different definition:

> "Institutions are the rules of the game in a society or, more formally, are the humanly devised constraints that shape human interaction. In consequence they structure incentives in human exchange, whether political, social or economic. Institutional change shapes the way societies evolve through time and hence is the key to understanding historical change" (North, 1990, p. 3).

The two ways of understanding "institutions" are complementary rather than mutually exclusive. While a degree of stability characterizes institutions, there is also the possibility of institutional change. Existing institutions may be modified, some institutions loose in importance or disappear, other institutions are strengthened and new institutions may emerge. Single actors and groups of actors have their agendas (ideological orientation) and may use their power position to facilitate or counteract specific kinds of institutional change. At issue is how institutions can deliberately be changed to become more supportive of sustainable development.

A "political economic system" can be understood as the total institutional framework or arrangements in a nation or a group of nations such as the European Union. Such an institutional framework consists of single institutions in dynamic interaction with other institutions. Two kinds of essential factors that explain social and institutional change can be discussed in relation to Table 2. The present kind of capitalism is largely made legitimate through the domination of neoclassical economics as economics paradigm and neoliberalism as ideology (left-hand column in the table). It should be made clear that neoclassical economics and neoliberalism are not totally separate but rather overlapping. Neoclassical economics is, as previously argued, specific in value or ideological terms and neoliberalism includes ideas about how to look upon economics and efficiency. The reason to regard economics paradigm and ideology as different factors has to do with the fact that the two kinds of perspectives are often considered as different discourses. Paradigm in economics is mainly discussed by economists while ideological orientation is often being regarded as being a matter for politicians.

Table 2 Paradigm and ideology as essential factors behind political economic system

	Mainstream	**Alternative (Example)**
Economics paradigm	Neoclassical economics	Institutional ecological economics
Ideological orientation	Extreme business and market ideology (neoliberalism)	Interpretation of sustainable development as ideology
Political economic system	Present kind of capitalism	Institutional arrangements compatible with alternative paradigm and ideology

Source: Modified after Söderbaum, 2000, p.84

Not only the mainstream perspectives but also alternative perspectives need to be considered in relation to sustainable development (right-hand column of Table 1). A different definition and conceptual framework for economics is suggested as well as an interpretation of sustainable development as ideology.

Potential elements of sustainability politics

In the present essay the role of paradigms in economics and ideological orientations among actors in society has been stressed. A dialogue about these issues is very much needed. The close to monopoly position of neoclassical theory and method at university departments of economics and elsewhere need, as a first step, be replaced by pluralism. The cognitive inertia of neoclassical economists emphasizing one thinking pattern at the expense of all other possibilities can be referred to as "narrative fixation" (Fullbrook, 2016).

The conceptual framework or language connected with a paradigm matters. Economic Man assumptions with their narrow focus on self-interest differ from Political Economic Person assumptions. The latter point in a potentially more active role of the individual as actor, not limited to self-interest. Ideological orientation enters as a concept suggesting that the individual is a responsible and accountable person.

Assumptions about organizations as profit-maximizing firms can be replaced by political-economic organization assumptions implying that there also exist organizations other than firms and that all kinds of organizations are guided by a mission which can be multidimensional and differ from the limited liability doctrine of joint stock companies. Thinking about the economy and society exclusively in (neoclassical) market terms is perhaps not such a good idea.

Closely related to neoclassical economics is neoliberalism, i.e. an extreme business and market ideology. The dominant role of neoclassical economics has contributed to the present dominance of neoliberalism as ideology at many arenas where political options are considered. Some steps can certainly be taken toward sustainable development through governmental manipulation (in neoclassical terms) of markets through taxes or by constructing "markets for pollution permits" etc. But neoclassical economists, as well as neoliberal politicians, generally believe so much in the beneficial aspects of unregulated markets that they are reluctant to implement measures that go against their fundamental beliefs.

A different conceptual framework in economics and a Green ideology will open the door for additional policy measures. Here I will point to a need for institutional change at the following levels:

- Laws and guidelines for universities, departments of economics in particular
- Laws and guidelines for pension funds, observing possibilities for so called divestments
- Laws and guidelines governing corporations, in particular transnational ones
- Laws and guidelines governing international institutions, United Nation organizations included
- The Bank of Sweden Prize in Economic Science in Memory of Alfred Nobel.

Such proposals certainly reflect my own experiences, scientific and ideological orientation as an actor and institutional-ecological economist. In a democratic society each citizen as actor will react or relate to the above guidelines in her way. I will only comment upon some of the suggestions:

- Concerning universities I think that ideas about value neutrality in social sciences have to be abandoned. Values are always with us even when making descriptive statements. Describing a phenomenon can be made in different ways.
- Concerning pension funds allocation of financial resources can be based on sustainability criteria rather than short term financial expectations.
- Too many transnational corporations are not performing well in relation to sustainable development. William Kapp's arguments about tendencies of "shifting costs of environmental disruption to third persons or to society" should be taken seriously. New legal forms of organization, such as the UK Common Interest Company have to be systematically considered. Such new initiatives are very much needed.
- International institutions such as the World Bank, International Monetary Fund and World Trade Organization and even UN organizations, such UNEP and UNDP, still rely very much on neoclassical theory and method. New approaches are needed.
- There are many reasons to abandon the Bank of Sweden Prize in Memory of Alfred Nobel. In recent times this prize has, as I see it, had a role mainly of protecting neoclassical theory. Ideology is involved which makes the prize comparable with the Nobel Peace Prize rather than anything else.

To summarise; in relation to climate change and other threats we need to consider scenarios of future political economic systems that in terms of social and environmental performance differ from the present kind of capitalism (Smith, 2016). The present ideas behind pluralism and democracy in relation to economics should also be stressed. When suggesting a partly new conceptual framework, this is not done as a proposal for "paradigm-shift" (Kuhn, 1970) in the sense of neoclassical theory and method being completely replaced by one other paradigm. While each economist may have her or his preferences, it appears more constructive to think in terms of "paradigm coexistence". In relation to a set of phenomena, one perspective may add to the understanding offered by another. Within the scope of coexistence there may still be shift in dominant paradigm, for example by the neoclassical paradigm losing ground to institutional theory.

References

Berger, John (1972) *Ways of Seeing*. London: Penguin.

Brown, Judy, Peter Söderbaum, Malgorzata Dereniowska (2017) *Positional Analysis for Sustainable Development. Reconsidering Policy, Economics and Accounting*. London: Routledge.

Connolly, William E., 1993. *The Terms of Political Discourse (Third edition)*. Blackwell, Oxford.

Von Egan-Krieger, Tanja (2014) *Die Illusion wertfreier Ökonomie. Eine Untersuchung der Normativität heterodoxer Theorien*. Franfurt am Main: Campus Verlag.

Fullbrook, Edward (2016) *Narrative Fixation in Economics*. Bristol: World Economics Association Books.

Harremoës, Poul, David Gee, Malcolm MacGarvin, Andy Stirling, Jane Keys, Brian Wynne, Sofia Guedes Vaz (Eds) (2002) *The Precautionary Principle in the 20th Century. Late lessons from early warnings*. London: Eartscan.

Kapp, K. William (1970) "Environmental Disruption: General Issues and Methodological Problems." *Social Science Information* (International Social Science Council), Vol. 4, No 9, pp. 15-32.

Kapp, K. William (1971 [1950]) *The Social Costs of Private Enterprise*. New York: Schocken Books.

Kapp, K. William (1976) "The Nature and Significance of Institutional Economics." *Kyklos*, Vol.29, No 2 (April), pp. 209-232.

Klein, Naomi (2009) *This Changes Everything. Capitalism vs. the Climate*. London: Allen Lane.

Korten, David C. (2001) *When Corporations Rule the World*. San Francisco: Berret- Koehler.

Kras, Eva (2007) *The Blockage. Rethinking Organizational Principles for the 21th Century*. Baltimore: American Literary Press.

Kuhn, Thomas S. (1970) *The Structure of Scientific Revolutions (Second edition)*. Chicago: University of Chicago Press.

Mankiw, N. Gregory and Mark P. Taylor (2011) *Economics*. Andover, Hampshire, UK: South-Western CENGAGE Learning.

Martin, Robert W. T. (2005) "Between Consensus and Conflict: Habermas, Post-Modern Agonism and the Early American Public Sphere." *Polito* Vol. 37, pp. 365-388, Accessed at: doi: 10.1057/palgrave.polity. 2300018

Mouffe, Chantal (2013) *Agonistics. Thinking the World Politically*. London: Verso.

Myrdal, Gunnar (1978) "Institutional Economics." *Journal of Economic Issues*, Vol. 12, No 4, pp. 771-783.

North, Douglass C. (1990) *Institutions, Institutional Change and Economic Performance*. New York: Cambridge University Press.

Raworth, Kate (2017) *Doughnut Economics. 7 Ways to Think Like a 21st Century Economist.* White River Junction, Vermont: Chelsea Green Publishing.

Shiva, Vandana (2006) *Earth Democracy. Justice, Sustainability and Peace.* London: Zed Books.

Smith, Richard (2016) *Green Capitalism. The God that Failed.* Bristol: World Economics Association Books.

Söderbaum, Peter (2000) *Ecological economics. A Political Economics Approach to Environment and Development.* London: Earthscan.

Söderbaum, Peter (2008) *Understanding Sustainability Economics. Towards Pluralism in Economics.* London: Earthscan.

Söderbaum, Peter (2018) *Economics, ideological orientation and democracy for sustainable development (Second edition).* Bristol: World Economic Association (WEA) Books.

United Nations (2015) *Sustainable Development Goals.* https://www.un.org/sustainable-development-goals/ (Accessed: 2018-11-29)

World Commission on Environment and Development (1987) *Our Common Future* (The Brundtland Commission). Oxford: Oxford University Press.

Chapter 14

Like blending chalk and cheese – the impact of standard economics in IPCC scenarios

Joachim H. Spangenberg & Lia Polotzek

There is no place where economics and ecosystem science, in particular climate science, are more intimately connected than in the IPCC climate scenarios, their Integrated Assessment Models (IAMs) and the conclusions derived from them. In official parlance, their results should be "policy relevant but not policy prescriptive". Nonetheless past policy discussions have shown that both the climate science part with its scenarios of droughts, floods and storms (e.g., IPCC, 2014a), and the economic part with cost calculations and suggestions for efficient policy strategies (e.g., IPCC, 2014b) have been influential. In the reports, and in particular in IAMs, both are merged, assuming that both disciplines provide adequate descriptions of the parts of reality they are in charge of analysing and understanding. But do they?

1. Background: dealing with complex systems

Since the early 2000s it is a consensus of climate scientists that the climate system is characterised by abrupt change, tipping points, irreversibility and the possibility of multiple stable states: it is a complex, dynamically evolving and highly non-linear, self-organising system:

- Complexity is characterised by time lags (e.g. the ocean delaying terrestrial warming, IPCC, 2013) and positive as well as negative feedback loops. A positive one is less arctic summer ice reducing the albedo and thus increasing warming (Kashiwase et al., 2017). In the same vein, thawing permafrost is releasing methane, a potent greenhouse gas (Knoblauch et al., 2018). A negative one is that increasing CO_2 concentration can stimulate plant growth, enhancing carbon fixation (Culotta, 1995). Some effects are context dependent, like the carbon fixation effect or the global warming contribution of clouds (McCoy et al., 2017).

- The evolution of the system has been felt with the weakening of the polar vortex (the system of circular winds usually keeping polar air in the polar region), resulting in deep-freeze conditions in Canada and the USA (Kretschmer et al., 2018). The weakening of the thermohaline circulation, a hypothetical shock scenario less than a decade ago (Dima and Lohmann, 2010; Spangenberg et al. 2012), is now being observed, and the mechanisms are increasingly better understood (Oltmanns et al., 2018). Another example is the now irreversible retreat of a huge sector of the West Antarctic

Ice Sheet (Showstack, 2014; Joughin et al., 2014) – only the speed of melting and with it, the speed of sea level rise is yet unknown.

- The dynamic of the system results from the fact that it is never in equilibrium. Driven by solar radiation, there is no standstill in the climate system; sudden transitions and bifurcations can appear anytime. What is perceived by some observers as an unchanging state and interpreted as equilibrium are instead dissipative patterns (Prigogine and Stengers, 1984; Lockwood, 2001). While such patterns can be relatively stable, they cannot be characterised as unchanging structures but are in fact maintained by a permanent throughput of matter and energy (a candle flame is such a dissipative pattern, and with our body cells replaced in months, years or decades, humans can be described as dissipative pattern as well).

So how do climate scientists deal with the challenge of having to make statements about a system with inherent uncertainty, i.e. a system with an essentially unpredictable behaviour? First of all, they use a wide variety of parameters, software and hardware variations reflecting different approaches and mechanism hypotheses, and diverse data sets. Before coming to a conclusion, the results of these models have to converge, divergence indicating open research questions (Knight et al., 2007). Secondly, climate scientists distinguish the scenarios they are developing from predictions, and thirdly they refrain from quantitative statements when describing the future; they use ordinal scales instead. Fourthly, they qualify their statements by linking each statement (in the summaries for policy makers) with the degree of confidence the scientific community places in it, based on published papers. And finally, they sometimes test "shock" assumptions, i.e. the result of trajectory-changing singular events, to explore the full range of plausible if not necessary probable development options, not least as probabilities cannot be quantified anyway (Spangenberg et al., 2012). Such shocks are hard to model and best represented in the scenario narratives the models should be embedded in (Alcamo, 2001).

This is a high level of caution, reflecting the uncertainties inherent in complex systems, and all the more remarkable as the climate system, despite its size, must be classified as of medium complexity (Allen, 2001; Spangenberg, 2015): after all, the individual parameters and objects (molecules) follow the laws of nature, and only their interplay creates the complexity. Compare this to the models and scenarios developed by biologists and ecologists, where an additional element has to be factored in: unlike molecules, organisms can adapt to a changing environment, in an often unpredictable way. This makes the system even more complex, and modelling it more difficult. It was the recognition of uncertainty and system dynamics by of Holling and May during the 1960s and 1970, which contributed to a stepwise paradigm shift in ecosystem science, away from the stable equilibrium paradigm which had persisted in the earlier decades. That time huge ecosystem models were built and some predictability expected, based on an ontology which perceived the real world as a "stable and infinitely resilient environment where resource flows could be controlled and nature would self-repair into equilibrium" (Folke, 2006, p. 253). The observations and the emphasis on non-linear dynamics, uncertainty, surprise, thresholds and regime shifts starting with Holling (e.g., Holling, 1973) and May (e.g., May, 1976) are today mainstream in the science of ecosystems (Folke, 2006). Consequently, biodiversity scientists are reluctant to make strong statements or predictions and prefer to talk about climate envelopes and how their effects are modified by, amongst others, landscape composition, microclimate, soil conditions and trophic interactions (Dominik et al. 2018), and in particular land use change (Chytrý et al., 2012). Only recently, with evidence for a global decline of insect population manifesting itself (Sánchez-Bayo and Wyckhuys, 2019) and the risks for human food supply resulting from of biodiversity loss now evident (Bélanger and Pilling, 2019), they have become more robust in their statements (IPBES, forthcoming; EEA, forthcoming).

Against this background the question arises what economists can learn from natural sciences regarding the models they use? A society and its economy are systems populated by elements (consumers, firms, etc.) which are neither passively pushed by the laws of nature, nor only reactively adapting to them. Instead they are actively choosing between options including the (often misguided) anticipation of future events. As a result, socio-economic systems represent the highest level of complexity (Allen, 2001; Bossel, 2000). How do economists deal with this complexity of the dynamically evolving, highly non-linear, self-organising system that constitutes our economy? How do they deal with the interaction of the systems, with how and to what degree the economy is changing the ecosystem, and the question to what degree the "age of environmental breakdown" (Laybourn-Langton et al., 2019) is changing the economy, its logic, its functions, structures and prospects? From Adam Smith's "invisible hand" argument to modern general equilibrium theory in the vein of Arrow-Debreu, most of the history of economic thought can be regarded as an attempt to make sense of a complex, non-linear system – the economy. The reminder of this paper is dedicated to show that current mainstream economic thinking is not adequate to deal with this complexity. Section 2 starts with the particularly striking case of economic modelling in the context of Integrated Assessment Models, questioning their capability to accommodate the uncertainty so typical for complex system development. One underlying reason for this weakness is the insufficient reflection on whether to conduct descriptive or normative science, as section 3 shows. That these shortcomings are policy relevant is illustrated in section 4, dealing with the IPCC 1.5°C Report. Section 5 takes a look at the reasons behind, identifying the unreflected approaches to both discounting and economic growth as weak points which must be addressed in any attempt to future-proof economics. Section 6 concludes with an outlook.

2. Precisely incorrect – modelling fiction, not science fiction

Economists combine the science-based climate scenarios (the biodiversity scenarios are rarely taken into account and the issue is dealt with in a more narrative way) with deterministic computable global equilibrium (CGE) models. One reason for doing so is that climate scenarios are most often presenting potential developments up to the midst and the turn of the century, i.e. for the next 30 and 80 years (e.g., IPCC, 2014a; 2018), and deterministic models are the only ones capable of making statements over such a long time period. But are these statements meaningful?

Expect the future as it was

Evolutionary, i.e. non-deterministic econometric models offer rather reproducible results for 15 to 20 years, and as a maximum for 30 years – after that, the results of model runs using the same starting parameters diverge too much for any robust statement to be made (Umweltbundesamt, 2018a; 2018b). The runs of CGE models, due to their deterministic character, are replicable over longer time spans – not all parameters are fully endogenised, and the fundamental structure of the economy they base their calculations on remains unchanged. However, what is the value of forecasts derived from such assumptions? Just imagine the world 80 years ago: World War II had not yet started, fascism reigned not only in Germany but all over Europe, there was an aggressive Japanese empire, Stalin's iron fist in the Soviet Union, the Baltic countries did not exist, Germany was bullying its neighbours, with France and Britain trying to appease it. The world outside America, Europe, China and Japan was mostly colonised. The Great Acceleration of the 1950s/1960s was still far off, the gold standard intact and institutions like the United Nations, the World Bank, IMF, GATT/WTO and international courts did not yet exist. If the world's most sophisticated economists had made a scenario for the next 80 years, up to 2019, it is hardly conceivable that their imagined future would have had any similarities with our present times. The same applies today: economic scenarios for 50, 80 or even more years are pure fiction, not even science fiction. In other words, neoclassical models with their three defining elements of methodological individualism, instrumentalism / utility maximisation intention and market equilibrium are *inherently deterministic* and thus incapable to appropriately reflect the unpredictable dynamics of disequilibrium processes (Arnsperger and Varoufakis, 2006). As deterministic models are representing a zero level of complexity (Allen, 2001), they are bound to fail in the real world (Spangenberg, 2015).

The inherent determinism has further implications: equilibrium models cannot predict tipping points (one reason for their failure in the financial crisis), and they are not capable to deal with uncertainty. Tipping points are non-linearities where small parameter changes can have large impacts on system structure and behaviour, affecting the physical and biological structures as well as kind and cost of the climate policy responses required (Cai et al., 2015). The regime shifts resulting from crossing critical tipping points cannot be modelled by structurally inert models (their differential equations have discontinuities there). Nor can triggering mechanisms be represented as they can be located on different scales of the system and spread to other levels (Rocha et al., 2018), a mechanism not foreseen in CGE models.

Uncertainty and ignorance

The real-world policy relevance of economic modelling is also undermined by the fact that strategic decisions are often confronted with risk, uncertainty and ignorance (Mayumi and Giampietro, 2001; van der Sluijs, 2012). In non-deterministic systems

- **risk** means knowing the potential impacts of a decision, and the probability that they occur,
- **uncertainty** means knowing the potential impacts, but having no idea about the probability they might occur, and
- **ignorance** means having no idea of either the potential impacts nor their probability.

These distinctions are important as the policy suggestions derived from the models are to be applied in a world characterised by the lack of full knowledge and a high level of uncertainty (Prigogine 1997), and the different levels of knowledge deficits call for different kinds of responses, as summarised in table 1.

Table 1 Knowledge and decision making

Situation	Kind of system	State of knowledge	Futures	Action required
Certainty	Determin-istic	"known" future system behaviour, probability =1	Prediction	Adaptation, preparation for the events to come
Risk	Dynamic	"known" impacts, "known" probabilities	Stochastic prediction	Preparing for adaptation, prevention to reduce known pressures, directly or by modifying their driving forces
Uncertainty	Self-organising and evolving systems	"known" impacts with "unknown" probabilities	Scenarios, vulner-ability analysis	Precautionary prevention to reduce potential pressures
Ignorance	Self-organising, evolving and un-determined systems	"unknown" impacts, "unknown" probabilities, e.g. surprises, shocks or "wild cards"	Scenario story lines, fiction	Precaution, action taken to anticipate, identify and reduce the impact of "surprises"

Source: EEA (2004), extended and modified

That their deterministic models cannot deal with real-world uncertainty, and that this hampers the relevance of their models, is not unknown to standard economists (e.g. Lange and Treich, 2008) – as a response, some have developed various classes of "fuzzy system models" (Lontzek et al., 2012), amongst them the dynamically stochastic general equilibrium models DSGEs (Syll, 2010). According to Yager and Filev (1994, p. xiii), the concept of fuzziness can directly attributed to L.A. Zadeh who intended to provide a tool to help in the modelling of complex phenomena, especially, but not restricted to those involving human agents. They claim that "By permitting a certain amount of imprecision in our models, we provide a robustness that allows us to model complex situations, which we might not otherwise be able to model and also provide a means for including the inherent vagueness of the human process of conceptualizing the external world", but that depends on how the "imprecision" is realised in what they call mathematical models. Usually cardinal parameters are replaced by probability distributions, forming a stochastic "cloud" around the initial parameter value. However, such "imprecisions" are not providing the "robustness that allows us to model complex situations" as Yager and Filev had hoped insofar as if they are stochastic probability distributions, they represent quantifiable risks (the everyday business of insurance companies), but not uncertainty, let alone ignorance. The results are predictable, and their probability is known – the models, including the DSGEs, are still deterministic, and due to the equilibrium assumptions ruling out system structure development, processes and developments are even assumed to be reversible.

Digging deeper

To understand why mainstream economics today is grappling to deal with complex systems and structural uncertainty, a brief look at the history of economic thought is helpful. In the more distant past, at least some economic scholars have incorporated structural uncertainty into their theories. Whereas in the writings of economists such as Mill, Keynes and even Hayek structural uncertainty played a certain role in explaining macro-structures of the economy, today the rational actor model has led to a reductionist containment of complexity as well as structural uncertainty. In the aftermath of the marginalist revolution and with its roots in Benthamian utilitarianism, the economic mainstream began to shift its attention away from understanding the causes of wealth, the functioning of markets and the macro-structure of the economy in order to focus on individual behaviour and choice as core units of analysis (Herfeld, 2013). Macroeconomic theorising is required to provide a microeconomic justification – exactly the opposite order of what it should be, taking into account the emerging properties on higher system levels.

However, the real birth of economics as a science of choice happened right after WWII with a more axiomatic formulation of rationality (Herfeld, 2013). This concept, which was initially put forward by von Neumann and Morgenstern, would set the standard for the treatment of the notion of uncertainty in post-WWII mainstream economics (Muthoo, 1999). They proposed conditions under which the maximisation of expected utility holds, which can be arrived at in case of given objective probabilities attached to the utilities that might be realised in a choice situation. Thus, it is a theory of risk and does not apply to situations where probabilities are not known (Gilboa, 2009). It was Savage (1954), who would suggested an approach how to realise expected utility maximisation in situations of structural uncertainty where no objective probabilities are known. Building on the concept of subjective probabilities and combining it with the idea of expected utility proposed by von Neumann and Morgenstern, Savage's theorem claims that decisions under uncertainty can in fact be converted into decisions under risk. According to Savage, the only difference is that it is not objective probabilities but merely subjective probabilities and thus beliefs of the decision maker herself, which are incorporated into the model. However, he warned that his theory of subjective probabilities "should only be applied in small worlds in which it is possible to 'look before you leap'" (Millner et al., 2013, p. 22). Nonetheless economists considered it possible to transform choices under structural uncertainty into choices under risk using Savage's theorem and applying it far beyond its domain of applicability. This way the expected utility framework combined with the assumption of rational agents and market equilibria led to complicated but not complex, rather neat and tidy economic models spanning over longer periods of time – precisely those deterministic CGE/DSGE models which are today combined with science-based climate scenarios. In a nutshell, currently economists apply their highly sophisticated tools to areas where they cannot provide meaningful results: the domain of legitimate applicability might be short term assessments, for time spans short enough to make the assumption plausible that major system changes have not occurred during the time period covered by an analysis. A future-proof economic theory will address these shortcomings the same way other sciences have done so, even it requires giving up highly valued tools and methods.

In the meantime, there is another problem associated with the reductionist approach of modern economics towards complex systems and to deriving recommendations regarding their management.

3. A fine line – descriptive or normative science?

Next to the inability to describe long-term developments and to take into account the structural uncertainty of complex systems, there is a more fundamental problem regarding current economic modelling manifesting itself in IAM/DSGE models. It consists of the fact that economic models are presented as being purely descriptive, while they actually carry quite some normative baggage. This becomes particularly relevant as the function of economics in society changed from depicting and explaining the reality of the economic system to serving as a tool to facilitate political decision making processes.

Through the rise of the rational choice paradigm and economics' development into a science of choice, it has become vague whether its approach to decisions is more of a descriptive or of a normative character. Usually, in neoclassical economics, expected utility theory is claimed to be used as a purely descriptive theory. Yet this claim is false as the idea of rational choice in conjunction with utilitarian assumptions is inherently tied to a specific concept of welfare and its normative assumptions (Muthoo, 1999). These circumstances have made it almost impossible for economists to draw a precise line between a descriptive and a normative approach, although few are aware of this challenge. This is dangerous as it disguises the outcomes of economic models as purely rational, whereas in fact they contain a plethora of underlying normative assumptions representing a specific world view (Spangenberg, 2016).

This has some very real and direct consequences when it comes to climate scenarios and the corresponding IAM/DSGE models: they are built in order to find solutions on how to achieve certain emission reduction targets (Kuhnhenn, 2018). The underlying economic models are optimisation models, which – under given boundary conditions – try to maximise the social utility function, most often represented by the GDP. Thus, in IAM/DSGE models, what is presented as the "optimal" outcome is more wealth in a national economy, in monetary terms (distribution plays no role). In such models, any measures leading to a reduction of GDP growth would not be regarded as an "optimum" as they would be regarded as "expensive" in and by the model. The consequence is that measures, which might lead to less production and consumption either cannot be depicted or are not used (Kuhnhenn, 2018). This is striking as changes of the consumption patterns and levels are a necessary condition for reaching the climate goals, as described above. In summary, it is our very standards of evaluation, which lead to deeply ideologically biased policy recommendations being presented as "objective" scientific insights, which has made economics the favourite legitimation science of neoliberal decision makers in politics and business.

4. The 1.5°C calamity

Until 2015 the most frequently cited and politically endorsed goal of climate policy was limiting global warming to 2°C. At the Paris Climate Conference, however, Small Island States and civil society lobbying brought about an unexpected change: the UNFCCC Conference of Parties agreed that the goal should be tightened to a warming level significantly below 2°C and at best not surpassing 1.5°C. Climate science was taken by surprise, but by taking a closer look at the discrepancy of impacts between 1.5° and 2°C came to conclusions strongly supportive to the 1.5°C maximum. Hansen et al. (2016) showed that evidence from paleoclimate data, climate modelling, and observations supported the assumption that 2 °C global warming could be dangerous for ice melt, sea level rise and superstorms. Two years later the IPCC confirmed the warning and pointed to a rapidly worsening situation when the temperature increase surpasses 1.5°C and reaches 2°C (IPCC, 2018). Climate science called for "rapid, far-reaching and unprecedented changes in all aspects of society" and highlighted the need for stringent policy reaction to avoid the potentially catastrophic effects which can occur when the 1.5°C threshold is surpassed, in particular as a result of the highly non-linear character of the climate system.

New targets require new policy strategies, and new economic assessments. The IPCC (2018) presents four scenarios, three the "usual suspects" (variants of the SRES scenarios used by the IPCC since 2002) and one explicitly ambitious sustainability scenario, including all kinds of technologies and policy interventions to reduce greenhouse gas emissions. The potential of these technologies is assessed by analysing the technical and biophysical limits and the cost of each measure (Smith et al., 2016). So do the scenarios foresee a stop the increase of greenhouse gas concentrations in the Earth's atmosphere sufficiently and in due time? Unfortunately not – all four socio-economic scenarios and storylines linking the physical scenarios to the socio-economic domain expect an overshoot of greenhouse gases and thus temperatures. These are intended to be temporary (the scenarios vary in the level and duration of overshoot).

However, in dynamically evolving, self-organising systems such as the environment and the society and its economy, systems changes emerging during the overshoot period are irreversible and initiate path dependent developments: you never cross the same river twice, and you never visit the same town twice. Thus it is obvious that an overshoot – temporary or permanent – is not acceptable once the lessons from complex systems theory are taken into account. Compensation later, in the second half of this century, by "negative emissions",[1] i.e. technologies extracting more CO_2 from the atmosphere

[1] The charm of this definition is that one only has to "think big" regarding the technologies to avoid a comprehensive phase out of fossil fuels.

than is still being emitted in the same time period offers no guaranty that it will not trigger changes beyond those accepted by setting the threshold.

Physics doesn't negotiate, so the task of economic and policy scenarios is to find ways how the limits identified can be kept, and this is where the IPCC report fails: in the suggestions how to do its own bidding. The technologies suggested in the scenarios to make good for a more or less large overshoot (without being able to quantify what this implies for avoiding tipping points) by providing "negative emissions" in the second half of the century include readily available and so far more experimental approaches; upscaling would be needed in most cases:

- One short-term available option is large-scale pyrolysis of organic matter, using the resulting carbon as soil stabiliser (known from the Amazon basin as "Terra Preta"). While it could indeed sequester significant amounts of carbon and reduce soil nitrous oxide emissions (Cayuela et al., 2014), there is a high probability that with it, aromatic and in particular polycyclic aromatic compounds (potentially toxic, carcinogenic and mutagenic) plus heavy metals would be inserted into the soil and accumulate there.

- The preferred longer-term option is Bio-Energy with Carbon Capture and Storage (BECCS), large-scale biomass cultivation as energy source, with subsequent carbon capture and underground storage. It is a combination of existing but separate technologies requiring integration, upscaling and commercialisation before effects can be expected. Both elements of that "rescue concept" have been criticised – a sustainable biomass base has limitations regarding energy provision (in particular in competition for land against food production), and large scale plantations would be damaging biodiversity (EEA, 2006; Spangenberg et al., 2014; Smith et al., 2016)., CCS has been questioned regarding its safety (storing carbon dioxide for centuries and longer), and the enormous energy consumption it entails (Stephens, 2013). Even the criticism articulated from an ecological and nature protection point of view in the last IPCC report has been ignored or at least side-lined (IPCC, 2014a), as has been the critical assessment by the Global Assessment of the Intergovernmental Platform on Biodiversity and Ecosystem Services, the biodiversity pendant of the climate-focussed IPCC (IPBES, forthcoming). Furthermore, the European Academies' Science Advisory Council recently concluded that "bioenergy with carbon capture and storage (BECCS) remains associated with substantial risks and uncertainties, both over its environmental impact and ability to achieve net removal of CO_2 from the atmosphere." (EASAC, 2019) Nevertheless, according to IPCC economists, "there is no alternative".

Essentially, the strategy recommended by the IPCC is one of shifting problems into the future, following the old and discredited motto "grow now, clean up later". The belief in technological solutions is a narrow and inadequate approach to changing the environmental performance of the economy, but – unfortunately – not unusual amongst standard economists. In essence it implies kicking the can down the road (or even in the long grass), by shifting the risk of failure to future generations. Believing in the possibility to revert the impacts of threshold transgressions by compensation measures later on is a clear indication that the authors have not understood the very character and the dynamics of complex evolving systems – or at least they base their suggestions on models which are unable to adequately deal with the complexity of the systems they are supposed to represent.

The technology fix corresponds with the supply-side policy focus dominating current economic theory and policy, leading to the neglect of demand side management. Demand editing and nudging are even denounced as paternalising, and rejected – but they are no more paternalising than the ban on toxics to be sold in the supermarket. Recent studies have shown that adding consumption sufficiency to production efficiency makes it significantly easier to reach the Paris targets (Mundaca et al., 2019; Wachsmuth and Duscha, 2019) and overcome rebound effects: sufficiency is necessary to make efficiency effective.

In a nutshell: the world views and established habits of standard economics misguide the policy recommendations derived, increase the risks through accepting a (temporary) overshoot, ignore irreversible changes happening in the meantime and rely on technology fixes to provide for "negative emissions" which are either proven to be environmental disasters (large-scale pyrolysis) or untested, coming too late and if coming, a threat to the web of life (BECCS). They rest on an overly optimistic belief in technology and ignore policy options not in line with their supply-side focussed economic ideology.

5. Money makes the world go down?

Economics in itself has been evolving since its inception, and the understanding of central terms like "value" has consistently changed over time (Baldissone, 2008). In this section we address two key traits of current economic thinking, the way the future is valued and the role of economic growth.

The discounting procedure

Whether from an economic point of view policy measures should be taken now or later depends on how the costs are calculated – this is supposed to be the core competence domain of economists. Some problems like how to calculate the value of a human life are unsolvable questions where insurance mathematics, human rights and value theories collide, but others could be solved if only economic orthodoxy would give way to more empirically based procedures. Much of the economic dispute is about the appropriate discount rate, about the implicit choice of the value allocated to the future. Besides the belief in problem-solving technologies and the reliance on ever-lasting economic growth (higher affluence in the future would diminish the relative importance of cost shifted to future generations), the key argument is that future consumption should be discounted simply because it takes place in the future and people generally prefer the present to the future (inherent discounting).

This is problematic in many ways. In particular, discounting the value of future consumption, gains and liabilities makes a very strong normative statement concerning the value of future generations and the freedom of choice left to them. A high discount rate decreases the estimated benefit of actions taken today designed to reduce greenhouse gas emissions (Moxnes, 2014), and makes it appear plausible to "kick the can down the road". For instance, Nicholas Stern chose a low average discount rate (he applied a stochastic approach whereby the discount rate varied with the expected outcomes) for his famous Stern Review and was criticised for this by William Nordhaus (who 2018 received the Bank of Sweden price commemorating Alfred Nobel for his work on the economics of global warming). Nordhaus criticised the Review for its use of a low discount rate and argued for taking today's marketplace real interest rates[2] (Stern, 2006; Nordhaus, 2007). On the opposite side, ecological economists like Clive Spash criticised the study for its use of cost benefit analysis including discounting the future and Ted Trainer for the narrowness of its economic perspective (Spash, 2007; Trainer, 2008). The market interest rate approach by Nordhaus also suffers from the fact that while indeed humans often prefer the present to the future, not all do that, all do it differently and not all objects are treated the same. For instance, Gowdy et al. (2013) find that in many cases discounting

[2] Times are changing: with market interest rate close to zero since a decade, Nordhaus' line of argumentation would now imply much more drastic measures to be taken than in Stern's initial approach.

makes no sense at all; in these cases policies could instead be ranked by inspecting policy consequences over time (Moxnes, 2014). Even where discounting is applicable, hyperbolic discounting is better reflecting human preferences than exponential discounting (Gowdy et al., 2013); it can be introduced into classical welfare functions as Gollier (2011) has shown. However, guided by their methodological assumptions, neoclassical economists rarely ask if the concept of exponential discounting is appropriate at all, if discounting as such makes sense, but if anything, they discuss the discount rate, i.e. the exponent in function defined as exponential curve.

So do the IPCC economists, and by assuming a discount rate of 4% they follow the Nordhaus pledge for higher discount rates, and thus a lower profitability of climate mitigation policies. Hyperbolic accounting, which would lower the profitability threshold for climate policy measures is not discussed at all, nor are far-reaching demand side policies for energy reduction (the sustainability scenario at least touches upon the issue in passing).

The growth obsession

Assuming higher affluence in the future is one of the justifications for discounting; it appears to be an assumption which cannot be questioned, a holy grail of standard economics. Together with the neglect of demand side policies this may be the reason why all four IPCC scenarios assume continuous economic growth, and transgress the 1.5°C threshold – which makes them dependent on the use of "negative emission" technologies. The scenarios assume economic growth rates between 0.6% and 1.7% for the affluent countries, and 1.1% to 2.8% for the global economy (Kuhnhenn, 2018). The assumption is that growth is exponential, implying an accelerating growth in absolute terms – in contradiction to empirical analyses which found that the vast majority of mature economies follows a linear, rather than an exponential growth path (Lange et al., 2018).

The IPCC offers no scenario exploring the effects of discontinuing economic growth, at least in the affluent countries, any time between now and 2100. Considering options like a policy of degrowth was rejected by IPCC economists as the results of such assumptions would be economically implausible. Thus assumptions of 80 years of growth and the risk of hothouse climate conditions are considered a realistic option, while the deep structural change necessary to limit climate damage isn't – at least not if it ends permanent economic growth.

However, since the 1972 publication of the "Limits to Growth" report to the Club of Rome it is well known that economic growth and resource consumption are closely linked, and that efficiency gains minus rebound effects can ease the burden, they do not eliminate the links (Meadows et al., 1972). Long-term decoupling of the overall physical throughput of an economy from its monetary growth has not been achieved anywhere, and given the limits of possible efficiency increases resulting from physics, technology and cost it is also implausible. Most economists agree these days that growth cannot be eternal, that there will be limitations, but consider them to be far away. They tend to ignore the rising trend of the Energy Return On (Energy) Investment EROI, i.e. the fact that the low hanging fruits in terms of energy sources and deposits of minerals and ores have been exploited, and that new resources tend to require more input for the same output, reducing the net production (Sers and Victor, 2018). This and other factors contributing to a secular stagnation of economic growth (Schmelzer, 2015) have been the arguments which led even some business-affiliated research institutes in Germany to support a "precaution-oriented post-growth position" (IÖW et al., 2018).

Sustainable development in the affluent countries, starting with a phase of degrowth (Victor, 2012; Eversberg and Schmelzer, 2018) and turning into a steady-state economy once the necessary reductions of resource consumption have been achieved (Daly, 1973) could contribute significantly to avoid a climate overshoot. Thus it is high time that the economic profession leaves the world of sterile models behind, which are incapable of capturing the dynamics of living and social systems, and takes a scientific turn towards fact-based argumentation and the search for evidence. The CGE/DSGE models are castles in the clouds, and the conclusions drawn from them are dangerous for humankind and the global environment.

6. Outlook

The discipline of economics has played and continues to play a major role in leading humankind towards the ultimate calamity. It is driven by world views and their ontologies, which are more based on Newton's mechanics than rooted in modern science's understanding of systems complexity (Spangenberg, 2016). To make good for its past harmful role, and to provide guidance for economic and other policies for a sustainable 21st century, economics must change. Change would have to be deep and broad; for instance, economics would have to change its ontology (the economy is a subsystem of society which in turn is embedded in the environmental systems), its epistemology (not least to accommodate ignorance and uncertainty), its anthropology (accepting human beings in their ambivalence instead working with the *homo oeconomicus* abstraction and its derivates) and its axiology (accepting diverse value systems to replace "economic rationality").

Part of adapting its ontology to modern scientific standards would be to overcome the mechanistic world view and replace it by one based on a physical economy of matter and energy flows, and abolish equilibrium thinking and modelling to replace it by evolutionary approaches (Spangenberg, 2018). This would help avoiding what Herman Daly (2000) has called "dumb mistakes", like for instance considering the collapse of agriculture (Bélanger and Pilling, 2019) as a minor problem, as agriculture only makes up for 2 to 4% of the GDP in affluent countries, without asking about what people will eat in that case. It would also offer the opportunity to better understand the physical processes of the environment impacting on the economy, for instance the role of tipping points and their relevance for cost-impact analyses of climate policies (Cai et al., 2015). In short, economics has to be reinvented if it is to become a force for leading us away from catastrophe – rather than toward it. Taking the science of complexity on board would be a first step in this direction.

Is this an illusory demand? Being a neoclassical economist has been the career and income base for scholars, consultants, administrators and politicians – they would hardly declare "*mea culpa, mea maxima culpa*" and leave their positions (they could hardly stay after their lack of real-world qualifications has been revealed). But this should happen as soon as possible, as even with environmental and climate evidence mounting, adaptation and precaution remain limited and restricted by the existing system logics and institutionally based motivational and incentive structures, rather than by externally defined "rational" motivations (Andersson and Keskitalo, 2018). Several schools of heterodox thought have developed the means for economic research and advice generation, like stock-flow consistent input-output models (Berg et al., 2015), system dynamics models like WORLD6, simulating potential future supply and supply deficiencies of a number of natural resources (Umweltbundesamt, 2018a), evolutionary models like the fully endogenised GINFORS (Umweltbundesamt, 2018b; Spangenberg et al., 2012) or cybernetic models based on orientation strategy simulating sustainable system behaviour in changing environments (Bossel, 2000).

So while there is an abundance of ideas and tools, it is still plausible to expect a continued dominance of neoclassical power holders, and an Economics 6.0 (if you count Physiocrats, classics, marginalists, Keynesians and neoclassicals as earlier incarnations of economics) will only become hegemonial once the representatives of the existing hegemony have disappeared, or standard economics has fallen into so complete disregard by decision makers that standard economists give up and leave the sinking ship. So far however, such a turn of the tide is not in sight as even without good prognoses, scenarios and policy recommendations for sustainability, standard economics has an important function in legitimising neoliberal policies.

References

Alcamo, J. (2001) "Scenarios as tools for international environmental assessments." EEA European Environment Agency Expert Corner Report "Prospects and Scenarios" No. 5. Luxembourg, Office for the Official Publications of the European Communities.

Allen, P. M. (2001) "The Dynamics of Knowledge and Ignorance: Learning the New Systems Science." In *Integrative Systems Approaches to Natural and Social Dynamics*. H. M. Matthies, J. Kriz. Berlin: Heidelberg / New York: Springer, pp 3-30.

Andersson, E., Keskitalo, E. C. H. (2018) "Adaptation to climate change? Why business-as-usual remains the logical choice in Swedish forestry." *Global Environmental Change*, 48, pp. 76-85.

Arnsperger, C., Varoufakis, Yanis (2006) "What is neoclassical economics?" *post-autistic economics review*, 38, pp. 2-12.

Baldissone, R. (2008) "Beyond economic fundament-alism." *real-world economics review*, 45, pp. 33-53.

Bélanger, J., Pilling, D. (eds.) (2019) "The State of the World's Biodiversity for Food and Agriculture." Assessments. FAO Food and Agriculture Organization of the United Nations, Commission on Genetic Resources for Food and Agriculture. Rome: FAO.

Berg, M., Hartley, B., Richters, O. (2015) "A stock-flow consistent input–output model with applications to energy price shocks, interest rates, and heat emissions." *New Journal of Physics*, 17(1), p. 015011.

Bossel, H. (2000) "Policy assessment and simulation of actor orientation for sustainable development." *Ecological Economics*, 35(3), pp. 337-355.

Cai, Y., Judd, K. L., Lenton, T. M., Lontzek, T. S., Narita, D. (2015) "Environmental tipping points significantly affect the cost-benefit assessment of climate policies." *Proceedings of the National Academy of Sciences*, 112(15), pp. 4606-4611.

Cayuela, M. L., van Zwieten, L., Singh, B. P., Jefferey, S., Sánchez-Monedero, A. M. (2014) "Biochar's role in mitigating soil nitrous emissions: A review and meta-analysis." *Agriculture Ecosystems Environment*, 191, pp. 5-16.

Chytrý, M., Wild, J., Pyšek, P., Jarošík, V., Dendoncker, N., Reginster, I., Pino, J., Maskell, L. C., Vilà, M., Kühn, I., Spangenberg, J. H., Settele, J. (2012) "Projecting trends in plant invasions in Europe under different scenarios of future land-use change." *Global Ecology and Biogeography*, 21(1), pp. 75-87.

Culotta, E. (1995) "Will plants profit from high CO_2?" *Science*, 268(5211), p. 654.

Daly, H. E. (2000) "When smart people make dumb mistakes." *Ecological Economics*, 34(1), pp. 1-3.

Daly, H. E. (1973) *Toward a Steady-State Economy*. San Francisco: W.H. Freeman.

Dima, M., Lohmann, G. (2010) "Evidence for two distinct modes of large-scale ocean circulation changes over the last century." *Journal of Climate*, 23(1), pp. 5-16.

Dominik, C., Seppelt, R., Horgan, F. G., Settele, J., Václavík, T. (2018) "Landscape composition, configuration, and trophic interactions shape arthropod communities in rice agroecosystems." *Journal of Applied Ecology*, 55(5), pp. 2461-2472.

EASAC European Academies' Science Advisory Council (2019) *Forest bioenergy, carbon capture and storage, and carbon dioxide removal: an update*. EASAC. Halle (Saale), Germany / Brussels, Belgium.

EEA European Environment Agency (2004)" A Framework for Evaluating Complex Scientific Evidence on Environment Factors in Disease Causation." Background Paper No. 3 from the European Environment Agency for the EU Environment and Health Strategy. EEA, Copenhagen.

EEA European Environment Agency (2006) "How much bioenergy can Europe produce without harming the environment?" EEA Report 7/2006. EEA. Copenhagen.

EEA European Environment Agency (forthcoming) "State of the Environment Report - SOER 2020" Integrated Assessment report. EEA. Copenhagen.

Eversberg, D., Schmelzer, M. (2018) "The degrowth spectrum: convergence and divergence within a diverse and conflictual alliance." *Environmental Values*, 27(3), pp. 245-267.

Folke, C. (2006) "Resilience: The emergence of a perspective for social–ecological systems analyses." *Global Environmental Change*, pp. 253–267.

Gilboa, I. (2009) *Theory of Decision under Uncertainty*. Cambridge.

Gollier, C. (2011) "Discounting and risk adjusting non-marginal investment projects." *European Review of Agricultural Economics*, 38(3), pp. 325-334.

Gowdy, J., Rosser Jr, J. B.Roy, L. (2013) "The evolution of hyperbolic discounting: Implications for truly social valuation of the future." *Journal of Economic Behavior & Organization*, 90, Supplement, pp. S94-S104.

Hansen, J., Sato, M., Hearty, P., Ruedy, R., Kelley, M., Masson-Delmotte, V., Russell, G., Tselioudis, G., Cao, J., Rignot, E., Velicogna, I., Kandiano, E., von Schuckmann, K., Kharecha, P., Legrande, A.N., Bauer, M., Lo, K.W. (2016) "Ice melt, sea level rise and superstorms: evidence from paleoclimate data, climate modeling, and modern observations that 2 °C global warming could be dangerous." *Atmospheric Chemistry and Physics* Discussion, (16), pp. 3761-3812.

Herfeld, C. (2013) "PhD thesis summary: The many faces of rational choice theory." *Erasmus Journal for Philosophy and Economics*, pp. 117-121.

Holling, C. S. (1973). "Resilience and stability of ecological systems." *Annual Review of Ecology and Systematics,* (4), pp. 1-23.IÖW Institut für ökologische Wirtschaftsforschung, RWI – Leibniz-Institut für Wirtschaftsforschung, Wuppertal Institut für Klima Umwelt Energie (2018). Gesellschaftliches Wohlergehen innerhalb planetarer Grenzen. Der Ansatz einer vorsorgeorientierten Postwachstumsposition. UBA Texte 89/2018. Umweltbundesamt. Berlin, Germany.

IPBES Intergovernmental Science-Policy Platform on Biodiversity and Ecosystem Services (forthcoming). *The Global Assessment.* IPBES, Bonn, Germany.

IPCC Intergovernmental Panel on Climate Change (2013) *Climate Change 2013: The Physical Science Basis*. Cambridge, New York: Cambridge University Press.

IPCC Intergovernmental Panel on Climate Change (2014a) *Climate Change 2014: Impacts, Adaptation, and Vulnerability*. New York, Nairobi, IMO, UNEP.

IPCC Intergovernmental Panel on Climate Change (2014b) *Climate Change 2014: Mitigation of Climate Change*. New York: Cambridge University Press.

IPCC Intergovernmental Panel on Climate Change (2018) "Global Warming of 1.5°C. An IPCC Special Report on the impacts of global warming of 1.5°C above pre-industrial levels and related global greenhouse gas emission pathways, in the context of strengthening the global response to the threat of climate change, sustainable development, and efforts to eradicate poverty." *IPCC Special Report 15*. Geneva, Switzerland, World Meteorological Organization.

Lange, A., Treich, N. (2008) "Uncertainty, learning and ambiguity in economic models on climate policy: some classical results and new directions." *Climatic Change*, 89(1-2), pp. 7-21.

Laybourn-Langton, L., Rankin, L., Baxter, D. (2019) *This is a crisis. Facing up to the age of environmental breakdown*. IPPR Institute for Public Policy Research, London

Joughin, I., Smith, B. E., Medley, B. (2014) "Marine ice sheet collapse potentially under way for the Thwaites Glacier Basin, West Antarctica." *Science*, 344(6185), pp. 735-738.

Kashiwase, H., Ohshima, K. I., Nihashi, S., Eicken, H. (2017) "Evidence for ice-ocean albedo feedback in the Arctic Ocean shifting to a seasonal ice zone." *Scientific Reports*, 7(1), p. 8170.

Knight, C. G., Knight, S. H. E., Massey, N., Aina, T., Christensen, C., Frame, D. J., Kettleborough, J. A., Martin, A., Pascoe, S., Sanderson, B., Stainforth, D. A., Allen, M. R. (2007) "Association of parameter, software and hardware variation with large scale behavior across 57,000 climate models." *Proceedings of the National Academy of Sciences of the USA*, 104(30), pp. 12259-12264.

Knoblauch, C., Beer, C., Liebner, S., Grigoriev, M. N.Pfeiffer, E.-M. (2018) "Methane production as key to the greenhouse gas budget of thawing permafrost." *Nature Climate Change*, 8(4), pp. 309-312.

Kretschmer, M., Coumou, D., Agel, L., Barlow, M., Tziperman, E., Cohen, J. (2018) "More-persistent weak stratospheric polar vortex states linked to cold extremes." *Bulletin of the American Meteorological Society*, 99(1), pp. 49-60.

Kuhnhenn, K. (2018) *Wachstum in der Klimawissenschaft: Ein blinder Fleck. Globale Szenarien aus wachstumskritischer Perspektive*. Heinrich-Böll-Stiftung, Berlin, Germany.

Lange, S., Pütz, P., Kopp, T. (2018) "Do mature economies grow exponentially?" *Ecological Economics*, 147, pp. 123-133.

Lockwood, J. G. (2001) "Abrupt and sudden climatic transitions and fluctuations: A review." *Journal of Climatology*, 21(9), pp. 1153-1179.

Lontzek, T. S., Cai, Y., Judd, K. L. (2012) "Tipping points in a dynamic stochastic IAM, RDCEP. " Working Paper 12-03. The Center for Robust Decision Making on Climate and Energy Policy. Chicago, IL, USA.

Mayumi, K., Giampietro, M. (2001) "The epistemological challenge of modelling sustainability: risk, uncertainty and ignorance." Paper presented at the European Society for Ecological Economics' Frontiers I conference, Cambridge University, UK, 4-7 July 2001.

van der Sluijs, J. P. (2012) "Uncertainty and dissent in climate risk assessment: a post-normal perspective." *Nature and Culture,* 7(2), pp. 174-195.

May, R. M., (ed.) (1976) *Theoretical Ecology - Principles and Applications*. Blackwell Scientific Publications, Oxford, UK.

Mayumi, K., Giampietro, M. (2001) "The epistemological challenge of modelling sustainability: risk, uncertainty and ignorance." Paper presented at the European Society for Ecological Economics' Frontiers I conference, Cambridge University, UK, 4-7 July 2001.

McCoy, D. T., Eastman, R., Hartmann, D. L., Wood, R. (2017) "The Change in Low Cloud Cover in a Warmed Climate Inferred from AIRS, MODIS, and ERA-Interim." *Journal of Climate*, 30(10), pp. 3609-3620.

Meadows, D. H., Meadows, D. L., Randers, J., Behrens, W. W. (1972) *Limits to Growth. A Report to the Club of Rome*. New York: Universe Books.

Millner, A., Dietz, S., Heal, G. (2013) "Scientific ambiguity and Climate Policy." *Environmental Resource Economics*, pp. 21-46.

Moxnes, E. (2014) "Discounting, climate and sustainability." *Ecological Economics*, 102, pp.158-166.

Mundaca, L., Ürge-Vorsatz, D., Wilson, C. (2019) "Demand-side approaches for limiting global warming to 1.5°C." *Energy Efficiency*, 12(2), pp. 343-362.

Muthoo, A. (1999) "Rationality in the Face of Uncertainty." In *The Elgar Companion to Consumer Research and Economic Psychology*, pp. 487-495. Cheltenham.

Nordhaus, W. D. (2007) "A review of the stern review on the economics of climate." *Journal of Economic Literature*, 45(3), pp. 686–702.

Oltmanns, M., Karstensen, J., Fischer, J. (2018) "Increased risk of a shutdown of ocean convection posed by warm North Atlantic summers." *Nature Climate Change*, 8(4), pp. 300-304.

Prigogine, I. (1997) *The end of certainty: Time, Chaos and the new Laws of Nature*. New York: The Free Press.

Prigogine, I., Stengers, Isabelle (1984) *Order out of Chaos: Man's New Dialogue with Nature*. London: Heinemann.

Rocha, J. C., Peterson, G., Bodin, Ö., Levin, S. (2018) "Cascading regime shifts within and across scales." *Science*, 362(6421), pp. 1379-1383.

Sánchez-Bayo, F., Wyckhuys, K. A. G. (2019) "Worldwide decline of the entomofauna: A review of its drivers." *Biological Conservation*, 232, pp. 8-27.

Savage, L.J. (1954) *The Foundations of Statistics*. New York: John Wiley.

Schmelzer, M. (2015) "The growth paradigm: History, hegemony, and the contested making of economic growthmanship." *Ecological Economics*, 118, pp. 262-271.

Sers, M. R., Victor, P. A. (2018) "The energy-emissions trap." *Ecological Economics*, 151, pp. 10-21.

Showstack, R. (2014) "Sector of West Antarctic Ice Sheet in 'irreversible retreat'." *Eos, Transactions American Geophysical Union*, 95(20), pp. 168-168.

Smith, P., Davis, S. J., Creutzig, F., Fuss, S., Minx, J., Gabrielle, B., Kato, E., Jackson, R. B., Cowie, A., Kriegler, E., van Vuuren, D. P., Rogelj, J., Ciais, P., Milne, J., Canadell, J. G., McCollum, D., Peters, G., Andrew, R., Krey, V., Shrestha, G., Friedlingstein, P., Gasser, T., Grubler, A., Heidug, W. K., Jonas, M., Jones, C. D., Kraxner, F., Littleton, E., Lowe, J., Moreira, J. R., Nakicenovic, N.,

Obersteiner, M., Patwardhan, A., Rogner, M., Rubin, E., Sharifi, A., Torvanger, A., Yamagata, Y., Edmonds, J., Yongsung, C. (2016) "Biophysical and economic limits to negative CO_2 emissions." *Nature Climate Change*, 6(1), pp. 42-50.

Spangenberg, J. H., Bondeau, A., Carter, T. R., Fronzek, S., Jaeger, J., Jylhä, K., Kühn, I., Omann, I., Paul, A., Reginster, I., Rounsevell, M., Schweiger, O., Stocker, A., Sykes, M. T., Settele, J. (2012) "Scenarios for investigating risks to biodiversity." *Global Ecology and Biogeography*, 21(1), pp. 5-18.

Spangenberg, J. H., von Haaren, C., Settele, J. (2014) "The Ecosystem Service Cascade: further developing the metaphor. Integrating societal processes to accommodate social processes and planning, and the case of bioenergy." *Ecological Economics*, 104, pp. 22–32.

Spangenberg, J. H. (2015) "Sustainability and the Challenge of Complex Systems." In J. C. Enders, M. Remig (eds.), *Theories of Sustainable Development.* Routledge, Abingdon, Oxford, UK, pp. 89-111.

Spangenberg, J. H. (2016) "The world we see shapes the world we create: how the underlying worldviews lead to different recommendations from environmental and ecological economics – the green economy example." *International Journal of Sustainable Development*, 19(2), pp. 127–146.

Spangenberg, J. H. (2018) "Behind the Scenarios: World Views, Ideologies, Philosophies. An analysis of hidden determinants and acceptance obstacles illustrated by the ALARM scenarios." *Sustainability*, 10(7), p. 2556.

Spash, C. L. (2007) "The economics of climate change impacts à la Stern: Novel and nuanced or rhetorically restricted." *Ecological Economics*, 63(4), pp. 706-713.

Stephens, J. C. (2013) *Time to Stop CCS Investments and End Government Subsidies of Fossil Fuels.* WIREs Climate Change.

Stern, N. H. (2006) *The Economics of Climate Change. The Stern Review.* Cambridge/New York: Cambridge University Press.

Syll, L.P. (2010) "What is (wrong with) economic theory?" *real-world economics review*, 55, pp. 23-57.

Trainer, T. (2008) "A short critique of the Stern Review." *real-world economics review*, 45, pp. 54-58.

Umweltbundesamt (2018a) The WORLD Model Development and The Integrated Assessment of the Global Natural Resources Supply. UBA Texte 100/2018, Federal Environment Agency, Berlin

Umweltbundesamt (2018b) System analysis for environmental policy – System thinking through system dynamic modelling and policy mixing as used in the SimRess project "Models, potential and long-term scenarios for resource efficiency" (SimRess). UBA Texte 49/2018, Federal Environment Agency, Berlin.

UNFCCC United Nations Framework Convention on Climate Change (2015) The Paris Agreement, UNFCCC doc no. FCCC/CP/2015/L.9/Rev.1. Paris, UNFCCC.

van der Sluijs, J. P. (2012) "Uncertainty and dissent in climate risk assessment: a post-normal perspective." *Nature and Culture*, 7(2), pp. 174-195.

Victor, P. A. (2012) "Growth, degrowth and climate change: A scenario analysis." *Ecological Economics*, 84, pp. 206-212.

Wachsmuth, J., Duscha, V. (2019) "Achievability of the Paris targets in the EU—the role of demand-side-driven mitigation in different types of scenarios." *Energy Efficiency*, 12(2), pp. 403-421.

Yager, R. R., Filev, D. P. (1994) *Essentials of fuzzy modeling and control*. New York: John Wiley.

Chapter 15

Of ecosystems and economies: re-connecting economics with reality

Clive L. Spash & Tone Smith

Introduction

The state of planet Earth is widely recognised as in jeopardy due to a range of environmental problems relating to a dominant economic system that extracts resources and uses energy on an unprecedented scale in human history. A long-running claim amongst mainstream economists, defenders of unregulated capitalism and those favouring a regulated productivist economy has been that human ingenuity can find substitutes for all resources and technology can solve all problems allowing humanity to change and adapt to anything. These arguments are made in almost total ignorance of how the economy interacts with ecosystems and impacts their structure and functioning, how dependent economies are on the flow of low entropy materials and energy and what are the basic limits to humans as biological animals. Indeed even ignorance itself is ignored and reduced down to risk and probabilities.

Yet, that economies must change is no longer in question. That they will change is also no longer even an issue. The question is what responses materialise as resources, energy supplies and functioning of ecosystems do change? The options being put forward are numerous, but most aim to preserve some form of high-technology, capital accumulating, growth economy embedded in price-making markets, including: green economy, climate economy, low carbon economy, circular economy, knowledge economy, bioeconomy. Yet, none of these addresses the causal mechanisms of the current crises, or structural issues facing social ecological transformation; they are concerned only with controlling for impacts and adapting to consequences, not with the bio-physical relations of the economy with non-human nature.

This article provides an overview of the relationships between economic systems and the environment, human society and non-human nature, ecology and economy. It brings together various literatures with the aim of introducing the reader to the importance of biophysical reality for the operation of real economies, and therefore also for economics. In the next section, we explain the problems facing standard economic approaches if they are to address environmental problems, but more generally their inability to even understand the social ecological crises due to a limited scope and direction. This is followed by outlining the place of economies in the context of their social and bio-physical structural relations, a basic general ontology. More specific detail is then added on the lessons that can be drawn from ecological understanding in terms of ecosystems, materials and energy. The final section draws out the implications of this understanding for social ecological transformation of the currently dominant economic systems and the type of economics required to help achieve that transformation.

What ails economics? Growth, development and the environment

Economists hold that their concern is an object of study called "the economy". An object most economists assume can be treated meaningfully without any consideration of the social ecological context in which it operates, the society from which it emerges or the biophysical reality on which it depends. This positon is challenged by the, now common, realisation that there are serious environmental problems looming, including: mass extinction of species, biodiversity loss, destruction of ecosystems structure and functioning, and pollution of land, air and water on all scales from local to global. Talk of limits to economic growth by Meadows et al. (1972) was denigrated by economists (e.g., Beckerman, 1974), but unfortunately their baseline scenario analysis has proven in line with real trends (Turner, 2012). Limits have now returned to the political agenda, conceptualised as planetary boundaries (Rockström, et al., 2009).

However, few economists pay any attention to the ultimate failure of economic growth as their guiding principle. Even those, like Tim Jackson, who do claim prosperity is possible without growth still defend the need for economic growth for "poorer countries". As Jackson (2009, p. 41) makes clear, a "key message" of his book on the topic is that: "There is no case to abandon growth universally. [...] It is in these poorer countries that growth really does make a difference". This positon totally conflicts with the post-development school that documents how equating development with growth has been an imperialist post-World War II policy promoted by the USA and implemented through various captured organisations, such as the IMF and the World Bank (Sachs, 2015 [1999]). Development policy has denigrated and destroyed the cultures of non-industrialised countries, livelihoods of the rural and materially poor and removed their autonomy. Sachs differentiates the materially poor into what can be described as living frugally, suffering deprivation and living under systems of economic scarcity. Traditional societies have economic systems of social provisioning that are structured on frugality and sufficiency. Interventions to "develop" their economic circumstances have typically resulted in expropriation and forms of primitive accumulation. Culture is destroyed along with sustainable livelihoods. Land is grabbed, resources exploited, agriculture is industrialised and the environment is polluted. The survivors add to the exponential growth in urban slum dwellers, more than a billion on conservative UN estimates a decade ago (Davis, 2006, p. 23). A class of people ready for exploitation as commodified labour due to their newly-created wage dependency and their new lives as those saved from "poverty" to live in the economy of material scarcity measured by money.

Economists have continued to promote the "growth=development" ideology of progress even as the consequences (e.g., human induced climate change and biodiversity loss) are realised to be increasingly severe and threatening to all. The standard economic response has been to extend markets and private property rights including attempts to make ecosystems into goods and services (Spash, 2015) and greenhouse gases into financially tradeable commodities (Spash, 2010). Economic growth has remained the primary concern, with environmental issues considered only if investments give a positive financial rate of return, economic growth and jobs (GCEC, 2014; Jaeger et al., 2011). The problem is not seen as ecological crises, but how business can realise and capture the economic value that ecosystems produce. The opportunities for profiting from environmental problems are a stimulating tonic for creators of new markets and financial instruments that "make Nature pay".

Despite their ever increasing type, number and scale, environmental problems are treated by most economists as isolated, individual instances of market failure. Their conceptualisation as "externalities" has been copied widely. The classic treatment, as in Coase and neoclassical environmental economics, is based on pollution being a minor problem between two contracting parties operating in an isolated system with no irreversibility, uncertainty, indeterminacy, unknowns, complexity or asymmetric information. Much is made of relaxing assumptions to take account of some of these things (one at a time *ceteris paribus*), but the basic "solutions" – unregulated markets and private property rights – remain, regardless of whether the simplest or most complex models are applied. This is a closed, self-referencing system of deductive thought. In short, it is a total fiction that bears no relationship to actual environmental problems operating in a complex open systems reality, and as a result it produces policies that fail. Neither is any attempt being made to identify real causal mechanisms. However, ideas have the power to motivate people and externality theory serves as a convenient fiction, suited to maintain economic theories of efficiency, the neoliberal ideology of "free" markets, and the supreme economic objective of technologically driven growth.

In reality, the creation of environmental degradation is nothing external to the economic system of industrial modernity, but rather an integral part of that system. In this system, success is the ability to pass on as many "costs" as possible to others, while exploiting all possibilities for gain at others' expense. As Kapp (1965; 1969; 1970; 1971; 1978 [1963]) pointed out long ago, this is an exercise in cost shifting,[1] and the output of a firm is dependent on its ability to shift part of its costs to other sectors of the economy or individuals. Cost shifting can be identified within the structure of the economic system made operational through the mechanism of market competition. While unable to recognise structure,

[1] As Kapp (1965, p. 1) stated: "The concept of social costs refers to a wide variety of harmful effects of productive activities which are not reflected in entrepreneurial cost accounts and, hence, tend to be neglected in private decision making. That is to say, social costs may be identified as those harmful effects of private action which, under given conditions and institutional arrangements, tend to be shifted to and borne by other sectors, third persons, or the economy as a whole."

neoclassical economists might have recognised that investment for profit does not entail social efficiency, anymore than does the individual aiming to maximise their utility. Consistent with their neoclassically designated roles, both the primary mainstream economic actors – firms and consumers – can act "optimally" by shifting costs onto others. On this basis, mainstream economics should regard environmental degradation, as well as other social costs, as endemic to the system, and not some minor aberration or instance of market failure to be fixed by adjusting a price at the margin (i.e. internalising externalities). Yet, they persist in their ideological commitment to "getting the prices right" to empower economic actors with "information" about how to allocate resources efficiently.

Attempts, supposedly justified by "new" welfare economics, to convert environmental degradation into social costs, estimated as monetary values, require the application of heroic assumptions, e.g., a monistic value theory with total and universal commensurability, utilitarian ethics, absence of lexicographic preferences. The Pareto criterion, which economists seem to assume is some uncontestable moral ethic, justifies making the rich richer while doing nothing for anyone else. Its Kaldor-Hicks adjustment means deliberate harm – even to the already worst-off – can be justified without any actual compensation. The application of cost-benefit analysis to global environmental change (e.g. enhanced greenhouse effect causing climate change) violates even this neoclassically-based welfare theory, not least by ignoring the requirements for maintaining money as a measuring rod of value (i.e. no substantive income changes).

After decades of criticism the arbitrary and unscientific economic analysis of environmental problems as "externalities" remains firmly in place. Worse still, the application of cost-benefit analysis to climate change due to greenhouse gas emissions has earned one economist, Stern, a place in the House of Lords, and another, Nordhaus, the highest international prize in economics. This despite Stern and his colleagues' work not addressing the basic issues they themselves identify (Spash, 2007a; 2014). In the case of Nordhaus he has persisted in producing numbers which even *The Economist* admitted were "massively simplified" (Spash, 2002a, p. 161), and he has always employed not only over-simplification but also numerous *ad hoc* assumptions and highly selective use of science, possible future impacts and economic scenarios (Spash, 2002b; 2007b).

Beyond the basic failures of mainstream economists, even to stay within the strictures of their own theories, there is a much larger failure of the economics profession in general, and that is a lack of relation to the natural world. Most economists are trained to ignore the existence of anything outside "the economy", as if this were a self-sustaining and singular type of system. In what follows, we will explain why there is no such isolated system as "the economy", nor any such universal type as "the economy", but only varieties of economies.

Relations between economy, society and nature: ontology

The basic relationship of an economy to the rest of reality is core to understanding the ecological crisis of modernity. Economics, in both orthodoxy and heterodoxy, largely fails to include the dependency of human society on nature. In the orthodoxy, resource and environmental economics uses neoclassical micro-economics and welfare theory, but is a marginalised and minor field within the economics profession. In the heterodoxy, outside of social ecological economics, there has been minimal attention to the environment: mainly amongst eco-socialists and eco-feminists, occasionally by institutionalists, and to a much lesser extent by a few post-Keynesians (Spash and Ryan, 2012). In general, economists treat the environment as an optional extra, an area for specialists, outside the central concerns of the profession, rather than of fundamental importance to understanding economic systems, their organisation, operation and reproduction.

The reality is that the modern economy is built on fossil fuels and mass throughput of low entropy resources. The standard picture of what constitutes "the economy" is narrowly framed around price-making markets and capital accumulation. There is no connection between the macroeconomic circular flow diagram, with its never ending cycle of goods and services flowing between firms and households, and the necessary resource inputs and waste outputs that make this system operative. If there were, the fallacy of such a model would be self-evident. As every military strategist knows, if you cut the resource supplies the economy soon collapses. Just as crucial, humans can die from accumulation of waste including their own excrement (a problem related to typhoid and hepatitis, documented for millions living in cities by Davis, 2006 pp.137-142 in a section entitled "living in shit" in a chapter on "Slum Ecology"). However, material provisioning and waste disposal have no place in modern economics where "the economy" is treated as a physically isolated system (i.e., with no material or energy exchange with any other system). Once this theoretical pretence is dropped, specifying the nature of the relationship of different types of economy to the environment becomes key.

That there are different types of economy is also something typically ignored by economists. Commonly the issue is to determine how "the economy" operates and what the relationships of "the economy" are that would maintain certain states (e.g. full employment, growth). That there are varieties of economic systems seems self-evident due to the potential variety of institutional arrangements for operating social provisioning systems, e.g. the extent of state planning, corporate control, ownership of the means of production, types of property rights, types of markets or no markets. Indeed, exactly what factors can vary across economies is contested. For example, a common neoliberal claim is that "free" market capitalism is the ultimate form of economy and the only way in which human societies, on the scale of the current population, can operate. Others contest that government intervention is essential. That there might be varieties of capitalism is one issue (e.g., Hall and Gingerich, 2009; Hall and Thelen, 2008). That there might be alternatives to capitalism has seemingly been pushed off most economists' research agendas. Yet this is a basic historical fact. That is, before capitalism there were other types of economic systems, other economies. Once both the possibility of and need for alternatives are accepted then questions arise as to the varieties of social structure, means of social provisioning and waste disposal, and relationships with nature and biophysical reality.

In general, the conceptualisation of the place of economies in relation to other structures is a matter of ontology. Clarifying the structural relationships and dependency amongst different structures has then been a matter of some debate (e.g. the social "embeddedness" of the economy, see Dale, 2010; Gemici, 2008). For those who have been attentive to the relationship between the economy and society, a popular interpretation of the rise of market capitalism is that "the economy" has taken over society. For example, Sachs (2015 [1999], p. 17) states that "the economy overshadows every other reality; the laws of economy dominate society and not the rules of society the economy". This line of reasoning can be traced back to Karl Polanyi and his ideas of the economy being embedded in society prior to capitalism (Polanyi, 1977b), and then the society becoming embedded in the market economy after its rise to power (e.g., Polanyi, 1977a, p. 9). However, such reasoning is contradictory and problematic because, as Polanyi recognises, no economy can exist without society and the form of an economy is emergent from and dependent upon social relationships. There are then no pure economic entities that can dominate the social, but rather different types of social economic relations. Market economies are still "embedded" in social relations, but they rely on very specific institutionalised forms.

What the emergence of economies from society emphasises is the necessity of social theory. That is, economics always entails a set of social understandings whether they are explicit or not. Economic policy recommendations that fail to pay any attention to social aspects are like planning a transport system by designing a car engine (Spash, 2017). Social relationships in the market economy are atomised to the individual, although this is contradicted by the necessity of non-market coordinating institutions (conventions, norms, rules and regulations, see Vatn, 2005) that make the market operational. In addition, the undermining of social relationships in market-based economies results in the necessity of government intervention to restabilise the systems and save capitalism from itself. This is Polanyi's "double movement". The need to give back to the exploited before chaos ensues or democracy becomes authoritarian, dictatorial and fascist.

The tendency to undermine the social relationships upon which the system depends is matched by the impact on the environment. Ecosystems functions and structure are not optional extras to be added as an afterthought. The quality of the environment is essential to human flourishing and survival. Humans are biological entities and as such need to maintain their metabolism and are subject to the needs and conditions – climate, temperature, nutrients, water, oxygen – of being such entities. The ability to create interventions that change the actualised environmental circumstances to human advantage does not change these structural limits, but rather works within them, e.g. houses maintain a certain necessary temperature. This is a major distinction that needs to be made clear because of the techno-optimist rhetoric that claims human imagination can achieve anything it conceives; something prevalent amongst a class of technocratic advocates of the controversial Anthropocene conceptualisation of social ecological crises (Baskin, 2015).

In this respect, a critical realist philosophy of science can help due to its depth ontology and understanding of stratification and emergence (see Collier, 1994). The depth ontology differentiates between the empirical (things sensed by humans) and actual (things that happen, not all which we sense), but also emphasises the role of an underlying structural aspect of reality. The relationship between, for example, the social, biological, chemical and physical is stratified and hierarchically ordered. Each stratum has its own causal mechanisms. What this philosophy of science explains is the asymmetric dependency of one set of mechanisms on another, but not in a reductionist or determinist sense. Higher strata have the properties of emergence, so they cannot be understood by reduction to the lower strata on which they depend, e.g. humans cannot be fully understood by reduction to the rules governing their biology. The structure of the natural world is slow to change or effectively (as far as humans are concerned) unchanging. Science has progressed by learning the rules, understanding the mechanisms of physics, chemistry and biology, and then technology has been developed by using these mechanisms for human ends.

The reason humanity faces limits is because it does not make the rules. However, in creating actual events and phenomena different mechanisms, from across the layers of nature, can be, and typically are, brought together. Thus, understanding concrete events and phenomena requires knowledge of the multiple mechanisms that cause them. Human economic and social systems impact on ecosystems, species, biological and physical entities, not by changing the mechanisms, but by using them, either intentionally or unintentionally. Of course, a class of humans now have the ability to destroy entire systems on Earth, which completely removes mechanisms and their potential.

So how should something like human induced climate change be understood from this perspective? The greenhouse effect is a phenomenon established by a set of physical and chemical mechanisms. A select minority of humanity have unintentionally used these mechanisms to such an extent that they are responsible for enhancing the greenhouse effect, leading to global warming in the absence of any counter-mechanisms. Geoengineering promises to develop the use of such counter-mechanisms, rather than stop using those of the greenhouse effect. However, why does this minority of humans use the greenhouse mechanisms on such a scale in the first place? This is because they live within fossil fuel based economies, and to stop using them would require changing the economic system. There has never been an industrial economy that was not based on fossil fuels. So a totally new type of economy is necessary, and because economies are dependent on social structure that would imply new social arrangements and new means of social provisioning. Thus, recognising human induced climate change as a serious structural problem, the preference of policy makers, corporations, industrialists, financiers, bankers and all those invested heavily in the fossil fuel economy is to maintain the system and hope for a technology that could provide a physical-chemical counter-mechanism. Yet, the enhancement of the greenhouse effect is just one of many ecological problems created by modern economies.

Linking economics to biophysical reality: ecosystems, entropy and values

In the 1970s, fundamental insights arose from ecology about modern human society and the operations of its economy under capital accumulation and mass consumerism. At the core of concerns was the disruption of ecosystems' structure and functions impacting on human and non-human life. Impacts were related to the expanding scale of human activity due to economic and population growth (e.g. land use change, appropriation of natural functions), technologically driven qualities of those activities (e.g., emissions from fossil fuels, radioactive waste from nuclear power, toxic waste from the creation of synthetic chemical substances), and their combined impact.

The interconnectivity of things was a major new understanding coming from ecology, based upon the developing concept of ecosystems. Nutrients, as essential to life, were linked to chemical cycles – carbon, hydrogen, nitrogen, oxygen, phosphorus and Sulphur – operating through ecosystems. As systems composed of physical-chemical-biological processes, ecosystems were recognised to provide a concept of the functioning of nature that combined the biotic and abiotic. The importance of the conversion of biomass into energy laid the foundations for studying ecosystems through energy flow analysis. For example, ecologists traced energy through agro-ecosystems to question the sustainability of the Green revolution in agriculture (Biswas and Biswas, 1976; Pimentel et al., 1973). Pollution had been treated as a local problem or something solved by dilution of matter into a large and accepting environment. Now the long range transport of air pollutants creating acidic deposition became a recognised phenomenon, as did the potential for bioaccumulation of chemicals (e.g. DDT, heavy metals). In all this new understanding, the centrality of ecosystems structure and functioning to life on planet Earth became evident, but also that the characteristics of ecosystems were not those of mechanistic science, i.e. stable, static, equilibrating, reversible.

Ecosystems change, irreversibility and strong uncertainty[2]

For a long time ecologists assumed ecosystems were largely closed systems dominated by internal recycling of

[2] This and the next subsection include some reworked text taken from the introductions to volumes one, three and four of the four volume collection on ecological economics by Spash (2009).

elements, self-regulating and deterministic, and stable with end points (e.g., climax communities). They also neglected human influence, externalised it and separated it off, as something outside their concerns. Holling (2009 [1986], p. 87; 1995) reinterpreted disturbance as part of ecosystem dynamics and described this as a cycle in four phases: (i) exploitation, where species get established; (ii) conservation, where a climax community is achieved and consolidated; (iii) release / creative destruction, where a disturbance destroys the structure; and (iv) reorganisation / renewal, where order and structure starts to reform incorporating released materials and energy. An ecosystem might dramatically change at stage (iii), thereby preventing reorganisation along the same path as before. That is, there is no guarantee that a system will keep going through the same cycle of succession and recreating the same structure and functions (e.g. an old growth forest might never reappear after a devastating forest fire, and instead might become a desert ecosystem). The resilience of a system is then defined in terms of maintaining certain structures and functions through change. This emphasises the boundary of stability, events far from equilibrium, high variability and adaptation to change (Holling, 2009 [1986], pp. 71). Economic growth emphasises "operational efficiency" and demands more from all systems leading to impacts on biophysical evolution (Holling, 2009 [1986], p. 92).

These developments in ecosystem theory led to awareness that the changing dynamic of systems may result in surprise as systems flip due to different attractors becoming dominant. Kay et al. (1999) developed the concept of a self-organising holarchic open system. Such a system shows spontaneous coherent behaviour but can suddenly change (i.e., show discontinuity) when reaching a "catastrophic" threshold. Learning from ecosystems dynamics is combined with thermodynamic theory and linked into the need for a new approach to science. The scientist is seen as providing narrative descriptions, based upon quantitative and qualitative understanding, rather than making deterministic predictions. Kay et al. (1994, pp. 737-740) recommend a process of management where science informs but decisions involve ethics, values and concerns, visions of the future and socio-political context.

Continuous human intervention creating disturbance to ecosystems structure and functioning is not some mechanistic engineering problem to be solved through controlled experimentation. Standard scientific epistemology is challenged due to complexity precluding reductionism, lack of control and inability to replicate relationships in open systems. "Not only is the science incomplete, the system itself is a moving target, evolving because of the impacts of management and the progressive expansion of the scale of human influences on the planet" (Walters and Holling, 2009 [1990], pp. 117-118). Ludwig, Hilborn and Walters (1993) note the failure of science to prevent resource overexploitation, collapse and extinction and see this as due to a lack of scientific consensus as to the causes. The recommendation is caution and, more specifically, attention to: human motivation, acting before scientific consensus, recognising scientists and their judgements are subject to political pressure, distrusting claims of sustainability (especially where problems of population growth and excessive resource use are ignored), and confronting uncertainty. Similar concerns lay behind the development of post-normal science and its recommendation to involve an extended peer community in science-policy, including laypersons (Funtowicz and Ravetz, 1991).

As can be seen from this brief overview, the literature on ecosystem dynamics emphasises surprise and strong uncertainty (i.e., ignorance and indeterminacy, see Spash, 2002c). However, economics remains mechanistic, quantitative, equilibrium seeking and so totally incompatible with understanding the reality of the ecosystems in which economies are embedded. As Holling et al. (1995) recognise, the result is that economists generally ignore ecological information, despite the accumulated body of evidence from natural, disturbed and managed ecosystems.

Rather than a more humble approach in human non-human relationships, the co-option of selected ecological concepts has been employed to support the opposite conclusion, that humans can create and control everything. For example, the idea that resilience is something mechanistic to be built into all systems as an inherently good quality, despite there being nothing that necessitates resilience in itself leading to sustainability, and it may even do the opposite, e.g. a resilient fossil fuel economy hurtling us headlong towards climatic disaster. Similarly, the use made of the ecological concept of adaptation can be seen as having undermined greenhouse gas mitigation especially once combined with economistic arguments about adaptation being more "cost-effective". The inappropriateness of such human hubris is further reinforced by the laws of physics.

Thermodynamics, entropy and economics

The marginalist revolution in economics during the 1870s, which led to the rise of neoclassical economics, borrowed heavily from mechanistic physics in terms of mathematical formalism and models (Mirowski, 1989). However, economics has managed to totally ignore the relevance of actual laws of physics, despite their importance for the social provisioning and reproduction of society. Economic growth predicated on material throughput creates vast amounts of waste. These wastes go into the environment and ecosystems with the implicit expectation of their harmless assimilation. The amount of energy remains the same from extraction to waste, as a direct consequence of the First Law of Thermodynamics, i.e. energy can neither be created nor destroyed. A similar law relates to matter and led to the idea of materials balance theory, that was briefly a topic of research in environmental economics (Kneese et al., 1970), that later developed into the field of industrial ecology. This means material that does not go into embodied capital will become waste and all the materials extracted from the environment will go back into the environment in equal mass.

Economic growth is dependent upon a specific form of energy, that is energy available for performing mechanical, chemical or thermal work. This useful energy is termed "exergy" to differentiate it from energy, which is neither created nor destroyed, because exergy is used up in all transformation processes (Ayres and Warr, 2009). Modern industrial society makes use of stored exergy in ores and fossil fuels. These sources are depleted and while the energy remains in the system it is no longer useful and so the exergy is reduced. The Second Law of Thermodynamics, or Entropy Law, in its classic form, states that energy changes quality from useful (low entropy) to less useful (high entropy) heading towards an equilibrium where all is evenly distributed (heat death of the universe). This process is irreversible and therefore associated with the phrase "times arrow". Creating concentrated forms of matter and energy (i.e., low entropy) is possible within a system, but only with energy added from another system; that is, overall in the combined system energy is still degraded, the Entropy Law remains in force. Georgescu-Roegen's (1971) major thesis, "*The Entropy Law and the Economic Process*", basically concluded that economic growth was infeasible over the long run and economic policy needed fundamental reform. His reasoning led to questioning human society from the size of population and the pressure placed upon systems, to the time allowed for change and the rate at which human systems impose change. Economic systems are then inseparable from ethical judgments both concerning others currently living and future generations. Herman Daly (1977a; 1977b) came to the conclusion that the best option in the face of the Entropy Law and critiques of growth was to aim for a steady-state economy.

The Entropy Law has been taken to imply absolute constraints on economic systems (Daly, 1977a; 1977b;

Georgescu-Roegen, 2009 [1975]). That is, energy use depletes stored exergy and dissipates minerals into "devil's dust" which can never be recovered (Marx cited by Daly, 1968). However, in theory the large amounts of energy input to the Earth system from the Sun can be used to create order and reverse dispersal. If human society relied upon solar energy and conserved the required amounts of ores to maintain man-made capital then a different type of economic system could be sustained over a long time horizon (Ayres, 1998). In fact, humans are not anywhere near meeting such requirements for a physically-sustainable system. We have no machines for filtering atomic particles from the atmosphere or oceans for reconstruction to replace essential ores, let alone ones which can do so while replacing all the materials they dissipate in the process and as they themselves decay. So in practice dissipation of ores and running down of useful energy sources (exergy), while creating all-pervasive pollution, are major problems posing ultimate limits. Indeed the rush to use these sources means the transition to a world which is of the physically sustainable type will be thrust upon future generations rather than achieved via a planned process. The great hope of the mainstream economic tradition is that prices will send signals to which producers will respond with substitution away from the increasingly unavailable resources. Yet, such economics is based on mechanistic equilibrium theories which bare little relationship to reality and cannot explain the evolution of technological change. Why mainstream economists, who have no theory to address past transitions, should predict a smooth future transition in the face of resource and exergy depletion, appears explicable more as a matter of blind faith than economic science.

In the absence of the means to re-concentrate dispersed ores a prudent approach would be to avoid their frivolous use. Of course what is frivolous, and whether a minority of humans should have a big all-consuming party while others starve, are value judgments of a most fundamental kind, not dictates from physical laws. Georgescu-Roegen (2009 [1975]) extrapolated from his interpretation of classical entropy as to the desirability of degrowth and avoiding luxury items constituted of metals which future generations would need for basic food production. Clearly physical laws only point to implications, they do not make ethical choices for us. Similarly the size of human population, type and scale of pressures placed upon systems, time allowed for change and rate of imposed change, are all matters for human judgment (if those responsible were able to stand back and use some).

One caveat to classical entropy is the neglect of self-organising systems arising to make use of available energy, i.e. organisation from disorder (Schneider and Kay, 1994). These systems include ecosystem functions but also geo-physical systems, such as climate regulation and ocean current circulation. Unfortunately some humans are destroying the ability of existing self-organising systems to operate. In addition, these systems fall outside the economic model of what is valued because they are not exchanged in market transactions. Ayres (1998; 2004) proposes starting to take account of what we are doing using measures of exergy, and others have suggested similar energy based approaches to measuring ecosystem health (Schneider and Kay, 1994).

Ecosystems' function, structure and value

The idea of stable equilibria is a fallacy. At the ecosystem level change is an ongoing reality and always has been, but human induced change is qualitatively and quantitatively different. Landscape modification, climate change and/or social developments all disturb ecosystem structure and function. The five main direct causes of biodiversity loss and degradation of ecosystems are: land use change, pollution, climate change, resource depletion and invasive alien species. All these factors are structurally part of current industrial economic systems with their focus on capital accumulation, appropriation of resources, global trade and innovative technology.

That humans are changing ecosystems is not in question, contestation is over the extent of human control, potential irreversibility and surprise, and consequences both bio-physically and in social, psychological and value terms. The idea that humans can recreate and restore ecosystems to their historical form (e.g. by invasive species removal) is popular enough. However, human inability and ignorance, plus the characteristics of the Holling cycle, imply that the outcomes are more likely to be novel ecosystems that are different from and cannot be restored to historic ones. Novel ecosystems may also arise from planned creation which can take on a variety of forms. For example, farming involves controlling non-human nature to establish specific ecological functions for human productivist ends. More recently the idea of promoting specific, typically singular, functions has moved to the planetary level as a means for survival e.g., carbon sequestration. Novel ecosystems may also arise from maintaining specific species or aesthetics because other aspects are simply ignored, and hence a new structure results. Then there is the whole area of compensation for loss, where totally different ecosystems, often in different locations, are created to justify destruction elsewhere.

How ecosystems functions are conceptualised and valued becomes a core concern. Regarding ecosystems as service providers facilitates regarding all change as good, because novelty can be described as supplying new ecosystem services. The central issue is how commensurable are new and old. The use of arguments to justify ecosystem destruction and re-creation is pervasive in the development of economic instruments for offsetting deliberately created damages, such as emissions trading (Spash, 2010), biodiversity offsetting (Spash, 2015), and species and ecosystem banking (Spash, 2011). Corporations and their financial backers, engaged internationally in resource extraction, have been particularly keen on seeing an "anything goes" policy, justified by commensuration of loss and gain. This has been supported by arguments that the worth of ecosystems can be converted into monetary values based on individual preferences (Spash, 2008).

An alternative is to focus explicitly on ecosystems functions, but this does not avoid commensuration and value judgements. A particular problem is where functional goals take priority over historical and compositional ones in ecosystem management. The contention is that ecosystem functions should be changed in novel ways to meet ecological crises, and traditional preservation goals should be dropped because they will prevent adaptation. Such logic is found in promotion of the bioeconomy, mainstream climate change mitigation and geoengineering. Desjardins, Donhauser and Barker (2019) identify a mechanistic approach to natural processes in such policy proposals, which also adopt a central aim of maintaining economic growth and industrial "development". Instead they argue for ecological integrity and value of place assessed through complex, multi-dimensional indices, rather than simple proxies. Such complex multidimensional evaluation severely restricts commensurability and means directly opposing economic and business logic based on bulldozing biodiversity and erasing ecosystems for monetary gain.

What cannot be avoided is the role of values and judgement. The aims of maintaining historical continuity, social-ecological relationships and a place for non-human autonomy sit uneasily with the values and institutions of price-making markets, love of money and capital accumulation. Contention over the values of modernity have always been evident when it comes to environmental concerns, and attempts to remove values for hegemonic conformity merely create the contradictions of new environmental pragmatism in the modern environmental movement. A place for "other values" is evident in the "rewilding" movement that includes a radical non-anthropocentric stance aimed at giving back autonomy to non-human nature (Gammon,

2018). This demands a reinterpretation of landscape and history, as well as the relations between humans and their environment, and thus challenges identities that are historically based (Drenthen, 2018).

Values are constitutive of human identity and reproduced (or not) through human practice. There are then real conflicts between the values of modernity promoted by industrialised technologically driven economies and other types of economies. Technology has become a force in itself that forecloses any notion of ends that would challenge what technology itself favours. As a hegemonic discourse it has real impacts on the world, motivating practices that eradicate human-independent entities from the surface of the Earth (Vetlesen, 2015, pp. 161-162).

Humans hold plural values that are in regular conflict. If economist wish to have any scientific credibility they can no longer continue the pretence that humans are preference utilitarians, or even purely consequentialists. In addition, the pretence that their work and its conceptualisations have no value implications and are merely factual, in some naïve objectivist sense, needs to be dropped as equally fallacious.

What type of economics and what type of economy?

That the global economy needs to change to avoid social ecological collapse, poses the problem of how and what sort of economics might help? There are three interrelated research questions. First, what is understood as being the current social ecological and economic reality and the causal mechanisms creating crises? Second, how can the current system be transformed, i.e. what are the barriers and enablers? Third, what is the goal of transformation, i.e. what kind of society is desirable?

What then is the point of the growth economy that modern economics tries so hard to sustain? Keynes advocated growth to avoid imminent social and economic collapse leading to international instability and war due to high unemployment (Spash and Schandl, 2009). Keynes (1930) outlined his vision in an article entitled "Economic possibilities for our grandchildren". He defined the economic problem as removing the struggle for meeting subsistence needs, a definable goal with an endpoint. His means of transformation was 100 years of economic growth (not an end in itself). Keynesians remain apologists for capitalism and the growth economy, although often growth for them this seems to have become an objective not a means. The ecological economist Tim Jackson (2009), as noted earlier, argues in line with Keynes, requiring the growth economy to transform society *before* there can be "prosperity without growth".

For Keynes the future goal was a leisure society sustained by the accumulated capital (ignoring maintenance requirements). Although he had doubts about this utopia when looking at the leisure class of his contemporaries. Worse, he recognised his transformative economic growth society would require empowering the worst of human values (i.e., greed, avarice, usury, the desire for ever more money) and people (i.e., those with "semi-criminal, semi-pathological propensities"). He had absolutely no answer as to what could be done after 100 years had been spent pretending "that fair is foul and foul is fair; for foul is useful and fair is not" (Keynes, 1930, p. 97). Under neoliberalism, the values Keynes apparently despised so much have been made into norms, supported by the institutions of private and public enterprise.

Concepts of "sufficiency" and "the good life for all" are challenges to how economic systems have been developing under Keynes growth imperative. Keynes recognised that affluence would not inform how "the art of life itself" should be conducted. Productivism makes life into labouring for a wage to survive and love of money into a virtue. While love of money results in people "which one hands over with a shudder to the specialists of mental disease" (Keynes, 1930, p. 97). Today, these are some of the most powerful people in the world.

The Polanyian double movement summarises the same tension between protecting and reacting against market capitalism (Polanyi, 1944, e.g., Chapter 11). Keynesian policy faces the dilemma of promoting this system, while also requiring major government intervention to control boom-bust cycles, and criticising and removing "market incentives" in the form of unemployment and bankruptcy. The welfare state was a necessary reaction to the social effects of unregulated market capitalism of the 19th Century and, more generally, the commodification of labour (Burawoy, 2015). Its gradual deconstruction by neoliberalism has led to a situation similar to the social and economic crises of the 1920s and 1930s, including the political encouragement of nationalism and fascism. So unsurprisingly, returns to the Keynesian policies of the "golden age" (1950 to 1973) are back on the agenda, but with the additional aim of trying to address the ecological crises. A currently prominent example is the Green New Deal. Principally the target is carbon emissions and the concern is how to finance policy initiatives while creating growth, jobs and more equality. Changing the conditions under which capitalism operates is what has made it resilient in the face of change, and this may be part of developing a future regulatory regime (Dannreuther and Petit, 2012). However, this plan for a new, Green, fully employed, productivist, capitalist, growth economy considers none of the causal mechanism that generate lifestyles of unsustainable consumption and involves no analysis of the structure of material and energy throughput of the existing or revamped capitalist system. Rather than the necessary radical change, a Gramscian "passive revolution" is offered, which reinforces and facilities the preservation of the hegemonic system.

Even those who strongly criticise growth can be found defending market capitalism. A prime example is the steady-state economy promoted by Herman Daly (1973; 1992). This recommends a monetary, price-making market, capitalist economy that operates in equilibrium at an "optimal" scale to stay within limits to avoid ecological disaster. Scale, while important, fails to address the issues highlighted in our coverage of ecosystems or the qualitative properties of pollutants (e.g. toxic waste, radiation, plastics, hormones). Social problems are limited to inequitable income distribution which fails to get to the heart of the social organisation of production. Most fundamentally there remains the contradiction of maintaining the social economic institutions of capital-accumulation while deconstructing economic growth. Indeed, in reply to the criticisms of Smith (2010), Daly (2010) has confirmed his preference for constrained markets over centralised planning with the aim of achieving allocative efficiency. He has long been an advocate for tradable permits markets, even for the allocation of rights to give birth (Daly, 1974). His apologia for capitalism is why some see the steady state as a Trojan horse for neoclassical economic thinking (Pirgmaier, 2017). Others believe they can adopt Daly as a mainstream economist (Auffhammer, 2009), which would clearly be difficult (Spash, 2013). However, there is much confusion as to what an alternative economics is all about with the two main ecological economics textbooks – (Common and Stagl, 2005; Daly and Farley, 2004) – both strongly supporting the basic validity of neoclassical economics.

The point of these critical reflections is that the structural and multiple causal mechanisms creating social ecological crises are not being addressed and cannot be addressed by neoclassical economics anymore than maintaining market capitalism will solve our problems. The major contribution of Daly, like his teacher Georgescu-Roegen, has been to emphasise the importance of biophysical reality for the operations of any economy. However, the move away from "growth=development", "growth removes poverty", "growth is necessary", and towards an economy without growth, requires more than income redistribution and limits on scale. Neither is this a simple matter of implementing market based policies or subsidising corporate development of Green technology. The core is how social provisioning is undertaken, within which institutional arrangements and for what ends.

In order to answer these questions, requires a research agenda that understands the social metabolism of an economy (see Gerber and Scheidel, 2018; Giampietro et al., 2009; Krausmann, 2017). That is, in the same way that the biological metabolism of a human necessarily needs inputs and outputs to maintain itself, so does society. Yet, society can be structured in different ways with different material and energy requirements. Prioritising reductions in material and energy throughput to sustain systems over a long time period means using simpler technologies and less automatised production systems that can be maintained by the users with readily available materials and without complex technical knowledge, i.e., appropriate technologies. The problem with Green economies, Green New Deals and Green revolutions is that they pay no attention to the structural relationships nor the requirements of the associated technologies, let alone the military interventions that maintain their supply chains. However, there is more to the structure of society than materials and energy. There are the values a society upholds and, through its practices, reproduces.

This is why there is no such thing as a value-neutral technology. The transformation of social practices by technology is clear to anyone who looks around them, from modes of transport to means of communication to work life. Technology additionally brings with it strong uncertainty (ignorance and indeterminacy), surprise, lock-in and social change. The values it entails relate not just to human relationships but also human to non-human and, in modernity the most neglected of all, non-human to non-human relationships. Technology is inherently anthropocentric and typically about human dominance over nature. Yet the rhetoric surrounding technology, innovation and growth is a better world for all.

The "Green revolution" in the 1960s, and the later push to use biotechnology and genetically modified organisms in food production, were undertaken in the name of "feeding the world". Yet, as Sen (1986) explained, famines have not occurred due to lack of food but due to lack of ability to pay, or actually pay high enough, in a monetary system of profit making. More commodification of nature, price-making markets, technology and capitalist growth do nothing to address this systemic problem, rather the exact opposite. A basic fact is that the number of undernourished people has remained at approximately 800 million since the mid-1990s (FAO, 1996; FIAN, 2018), although food production has been high enough to feed the whole world. The aim of *sufficient* food to feed the world is fundamentally at odds with the current systems that create excess and waste for profit, while others starve. From Western obesity to third world starvation, no one gets a good life.

Economics, to be of use for the future, must address how to meet basic needs through social provisioning, not how to create markets for profit making. Billions suffer deprivation of food, water, shelter and sanitation. The variety of economies that might operate to address these issues is not even on the research agenda. Instead a one-size-fits-all approach is backed by simple quantitative minimum standards that reduce the human condition to a common metric that ignores culture and meaning (Sachs, 2015 [1999], pp. 9–10). In contrast, needs can be associated with contextual satisfiers that are culturally specific and signify the diversity and difference that gives meaning to people's lives (Rauschmayer and Omann, 2017). At the same time, that needs can be met by different satisfiers allows analysis and creation of alternative economies for social provisioning.

Concluding remarks

To suggest ways out of the current social ecological crisis, we need an economics that can lead us away from catastrophe rather than towards it. Such an economics needs to understand both how the current economy is working and impacting on ecosystems, how ecosystems work and the basic structural mechanisms of the natural world, as well as understanding potentials that could be built on to create new and different kinds of economic systems. Current projects of Green growth or a Green New Deal, unfortunately, do not live up to these criteria because they fail to conceptualise nature and environmental problems in their own terms.

John Stuart Mill believed that economics, as political economy, needed to be more than abstract theory and should be practical. In order to achieve that end, and contrary to the later development of economics, his *Principles*

> "treated Political Economy not as a thing by itself, but as a fragment of a greater whole; a branch of Social Philosophy, so interlinked with all the other branches, that its conclusions, even in its own peculiar province, are only true conditionally, subject to interference and counteraction from causes not directly within its scope: while to the character of a practical guide it has no pretension, apart from other classes of considerations" (Mill, 1874, pp. 236).

More than being this sort of interdisciplinary social science, we argue for economics to also connect to the natural sciences in order to understand the basic requirements for social provisioning and the reproduction of society.

References

Auffhammer, M. (2009) "The state of environmental and resource economics: A google scholar perspective." *Review of Environmental Economics and Policy*, 3 (2), pp. 251-269.

Ayres, R.U. (1998) "Eco-thermodynamics: economics and the second law." *Ecological Economics*, 26 (2), pp. 189-209.

Ayres, R.U. (2004) "On the life cycle metaphor: where ecology and economics diverge." *Ecological Economics*, 48 (4), pp. 425-438.

Ayres, R.U. and Warr, B. (2009) "Accounting for growth: The role of physical work." In: Spash, C.L. (ed.). *Ecological Economics: Critical Concepts in the Environment*. Abingdon: Routledge, pp. 427-459.

Baskin, J. (2015) "Paradigm dressed as epoch: The ideology of the anthropocene." *Environmental Values*, 24 (1), pp. 9-29.

Beckerman, W. (1974) *In Defence of Economic Growth*. London: Jonathan Cape.

Biswas, A.K. and Biswas, M.R. (1976) "Energy and food production." *Agro-Ecosystems*, 2 (3), pp. 195-210.

Collier, A. (1994) "Stratification and Emergence." In *Critical Realism: An Introduction to Roy Bhaskar's Philosophy*. London: Verso, pp. 107-134.

Common, M.S. and Stagl, S. (2005) *Ecological Economics: An Introduction*. Cambridge: Cambridge University press.

Dale, G. (2010) "'Disembedded' and 'always embedded' economies." In: Dale, G. (ed.). *Karl Polanyi: The Limits of the Market*. Cambridge: Polity Press, pp. 188-206.

Daly, H.E. (1968) "On economics as a life science." *Journal of Political Economy*, 76, pp. 392-406.

Daly, H.E. (1973) "*Towards a steady-state economy*." San Francisco Freeman.

Daly, H.E. (1974) "Economics of steady state." *American Economic Review*, 64 (2), pp. 15-21.

Daly, H.E. (1977a) *Steady-State Economics*. San Francisco, California: W H Freeman.

Daly, H.E. (1977b) "Steady state and thermodynamics." *BioScience*, 27 (12), pp. 770-771.

Daly, H.E. (1992) *Steady-State Economics: Second Edition with New Essays*. London: Earthscan.

Daly, H.E. (2010) "The operative word here is 'somehow'." *Real-World Economics Review*, September (54): 103. Available from: http://www.paecon.net/PAEReview/issue54/Daly54.pdf. Accessed:12 jan 2013

Daly, H.E. and Farley, J. (2004) *Ecological Economics: Principles and Applications*. Washington: Island Press.

Dannreuther, C. and Petit, P.P. (2012) "Introduction to the special issue on the regulation approach and the contemporary crisis." *Capital & Class*, 37 (1), pp. 3-4.

Davis, M. (2006) *Planet of Slums*. London: Verso.

Desjardins, E., Donhauser, J. and Barker, G. (2019) "Ecological historicity, functional goals and novelty in the anthropocence." *Environmental Values*, 28 (3), forthcoming.

Drenthen, M. (2018) "Rewilding in layered landscapes as a challenge to place identity." *Environmental Values*, 27 (4), pp. 405-425.

FAO (1996) "Rome Declaration on World Food Security." World Food Summit. Rome: Food and Agriculture Organisation.

FIAN (2018) FIAN International Annual Report 2017. Heidelberg: FoodFirst Information and Action Network (FIAN), 24.

Gammon, A.R. (2018) "The many meanings of rewilding: An introduction and the case for a broad conceptualisation." *Environmental Values*, 27 (4), pp. 331-350.

GCEC. (2014) "Better Growth Better Climate: The New Climate Economy Report; The Synthesis Report." In: Calderon, F., Oppenheim, J. and Stern, N.H. (eds). Washington, D.C.: The Global Commission on the Economy and Climate, pp. 1-72.

Gemici, K. (2008) "Karl Polanyi and the antinomies of embeddedness." *Socio-Economic Review*, 6 (1), pp. 5-33.

Georgescu-Roegen, N. (1971) *The Entropy Law and the Economic Process*. Cambridge, Massachusetts: Harvard University Press.

Georgescu-Roegen, N. (2009 [1975]) "Energy and Economic Myths." In: Spash, C.L. (ed.). *Ecological Economics: Critical Concepts in the Environment, 4 Volumes*. London: Routledge, pp. 328-373.

Gerber, J.-F. and Scheidel, A. (2018) "In search of substantive economics: Comparing today's two major socio-metabolic approaches to the economy – MEFA and MuSIASEM." *Ecological Economics*, 144 (February), pp. 186-194.

Giampietro, M., Mayumi, K. and Ramos-Martin, J. (2009) "Multi-scale integrated analysis of societal and ecosystem metabolism (MuSIASEM): Theoretical concepts and basic rationale." *Energy*, 34 (3), pp. 313-322.

Hall, P.A. and Gingerich, D.W. (2009) "Varieties of capitalism and institutional complementarities in the political economy: An empirical analysis." *British Journal of Political Science*, 39 (July), pp. 449-482.

Hall, P.A. and Thelen, K. (2008) "Institutional change in varieties of capitalism." *Socio-Economic Review*, 7 (1), pp. 7-34.

Holling, C.S. (2009 [1986]) "The resilience of terrestrial ecosystems: local surprise and global change." In: Spash, C.L. (ed.). *Ecological Economics*. London: Routledge, pp. 64-103.

Holling, C.S., Schindler, D.W., Walker, B.W. and Roughgarden, J. (1995) "Biodiversity in the functioning of ecosystems: An ecological synthesis." In: Perrings, C., Mäler, K.-G., Folke, C., Holling, C.S. and Jansson, B.-O. (Eds). *Biodiversity Loss: Economics and Ecological Issues*. Cambridge: Cambridge University Press, pp. 44-83.

Jackson, T. (2009) *Prosperity without Growth: Economics for a Finite Planet*. London: Earthscan.

Jaeger, C.C., Paroussos, L., Mangalagiu, D., Kupers, R., Mandel, A. and Tàbara, J.D. (2011) "A New Growth Path for Europe: Generating Prosperity and Jobs in the Low-Carbon Economy: Synthesis Report." Potsdam: European Climate Forum, 27.

Kapp, K.W. (1965) "Social Costs in Economic Development." In: Sicat, G.P. (ed.). *Economics and Development*. Quezon City: University of Philippines Press.

Kapp, K.W. (1969) "On the nature and significance of social costs." *Kyklos*, 22 (2), pp. 334-347.

Kapp, K.W. (1970) "Environmental disruption and social costs: Challenge to economics." *Kyklos*, 23 (4), pp. 833-848.

Kapp, K.W. (1971) *The Social Costs of Private Enterprise*. New York: Schocken Books.

Kapp, K.W. (1978 [1963]) *The Social Costs of Business Enterprise*. Nottingham: Spokesman.

Kay, J.J., Regier, H.A., Boyle, M. and Francis, G. (1999) "An ecosystem approach for sustainability: Addressing the challenge of complexity." *Futures*, 31 (7), pp. 721-742.

Keynes, J.M. (1930) "Economic possibilities for our grandchildren." *Nation and Athenaeum*, 48 (11 and 18), pp. 36-37, 96-98.

Kneese, A.V., Ayres, R.U. and d'Arge, R.C. (1970) *Economics and the Environment: A Materials Balance Approach*. Washington, District of Columbia: Resources for the Future.

Krausmann, F. (2017) "Social Metabolism." In: Spash, C.L. (ed.). *Routledge Handbook of Ecological Economics: Nature and Society*. Abingdon: Routledge, 108-118.

Meadows, D.H., Meadows, D.L., Randers, J. and Behrens III, W.W. (1972) *The Limits to Growth*. New York: Universe Books.

Mill, J.S. (1874) *Autobiography*. London: Longmans, Green, Eeader and Dyer.

Mirowski, P. (1989) *More Heat than Light: Economics as Social Physics, Physics as Nature's Economics*. Cambridge: Cambridge University Press.

Pimentel, D., Hurd, L.E., Bellotti, A.C., Forster, M.J., Oka, I.N., Sholes, O.D. and Whitman, R.J. (1973) "Food production and the energy crisis." *Science*, 182 (4111), pp. 443-449.

Pirgmaier, E. (2017) "The neoclassical Trojan horse of steady-state economics." *Ecological Economics*, 133 (March), pp. 52-61.

Polanyi, K. (1944) *The Great Transformation*. New York/Toronto: Rinehart & Company Inc.

Polanyi, K. (1977a) "The Economistic Fallacy." In: Pearson, H.W. (ed.) *The Livelihood of Man*. New York: Academic Press, pp. 5-17.

Polanyi, K. (1977b) "The Economy Embedded in Society." In: Pearson, H.W. (ed.) *The Livelihood of Man*. New York: Academic Press, pp. 47-56.

Rauschmayer, F. and Omann, I. (2017) "Needs as a Central Element of Sustainable Development." In: Spash, C.L. (ed.) *Routledge Handbook of Ecological Economics: Nature and Society*. Abingdon: Routledge, pp. 246-255.

Rockström, J., Steffen, W., Noone, K., Persson, Å., Chapin, F.S., Lambin, E., Lenton, T.M., Scheffer, M., Folke, C., Schellnhuber, H., Nykvist, B., De Wit, C.A., Hughes, T., van der Leeuw, S., Rodhe, H., Sörlin, S., Snyder, P.K., Costanza, R., Svedin, U., Falkenmark, M., Karlberg, L., Corell, R.W., Fabry, V.J., Hansen, J., Walker, B., Liverman, D., Richardson, K., Crutzen, P. and Foley, J. (2009) "Planetary boundaries: exploring the safe operating space for humanity." *Ecology & Society*, 14 (2), pp. 1-32.

Sachs, W. (2015 [1999]) *Planet Dialectics: Explorations in Environment and Development*. London: Zed Books.

Schneider, E.D. and Kay, J.J. (1994) "Life as a manifestation of the 2nd Law of Thermodynamics." *Mathematical and Computer Modelling*, 19 (6-8), pp. 25-48.

Sen, A. (1986) *Poverty and Famines: An Essay on Entitlement and Deprivation*. Oxford, England: Clarendon Press.

Smith, R. (2010) "Beyond growth or beyond capitalism?" *Real-World Economics Review*, June (53), pp.28-42. Available from: http://www.paecon.net/PAEReview/issue53/Smith53.pdf. Accessed:12th January 2013

Spash, C.L. (2002a) "Calculating the Costs and Benefits of Greenhouse Gas Control." In: Spash, C.L. (ed.). *Greenhouse Economics: Value and Ethics*. London: Routledge, pp. 153-183.

Spash, C.L. (2002b) "Loading the dice?: Values, opinions and ethics." In: Spash, C.L. (ed.). *Greenhouse Economics: Value and Ethics*. London: Routledge, pp. 184-200.

Spash, C.L. (2002c) "Strong Uncertainty: Ignorance and Indeterminacy." In: Spash, C.L. (ed.). *Greenhouse Economics: Value and Ethics*. London: Routledge, 120-152.

Spash, C.L. (2007a) "The economics of climate change impacts à la Stern: Novel and nuanced or rhetorically restricted?" *Ecological Economics*, 63 (4), pp.706-713.

Spash, C.L. (2007b) "Problems in economic assessments of climate change with attention to the USA." In: Erickson, J. and Gowdy, J. (Eds) *Frontiers in Environmental Valuation and Policy*. Cheltenham, UK/Northampton, MA, USA: Edward Elgar Publishing Ltd.

Spash, C.L. (2008) "How much is that ecosystem in the window? The one with the bio-diverse trail." *Environmental Values*, 17 (2), pp. 259-284.

Spash, C.L. (ed.) (2009) *Ecological Economics: Critical Concepts in the Environment, 4 Volumes*. London: Routledge.

Spash, C.L. (2010) "The brave new world of carbon trading." *New Political Economy*, 15 (2), pp. 169-195.

Spash, C.L. (2011) "Terrible economics, ecosystems and banking." *Environmental Values*, 20 (2), pp. 141-145.

Spash, C.L. (2013) "Influencing the perception of what and who is important in ecological economics." *Ecological Economics*, 89 (May), pp. 204-209.

Spash, C.L. (2014) "Better Growth, Helping the Paris COP-out?: Fallacies and Omissions of the New Climate Economy Report." Vienna: Institute for Environment and Regional Development.

Spash, C.L. (2015) "Bulldozing biodiversity: The economics of offsets and trading-in Nature." *Biological Conservation*, 192 (December), pp. 541-551.

Spash, C.L. (2017) "Social Ecological Economics." In: Spash, C.L. (ed.) *Routledge Handbook of Ecological Economics: Nature and Society*. Abingdon and New York: Routledge, pp. 3-16.

Spash, C.L. and Ryan, A. (2012) "Economic schools of thought on the environment: Investigating unity and division." *Cambridge Journal of Economics*, 36 (5), pp. 1091-1121.

Spash, C.L. and Schandl, H. (2009) "Challenges for Post Keynesian Growth Theory: Utopia meets Environmental and Social Reality. " In: Holt, R.P.F., Spash, C.L. and Pressman, S. (eds). *Post Keynesian and Ecological Economics: Confronting Environmental Issues*. Cheltenham: Edward Elgar, pp. 47-76.

Turner, G.M. (2012) "On the cusp of global collapse? Updated comparison of the Limits to Growth with historical data." *Gaia-Ecological Perspectives for Science and Society*, 21 (2), pp. 116-124.

Vatn, A. (2005) *Institutions and the Environment*. Cheltenham: Edward Elgar.

Walters, C.J. and Holling, C.S. (2009 [1990]) "Large-scale management experiments and learning by doing." In: Spash, C.L. (ed.) *Ecological Economics*. London: Routledge, pp. 104-119.

Chapter 16

How to achieve the Sustainable Development Goals within planetary boundaries by 2050

Per Espen Stoknes

Rising to the grand challenge

What will it take to bring about human prosperity and equity within a safe biosphere? If the world is serious about the Sustainable Development Goals (SDGs), and thus the need for a truly integrated prosperous and peaceful people-planet trajectory for development, what will it take to succeed? Is it at all possible to transition the world to global sustainable development as it is now defined: attaining the SDGs within Earth's planetary boundaries (PBs) – through conventional means of economic development? What potential trade-offs and synergies do societies face when taking a truly systemic approach to the SDGs? And, most importantly, what are the transformational requirements to succeed in attaining human prosperity within a safe operating space on Earth?

This paper will provide an adapted excerpt from a research project[1] aiming to answer the above questions. The project, named SDGinPB for short, developed a transparent, integrated and easily understandable modelling framework, which we call Earth3 (See Box 1) to give science-based answers.

Pathway analysis for achieving SDGs within PBs

The Earth3 model calculates the effects on the 17 SDGs of major socio-economic developments for seven regions of the world, and assesses the status of global environmental pressures on the nine PBs. In essence, it is a tool to answer the question: will given policies help the world move in an inclusive direction while staying within Earth's safe operating space?

[1] The full report is available here: Randers, J., Rockström, J., Stoknes, P. E., Goluke, U., Collste, D., & Cornell, S. (2018a). *Transformation is Feasible - How to Achieve the 17 SDGs within planetary boundaries.* BI, Stockholm Resilience Center. Retrieved from https://www.stockholmresilience.org/ So far, two research articles have been released in pre-print version, while in peer-review for publication Randers, J., Rockström, J., Stoknes, P. E., Goluke, U., Collste, D., & Cornell, S. (2018b). *Achieving the 17 sustainable development goals within 9 planetary boundaries,* https://doi.org/10.31223/osf.io/xwevb and Collste, D., Randers, J., Goluke, U., Stoknes, P. E., Cornell, S., & Rockström, J. (2018) *The Empirical Bases for the Earth3 Model: Technical Notes on the Sustainable Development Goals and Planetary Boundaries* https://doi.org/10.31223/osf.io/ephsf. Earth3 includes a simple system dynamics model, ESCIMO (Earth System Climate Interpretable Model) which is able to reproduce the main output from more complex climate models. ESCIMO represents the main causal mechanisms at work in the Earth system and is able to reproduce the broad outline of climate history from 1850 to 2015. See Randers, J., Golüke, U., Wenstøp, F., & Wenstøp, S. (2016) "A user-friendly earth system model of low complexity: the ESCIMO system dynamics model of global warming towards 2100" *Earth System Dynamics*, 7(4), 831–850. https://doi.org/10.5194/esd-7-831-2016 .

Figure 1 What pathways can lead to achieving the SDGs within planetary boundaries in 2050?

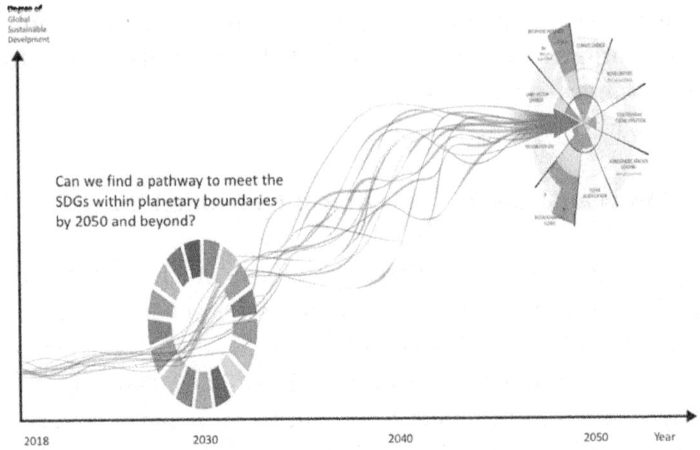

Box 1: About the Earth3 model

Earth3 is a Global Systems Model linking socio-economic and biophysical processes. It builds on more than 100,000 historic and new data points, from existing databases all over the world.

Earth3 first calculates the main socio-economic developments (GDP, population, distribution, energy use, etc.). Then it calculates estimates of how many of the 17 SDGs can be achieved by adopting certain policies in seven regions of the world. It also gives estimates of the status of global pressures on nine planetary boundaries for different world-development trajectories to 2030 and 2050.

Our modelling approach is described in Randers et al. (2018b).

Data sources are further described in appendix 2 of the "Transformation is Feasible" report and in Collste (2018): "The empirical basis for Earth3 model system". How we model the Earth's biophysical systems, including 15 identified tipping points, are described in Randers (2016) "A user-friendly earth system model of low complexity: the ESCIMO system dynamics model of global warming towards 2100."

The SDGinPB project has focused on calculating the effects of policy actions needed for meeting the globally agreed development goals within the safe operating space of a stable planet. Earth's safe operating space is defined through the nine planetary boundaries boundaries – global quantifications of human-caused environmental changes, where continued pressure risks destabilizing the long-term dynamics of the Earth system (see figure 2).

The grand ambition quantified: the SDG success score and PB safety margin

Our guiding question is: **how can the world succeed in achieving the Sustainable Development Goals within planetary boundaries?**

New studies show that currently no country meets the basic needs for its citizens at a globally sustainable level of resource use.[2]

[2] O'Neill, D. W., Fanning, A. L., Lamb, W. F., & Steinberger, J. K. (2018). A good life for all within planetary boundaries. *Nature Sustainability*, *1*(2), 88–95. https://doi.org/10.1038/s41893-018-0021-4

Figure 2 Nine planetary boundaries (PB) from Rockström et al. (2009) and Steffen et al. (2015). The dotted area represents the safe operating space. The greater the human-caused perturbation, the greater the risk of large-scale abrupt, and irreversible Earth system changes.

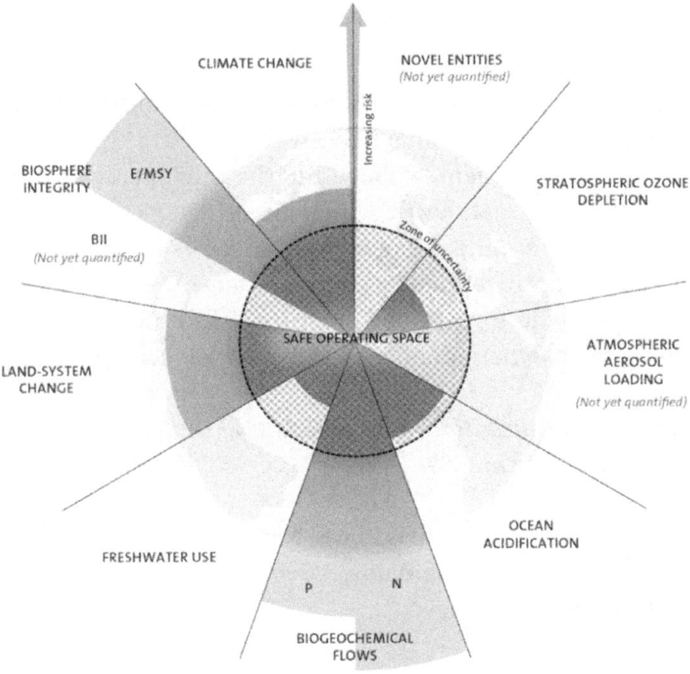

To study the whole world's progress on the SDGs into the future, we calculate the number of SDG achieved every year, the "SDG success score". The SDG success score thus goes from 0 to 17.[3] This is done for each of seven regions in the world as well as aggregated for the whole world weighted by population. We also calculate how this progress impacts the Earth's safety margin. To see if any SDG achievement is inside the planetary boundaries, we calculate how this progress impacts the Earth's safety margin over time. Earth's safety margin goes from 0 to 9, in steps of 0.5. If all PBs are in the safe zone (green), the safety margin is nine. If all PBs are violated (high risk = red), the safety margin is zero. Which would give a high probability of irreversible decline in Earth's life supporting systems and possibly societal collapse.

We assume that most of humanity would agree that a SDG success score of 17 with a PB safety margin of 9 is where we all want to be, whatever the population size is. The vision of 9 billion people living well on one planet,[4] can now be expressed more concretely as 9 billion people achieving 17 SDGs with Earth's 9 life-supporting systems in a safe state.

---

[3] We grade the SDG achievement in a simple way: An achieved goal (green) means 1 point. A goal that has passed the half-way target is 0,5 point (yellow). A red, i.e. not achieved goal is 0 points. See table 5.5 in appendix 1 of the *Transformation is Feasible* report for details on goals, chosen indicators and thresholds. Targets and thresholds are as far as possible based on Sachs et al. (2016). *SDG index & dashboards: A global report.* Bertelsmann Stiftung.

[4] World Business Council for Sustainable Development. (2011). *WBCSD - Vision 2050*. Retrieved Aug 27, 2018, from https://www.wbcsd.org/Overview/About-us/Vision2050/Resources/Vision2050-Road-Map

Figure 3 The UN 17 Sustainable Development Goals (SDGs), implemented by all the world's countries in 2016.

The SDGinPB project approach

In short, the project answers the research question by analysing the developments in all 17 SDGs, the 9 planetary boundaries across 7 regions of the world to 2050. To our knowledge, this is the world's first study to see if all SDGs can be reached within the PBs based on an integrated Global System Model.

The main types of input to our modelling approach are socio-economic data from 1980 to 2015 for all the world's countries. These include economic growth rates, population, education, health data, resource use and more aggregated into the regions. We use the most suitable publicly available databases to establish the historical trends.[5]

---

[5] One such central database is Feenstra, Robert C., Robert Inklaar and Marcel P. Timmer (2015), "The Next Generation of the Penn World Table"

The Earth3 model includes parameters that can reflect policy levers in many areas. The parameters can be seen as a "policy dashboard" for running the world model to 2050. There are levers per region to influence the expected a) Growth rates, b) Jobs, poverty and inequality levels, c) Energy use and composition, d) Food- and agriculture productivity, and finally e) Education, health and gender variables.

Based on this input, the Earth3 model can then calculate the SDG Success Score for each region and the Earth's common safety margin based on the state of the planetary boundaries.

The project explored four possible and plausible pathways to 2050. The exploration consisted of four scenario simulations of how the world can respond to the grand challenge, with each scenario giving both a regional and a world SDG as well as a global PB score. The four scenarios are all based on the same historic facts but are shaped by different policy and investment choices made in the coming decade(s).

We do not assign probability to the scenarios, which means they are not predictions. Some people may consider the first, business-as-usual scenario most likely and the fourth transformational scenario very unlikely. Others the opposite. We hope such foresight analysis will stimulate debate and create understanding about the long-term view on the SDGs, the synergies between them, and how they are systemically related. But based

---

American Economic Review, 105(10), 3150-3182, available for download at www.ggdc.net/pwt For more on the databases, see Collste (2018), ibid.

on historic and current trends, using the best socio-economic and biophysical data available, the modelling clearly shows that only the most transformational scenario points to a sustained higher and inclusive human wellbeing. This fourth, "Smarter" scenario achieves most of the SDGs while staying within most of the PBs. Thus, the results from our analysis show that only one out of the four pathways actually rises to the grand challenge.

Summary of four scenarios to 2050:

Through the scenarios we tested four different answers to the same research question: *"How can the world achieve the Sustainable Development Goals within planetary boundaries?"* The first answer comes from modelling how far the world will get by following business as usual to 2050. The second from simulating how far the world could get with faster economic growth. The third by pushing known policies harder toward sustainability. The fourth by calculating the scale of key transformational actions actually needed to get there. Below I give a short summary of the three first scenarios, and then a more detailed description of the fourth.

Scenario 1) *"Same":* how far will business as usual take the world to 2050?

This baseline scenario explores a future where the ***Same*** policies and actions are applied at the same pace into the future. Governments and industry will respond to technology, inequality and climate change in the conventional ways that the world has done over the last three decades. Despite rapid technological changes, digitalisation in particular, the data from the last decades shows that most rates of socio-economic change are slow.

Figure 4 SDG success score per region in the **Same** scenario. Regional SDG scores for 2010, 2030 and 2050 are shown.

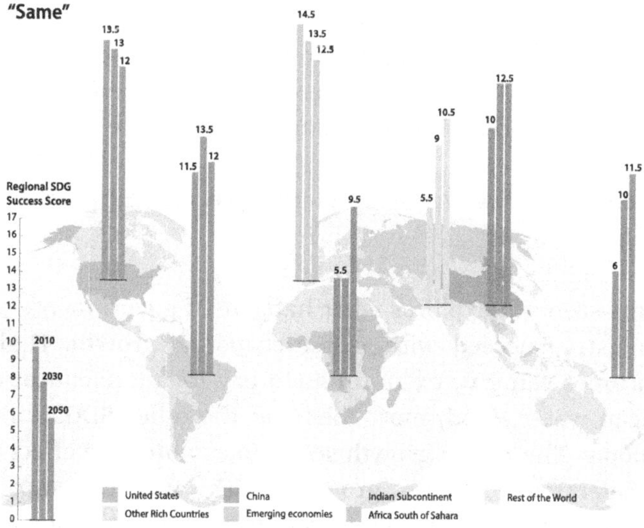

In a more-of-the-***Same*** world, there is even more *talk* about sustainability and SDGs, but in practice nations still continue to change at the very same pace. But this pace of progress proves insufficient to deliver on the SDG targets by 2030 nor 2050. The good news is that poverty and hunger is finally eradicated by 2050! However, the economy's large resource use and waste flows leads to more planetary boundaries in the red zone. This leaves many of Earth's life-supporting systems in a high-risk of irreversible decline, and people's prospects for wellbeing, particularly the poor, bleaker by 2050. In total, the world's SDG score only improves from 9 in 2015 to 11 in 2050. The main two reasons are that inequality continues to grow both within and between countries, and the second that total human footprints are too high. Planetary boundaries in high risk zones along with failing achievement on SDGs 13-15, pull scores downward to 2050. By responding to our new problems in the ***Same***, conventional ways, most people on Earth end up in a more precarious situation in 2050 than we are in 2018.

Scenario 2) ***Faster***: will accelerating economic growth help?

This scenario explores what happens if governments and industry succeed with faster economic growth. Higher incomes can give extra funds to pay for more education, clean water, food, more jobs and the other SDGs for all people. The ***Faster*** growth scenario explores the effects of accelerated economic growth all the way to 2050. To achieve this, governments ramp up conventional policy tools, such as increasing trade, innovations and investments, keeping corporate taxes and interest rates low. We model growth rates that are +1% higher in GDP per person than the historic trend, which makes the global economy significantly larger by 2050. In this way, higher incomes are available to solve the world's problems. But this approach only delivers a little bit better on the weighted SDG success score, from 8.8 in 2015 to 11.7 in 2050. Indeed, the planetary boundaries are more severely violated than in the ***Same*** scenario. Many people get very wealthy, but societies suffer even more destabilising inequality, and humanity as a whole undermines Earth's safe operating space by overexploiting nature's life-supporting systems. Earth's safety margin is down from 8 in 1980 to 3 in 2050 (see figure 10 below).

Scenario 3) *Harder*: what if both governments and industry try even harder to deliver on SDGs?

In this scenario, we explore where working harder for sustainability on all fronts will lead. The world's decision-makers focus real attention and energy on achievement of the SDGs. They allocate more funds to pay for more education, clean water, food, more jobs and the other SDGs for all people. In this way, governments strengthen their conventional policy tools, starting in 2018 and soon do on average 30% more rapid progress on SDG-achievement than they did in the 1990–2015 period. More workforces and funds are redirected to projects that help achieve SDGs and/or reduce the pressure on PBs. But by delivering on the SDGs one by one in a piecemeal way, department by department and ministry by ministry, there are many trade-offs. And by 2040 the planetary boundaries are still under strong pressure, which leads to flat SDGs scores from 2030 to 2050. Many regions still struggle with destabilising inequality that undermines the sustainability policies. The ***Harder*** pathway leads humanity to still undermine Earth's life-supporting systems, even if less so than in ***Same*** or ***Faster***.

Scenario 4) *Smarter*: what if governments and industry actually choose transformational actions?

This scenario explores five bold transformations in our societies and economies to see whether these can bring the human world to a desired future on Earth. This is a challenge-and-response scenario which describes the extent of what is needed to "hit target". Rather than repeating the ***Same*** conventional solutions, growing ***Faster*** or trying ***Harder***, this scenario explores what would happen if five bold, extraordinary actions were rapidly executed by decision- and policymakers in all regions. The ***Smarter*** scenario assumes that the world's countries and their leaders together, pushed by their citizens, become aware of the massive scale of the challenge ahead, and that the changes in mind-sets spread worldwide. Then, they rise to the challenge by implementing five turnarounds:

1. Rapid renewable energy growth – sufficient to halve carbon emissions every decade from 2020.
2. Accelerated productivity in food chains – improving resource productivity by +1%/year.
3. New development models in the poorer countries – which follow relevant models such as China, Scandinavia, Ethiopia or Costa Rica.
4. Active inequality reduction – ensuring that the richest 10% take no more than 40% of income.
5. Investment in education for all, gender equality, health, family planning – stabilising the world's population.

Despite initial criticisms for being too radical during the 2020s, in the ***Smarter*** scenario these five transformative efforts are then widely adopted, accelerated, and scaled over the coming decades. The scenario model runs indicate that these actions together create synergies that are capable of attaining (nearly) all SDGs while staying within (nearly all) planetary boundaries.

What are the transformational requirements to succeed?

As the focus has been to explore a scenario which leads to transformation, I will in the following deepen the description of this "Smarter" pathway to 2050. The 5 requirements to succeed mentioned above are all systemic interventions that hold the promise to achieve multiple of the SDGs simultaneously.

Smarter progress by focusing on human wellbeing
In this scenario, during the 2020s, the most governments increasingly accept that maximizing GDP as a first priority is not the best way to achieve sustainable human wellbeing. The EU's work on "Beyond GDP"[6], OECD's work on wellbeing[7] and China's work on *héxié shèhuì*[8] (a harmonious society) and the China dream, all reflect a deeper value shift in many societies. In the ***Smarter*** world, the broader objectives of society are seen as better

---

[6] EU Commission, (2013), *Progress on 'GDP and beyond' actions*, Commission Staff Working Document
http://ec.europa.eu/environment/enveco/pdf/SWD_2013_303.pdf
[7] OECD Statistics, (2018), *Measuring Well-being and Progress: Well-being Research*
http://www.oecd.org/statistics/measuring-well-being-and-progress.htm
[8] Joshua J. (2017) *China's Economic Growth: Towards Sustainable Economic Development and Social Justice.* Palgrave Macmillan, London

defined by the first generally agreed sustainability framework signed up to by the majority of the world's nations back in 2016: achieving all SDGs without endangering planetary ecosystem. This proved – in seen retrospect from 2050 – a paradigm shift for global development, when the move away from a sectorial approach to dealing separately with social, economic and environmental issues to a model of mutual leverage.

Recognizing the deep transformation required, more governments explicitly shift their long term purpose to *maximizing wellbeing and human capabilities*. They acknowledge that a conventional market-based growth approach has weak incentives to achieve SDGs; and none at all for protecting the commons, social welfare or any other non-economic values. In national decision-making, the wellbeing measure is accordingly widened from the previous function of just consumption, to also include the SDGs.

The "Smarter" pathway to 2050: How five transformational policies are rolled out

Rather than pushing for faster growth or working incrementally harder at each SDG separately, the ***Smarter*** approach happens through a systems transformation where five main policy initiatives start to create synergies and deliver on multiple SDGs at the same time.

1) Smarter Energy: accelerated renewables growth

A worldwide *rapid electrification* in power, transport, as well as heating and cooling, is rolled out. This happens by scaling up mainly solar and wind power, energy efficiency, distributed energy storage, electric vehicles, heat pumps and necessary distribution infrastructure, all digitized and integrated in smart grids to replace fossil fuels. Nearly all investments in fossil fuels (a historical average of 1.5%-2% of GDP per year[9]) are shifted to renewables and power infrastructure during the 2020s. The higher investments are driven both by a combination of demand-side pull as renewables start delivering higher profitability than fossils, with a government push through tougher regulations. This results in a doubling of the annual growth rates in wind, solar and other renewables during the 2020s.

Most countries also put in place bans on new fossil-fuel investments, including announcements during the 2020s of coming bans on sales of new fossil-fuel cars. Most regions adopt some form of the "Carbon Law"[10]: That means halving carbon emissions every decade, starting in 2020. This rapidly reduces global carbon emissions and at the same time eliminates human suffering by spreading affordable electricity to cities, slums and remote areas. In this field China takes the global lead, with strong policies for transforming coal-reliance to low-cost distributed renewables and electric mobility that make it profitable for other countries to follow. The direct use of fossil fuels and in buildings are replaced with electrification and smart system redesign. In this ***Smarter*** scenario, global carbon emissions fall from over 30 $GtCO_2$ in 2015 to 20 in 2030, 10 in 2040, and just 6 in 2050.

---

[9] IEA / OECD (2018) World Energy Investment 2018, , accessed 1 Aug. 2018 from https://www.iea.org/wei2018/

[10] Rockström, J., Gaffney, O., Rogelj, J., Meinshausen, M., Nakicenovic, N., & Schellnhuber, H. J. (2017). A roadmap for rapid decarbonization. *Science*, 355(6331), pp. 1269–1271. https://doi.org/10.1126/science.aah3443

The effect of this energy system transformation is that it starts to wean the world off fossil fuels and hits the nail on the clean energy goal (SDG7). Giving most people access to safe and clean energy creates a functioning energy democracy, which improves the development of many other SDGs (1, 2, 6, 8, 9, 11-13). There is, finally, "power to the people". It provides better access to light, education, clean water and communications. In addition to reducing climate change (13) it also helps fight poverty (1) and make more jobs (8). It makes innovations and infrastructure (9) more available, reduces food-waste and hunger by access to refrigeration and logistics. It helps making city air cleaner (11) by replacing combustion. In sum, universal access to cheap and clean electricity changes everything!

2) Smarter Food: accelerated shift to sustainable food chains

In the ***Smarter*** scenario, the world also accelerates the transformation to sustainable agriculture, linking production to better logistics that drives down food waste, as well as fertilizer and pesticide overuse. People shift their diets to more plant-rich foods which lowers the share of meat per person (particularly in richer countries).[11] The food system gets more direct links between food producers and consumers, i.e. direct delivery of easily available, affordable and nutritious foods that people actually need and want. This brings down food waste along the entire food chain, from soil to table.

---

[11] See p. 39 ff. in Hawken, P. (Ed.). (2017). *Drawdown: the most comprehensive plan ever proposed to reverse global warming*. New York, New York: Penguin Books.

New technology builds on the rapid development of digitalization of agriculture, cheap sensors, satellite monitoring and the Internet of Things to make real-time big data available to monitor the state of each field, river, crop and shop. Through better water management, total water use is brought within planetary boundaries. Intelligence embedded in water pipelines helps stop water loss from leakages and secures good water management in all river basins. It makes fresh-water pricing more accurate and feasible, giving incentives for better water efficiency. Biogas and composting replace landfills and surface run-off to the oceans, creating the capacity to recapture nitrogen and phosphorus and circulate these nutrients within bioregions.

These kinds of both low-tech and high-tech solutions enable increasingly regenerative agriculture to produce more food without any further land expansion. The release of bioactive nitrogen starts to decline. Climate-smart agriculture becomes a net carbon sink and draws down over one billion tons of carbon into the soil per year from 2040.

A less waste-full and more productive food system will also increase people's health as they get more nourishing and affordable food. With recycling of nutrients, it also improves clean water (SDG6), responsible consumption (SDG12), and reduces the pressure on climate change, life on land and life below water (SDG 13-15). In sum, all these improvements lower the footprint of the entire food chain by an extra 1% per year, relative to the ***Same scenario***.

3) Smarter Growth in poor countries: rolling out new development models

A higher growth rate is achieved in the world's poorest countries by increasing investment, strengthening institutions and allowing favorable trade arrangements in the early stages of industry development. The liberal market ideal is supplemented with various planned developments where certain industries that are of national interest are cultivated – inspired by role models of countries such as China, South Korea, Ethiopia, Scandinavia and Costa Rica. First Japan, then South Korea, Singapore and China managed to quadruple the GDP per person over thirty years. As other poor countries repeat these feats, they start providing each citizen with a reasonable standard of living. China has achieved an unprecedented duration of sustained economic growth and lifted hundreds of millions of people out of poverty in the process. (See box 2, on "the Chinese model".)

The Chinese model is preferred by many such countries over the Washington Consensus, which prescribes policies such as macroeconomic stabilisation, rapid economic opening with respect to both trade, finance and investment, and the expansion of market forces within the domestic economy. During the 2020s many of the world's poorer countries thus roll out forward-looking protectionist policies too, to raise standards of living by allowing their economies to catch up, and protect infant industries, without full immediate exposure to competition with advanced global industries in their home market in the beginning stages.[12] The effects in these countries are more rapid inclusive economic growth that lifts many millions more out of poverty quicker, and also delivers on hunger, jobs growth, clean water, better health, education, infrastructure (SDGs 2, 3, 4, 8, 9).

Box 2: What do we mean with "the Chinese model"?

"The Chinese model" is often equated with authoritarian capitalism – single-party rule combined with extensive state ownership and control over the economy. Others call it a political meritocracy[13] in contrast to democracy.

Rather than just authoritarianism dominating over markets and people, it seems there were many factors stimulating China's dynamism in the latest decades. Key factors were the introduction of some democratic qualities through bureaucratic reforms according to *long-term plans*, and Beijing's willingness to allow and direct local improvisation. In her research, Yuen Yuen Ang found that under Deng's rule: "Instead of trying to command their way to rapid industrialisation and growth, reformers focused on creating the right conditions for lower-level officials to kick-start development in their own communities using local resources."[14] Instead of only top-down commands, the country often leveraged local knowledge and resources, promoted diversity, and motivated and incentivized

---

[12] Reinert, E. S. (2008). *How rich countries got rich... and why poor countries stay poor*. London: Constable.
[13] Bell, D. (2015). *The China model: political meritocracy and the limits of democracy*. Princeton, New Jersey: Princeton University Press.
[14] Yuen Yuen Ang,(2018), "The Real China Model", *Foreign Affairs*, accessed 1 Aug 2018 https://www.foreignaffairs.com/articles/asia/2018-06-29/real-china-model

people to step up efforts and share ideas.

In short, with "the Chinese model", we refer in this report to the characteristics and conditions under which certain newer historic experiences in China – and in several other countries such as Ethiopia and Costa Rica – may have high relevance and serve as inspiration for the development of other countries. No such model is a perfect ideal to copy-paste; and each should be seen in the light of the other transformational strategies (further rapid transition from coal to renewables, reduction of inequality, etc). The rapid, intentional and positive change these models have delivered substantiate our claim that this kind of transformation is possible and shows how it can be done.

4) Smarter inequality reduction

Increasingly both rich and poor countries face the need to reduce growing unemployment and inequity. During the early years of the 2020s there is a series of political crises which are fed by broad protests and discontent among the public about the extreme unfairness of wealth inequality. A push for fairer wages and more progressive taxation succeeds at redistributing total output. Many developing countries intensify the domestic resource mobilization by improving their tax systems. As a result, there are funds for better service delivery and development for the majority.[15] There is, also in richer countries, growing accept of the recommendations from IMF[16] and OECD[17] to reduce inequality to enhance growth

[15] Lundstøl, O. (2018) 'Tax in Development: Towards a Strategic Aid Approach' *ICTD Working Paper* 77.
[16] IMF, 2017, The IMF's Work on Inequality,

and wellbeing. By shortening the work-year for everyone, it becomes possible to create and share more jobs, even in regions and sectors where there is low or no per capita GDP growth.

By 2025, there is broad and growing recognition among voters that it is of interest to national stability to ensure that the 10% richest take no more than 40% of income. Downward redistribution of wealth, work, and incomes through policies such as higher unemployment benefits and a shorter working year is the best way for businesses and banks to guarantee a stable economic future in the developed world, because it will put more money into the pockets of the poor. It allows the less well-off to spend more, which also improves conditions for business, investors, and the banking sector.

The funds raised by progressive taxation of income and wealth are also used to stimulate well-being by delivering on SDG achievement (particularly on health, education, infrastructure, sustainable cities and responsible consumption, i.e.SDGs 3, 4, 9, 11 and 12). Extensive downward redistribution efforts through more progressive taxation and unemployment benefits are stepped up in most countries during the 2020s. The historic trend of a falling median incomes since the 1980s is reversed starting in the 2020s. This proves conducive to regain more trust in government and stability in

---

https://blogs.imf.org/2017/02/22/the-imfs-work-on-inequality-bridging-research-and-reality /

[17] OECD, 2012, Reducing income inequality while boosting economic growth: Can it be done? https://www.oecd.org/eco/growth/49421421.pdf

politics, which strengthens institutions (16) and partnerships for the goals across national borders (17).

5) Smarter investment: education for all, gender equality, health, family planning

Figure 5 In Smarter, in an overall richer world by 2050, the regions gradually succeed in reducing income inequality to the before-1990 levels, at which the top 10% richest take <40% of total incomes.

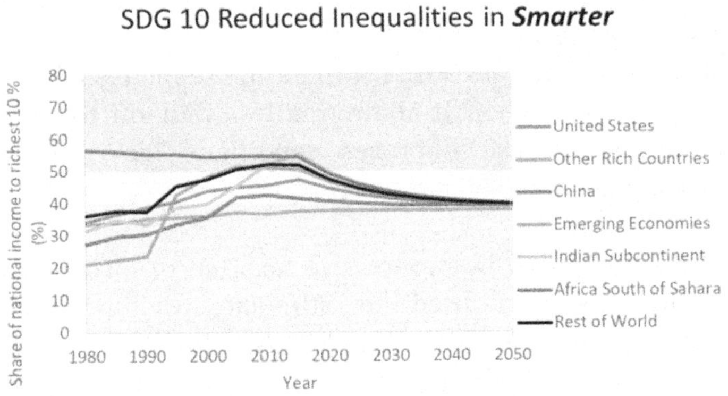

Global funds that focus on education, especially for all women, are strengthened. This gives women broader opportunities for autonomy and work. In addition, better family planning and urbanization give women more freedom to choose the kind of life they want. The more female leaders the world gets, the more women become empowered to take positions of leadership, a self-reinforcing loop.

While women worldwide were closing the gender gap before 2018 in critical sectors such as health and education, significant gender inequality persists in the workforce and in politics. The rate of progress for women starts slow, too. Between 2006 and 2016, the proportion of female leaders increased by only 2%.[18] But when women are better represented in leadership roles, more women are hired across the board. This picks up speed from 2025 and onwards, when the world recognizes that to encourage more female leadership is one of the levers for increasing gender equality in the entire workforce. Results speak for themselves, and by the 2030s it is becoming increasingly clear that a good gender balance is much smarter and more profitable[19] (SDGs 5, 8, 16) than the conventional male-dominated networks.

This also results in women choosing freely to have lower average birth rates. In many countries, these five factors (education, urbanization, job opportunities, family-planning and reproductive health) combine to give better wellbeing for both women and children.

Smarter, but there's still widespread resistance to transformation...

Regardless of good progress and smarter policies, environmental stresses – air pollution, water, heatwaves, wildfires – have been building up and worsening for many decades. These cause more urban crises and waves of migration, in the decades up to 2040 relative to 2015,

---

[18] World Economic Forum, 2.nov. 2017 https://www.weforum.org/agenda/2017/11/women-leaders-key-to-workplace-equality

[19] Noland & Moran, 2016, Study: Firms with More Women in the C-Suite Are More Profitable, HBR, https://hbr.org/2016/02/study-firms-with-more-women-in-the-c-suite-are-more-profitable

and contribute to conflicts, and sometimes civil wars. These put severe pressure on many fragile institutional structures. Political crises, corruption and distrust of interventionist government cause an outspoken opposition to the active planning and government roles key to rolling out the transformative actions. The increased progressive taxation to reduce inequality is also a hotly contested topic for decades.

	Scenario SMARTER overview
Main policies (2020–2040)	Accelerated renewable energy growth, sufficient to halve carbon emissions every decade from 2020. Accelerated productivity in food chains, improving by extra +1%/year. New development models in the poorer countries, following models such as South Korea, China, Ethiopia or Costa Rica. Active inequality reduction, ensuring that the richest 10% take no more than 40% of income. Investment in education for all, gender equality, health, family planning, stabilising the world's population.
Unintended obstacles & challenges (2025–2050)	Distrust of central government roles; More nationalism that discredits global cooperation; Ideological opposition to downward redistribution, particularly in Anglosphere; Market fundamentalism that opposes government work on market design.
Outcomes & consequences (2050 →)	World SDG success score of 14 (out of 17); Safety margin of 7: Zero PBs are in high-risk zones, and four in yellow: Global warming, forest degradation, air pollution, toxics (all at 0.5 pts).

Yet, since there are better redistribution measures in place that large majority groups benefit from, a stronger taxation base, and an international commitment to peace and partnership that remains resilient and responds rapidly, the worst crises are dealt with before descending into full collapse of large cities or more failed states.

Despite such obstacles, there is growing acceptance of the evidence that effective and smarter – rather than just bloated and bigger – governments are a huge boon to all market economies, of whatever culture and ideology. And there is evidence that their functions are indispensable for social stability by reducing inequity, in addition to maintaining law and order. Overall, there is also a growing willingness to invest public funds in repairs and rebuilding of infrastructure. Government oversight also helps to account for the material flows through the economy in real-time. The data is used for optimising a circular economy and resource productivity so that total resource use does not threaten the safe operating space of both local ecosystems and global natural commons.

Increasingly, the financial sector and capital markets start connecting corporate activity to positive Environment, Social and Governance criteria (ESG) for investment. From a feeble start around 2015, ever more of the world's funds start to actively *apply* the Principles of Responsible Investment (PRI). The strength of this connection is priced into security and capital markets, and investors include these considerations in their day-to-day operations. Rather than just subscribing to PRI but doing little in practice (as in ***Same***), in the ***Smarter*** scenario the talk becomes reality, and more than half of the world's wealth gets invested in line with effective PRI and ESG guidelines. This has a large direct effect on how businesses join governments' concern for achieving the SDGs inside Earth's safe operating space.

Figure 6 In this scenario, the poorest catch up earlier than in Same, Faster or Harder. This is the result of the five transformational actions having systemic effects on several SDGs The world's total success score also goes higher.

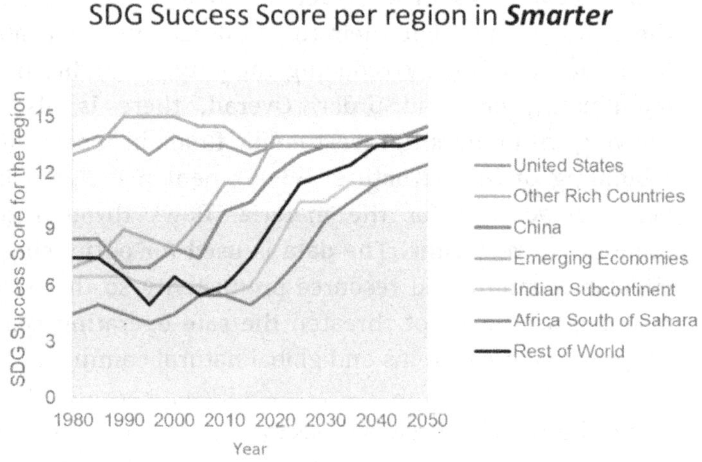

Smarter delivers results from 2030 to 2050

As 2050 approaches, the ***Smarter*** world sees a rapid trend to job-sharing, fairer wages and extensive redistribution reversing the trend of worsening inequity from the three decades that led up to 2020. More regions also manage to provide equitable access to natural resources and ecosystem services. Both these give a net positive impact on GDP/capita growth, the first by stimulating demand, the second by better resource use.

Population stabilizes more quickly as more women get radically better opportunities for education, jobs, economic autonomy, reproductive health and security, particularly in cities.

Among investors and private companies, there is a rapidly growing realization that business cannot succeed in societies that fail. The corollary is also true: societies striving for sustainability require the many opportunities that business can provide. The rationale for broad business engagement with the SDGs could not be plainer[20]: the smarter solution is when government and businesses mutually reinforce markets and regulations to deliver on the goals that they are (or should be) designed for. The new conventional wisdom in the 2030s is that delivering on the SDGs can happen only if business, governments, and civil society work together; and this is rapidly put into action in transformative ways.

By 2050, most regions of the world are delivering on nearly all SDGs. Both India and Africa South of Sahara has shown tremendous progress (from a 5.5 regional SDG-success score in 2010 to 12.5 in 2050, see figure 7). This ***Smarter*** pathway seems to point the world's economy further toward a prosperous direction within the Earth's safe operating space by 2100.

---

[20] UN Global Compact, and DNV GL *Spaceship Earth*, 2016, p.10.

Figure 7 SDG success score per region in the Smarter scenario. Regional SDG scores for 2010, 2030 and 2050 are shown.

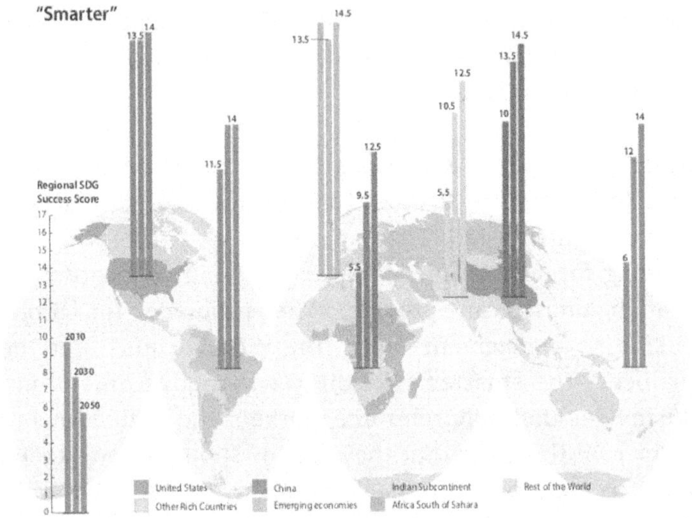

Conclusion: so, will we achieve the SDGs within PBs by 2050?

The results show that while scenarios "Same", "Faster" and "Harder" can somewhat improve on the world's SDG achievement, they tend to do that at high cost to the stability and risk level of Earth's life supporting systems.

In summary, the analysis shows that – of the four scenarios – only the five actions in the ***Smarter*** scenario can keep developing the world economy in an inclusive manner while staying within planetary boundaries. This type of transformative development seems to be able to secure a safe operating space for all of humanity before mid-century so that the world's societies can continue to flourish into the future beyond 2050 with safe(r) life-supporting systems on Earth.

Figure 8 The world's SDG Success Score for each scenario. The score is calculated as the sum of the regional success indices, weighted by population, for each scenario.

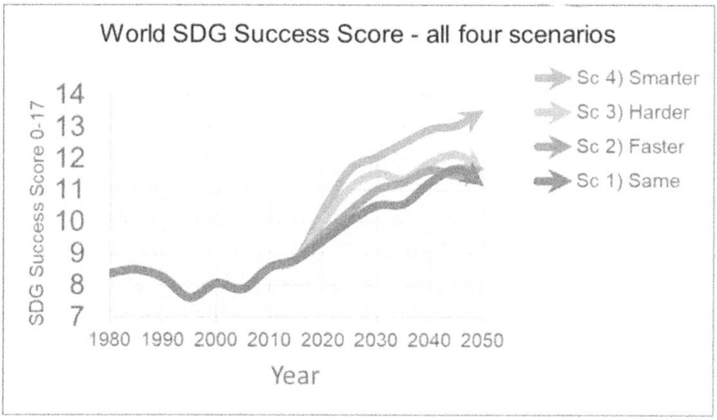

The set of four scenarios together shows the necessity to consider global transformations if we're serious about attaining the SDGs within a stable Earth system.

Figure 9 The four scenarios' impact on the Earth's safety margin. The safety margin is determined as the number of Planetary Boundaries that are within their safe operating space.

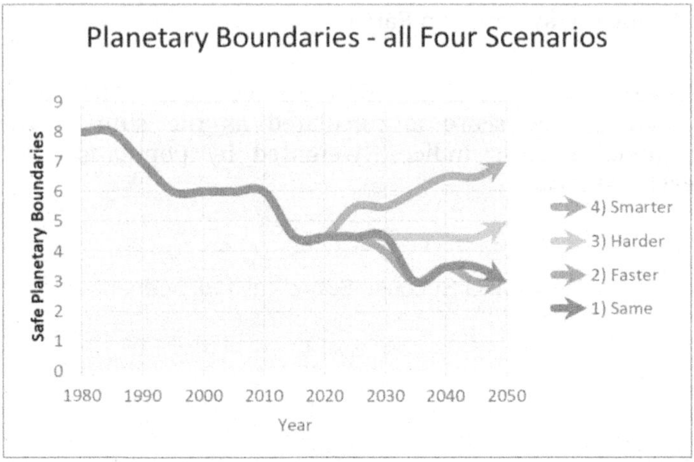

Figure 10 Achieving the SDGs within PBs: Only one of the four scenarios achieves the grand challenge of improving the world's SDG Success Score, without eroding Earth's Safety Margin.

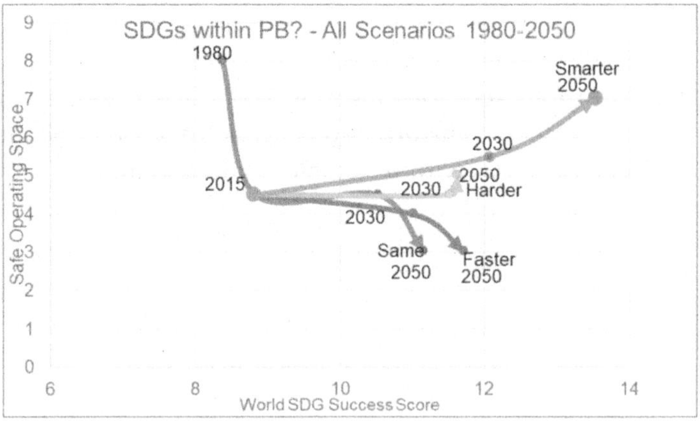

Our second caveat: the Earth3 model system is not a comprehensive model of all Earth systems. It does have representations of the complex environmental feedbacks between physical and living systems. Some of these feedbacks themselves are changing, such as greenhouse gas emissions from permafrost, ice sheet melting, forest dieback and more. If Earth3 underestimates these effects, then it could mean that failure to meet the SDGs within PBs may have even larger negative environmental social and economic impacts than illustrated in the scenarios, and thus lead to failure in reaching the SDGs in 2050 at all.

Chapter 17

The simpler way: envisioning a sustainable society in an age of limits

Ted Trainer & Samuel Alexander

1. Overview

Although it is widely understood that the global economy is currently unsustainable, few people seem to understand how far beyond sustainable limits it is. Fewer still seem to grasp what it would take for the global population to live within the sustainable carrying capacity of the planet. The most developed regions of the world have greatly exceeded levels of resource use and ecological impact that could be maintained or spread to all people, and as a result we are rapidly moving towards catastrophic planetary breakdown (WWF, 2018; Steffan et al, 2015; Ehrlich, 2013). The alarming global problems threatening humanity today include climate change; the pollution of air, soils, rivers, and oceans; deforestation and severe biodiversity loss; resource and topsoil depletion; and conflicts over access to scarce resources. From a social perspective, billions of people, mostly in the Third World, still suffer under conditions of deprivation (Hickel, 2017), while quality of life deteriorates even for many living high-impact, consumerist lifestyles in the richest countries (Lane, 2000). The global economic system and the cultures of consumption it celebrates are failing both people and planet.

It will be argued that these interconnected problems cannot be solved unless we move toward an economics of sufficiency, which means, among other things, that rates of production, consumption and resource use in rich countries must be dramatically reduced, probably by 80-90% or more (see Section 2 below). These reductions cannot be achieved merely by "greening production" through piecemeal efficiency improvements or technological innovation, and they cannot be achieved in a society that is governed by market forces or driven by the insatiable quest for material affluence or the limitless increase in GDP. In other words, a sustainable economy must be a post-growth and post-capitalist economy. But what does that actually mean?

We contend that the only way sustainability can be achieved is through a radical transition to a "Simpler Way" society which would be defined by low but sufficient material living standards, renewable energy, highly localised or "bioregional" production, egalitarian approaches to wealth distribution, participatory democratic practices of self-governance, and major degrowth to a zero-growth economy (Trainer, 2010; Alexander, 2015). We do not argue that this transition is likely, only that it signifies the only way for human civilisation to operate viably on our finite planet.

Fortunately, there are many social and ecological reasons to believe that this transition is in the interests of humanity, both rich and poor, but this requires reimagining the good life beyond consumer culture and embracing ways of living that are outwardly simple, but inwardly rich. Although we will surely raise more questions than we can answer, this short article seeks to provide more detail on this vision of a just and sustainable society and briefly highlight the implications this perspective has on questions of transition strategy.

2. The global predicament

It is important to indicate briefly why the reductions in resource and energy use must be dramatic and far greater than most people appreciate. Only then can an adequate response be formulated. Consider the following outline of the global predicament from a "limits to growth" perspective.

The commonly cited "Ecological Footprint" index shows that to provide the average Australian with food, settlement area, water and energy takes about 7gha of productive land (Global Footprint Network, 2018). If by 2050 the expected 9.8 billion people were to have risen to the present "living standard" in Australia, and the planet's amount of productive land remains the same as it is today (which is doubtful), then the amount available per capita would be in the order of 1.25 gha. In other words, Australians today are making ecological demands per capita that are about six times what would be possible for all people to make. It follows that the impacts of affluent living standards would need to be reduced by more than 80% to meet the sustainability criteria of the Ecological Footprint analysis.[1]

Wiedmann et al. (2015a) state an even more striking conclusion: the per person consumption of the ten highest iron ore and aluminum consuming nations is around 80 times that of all the rest. On a finite planet facing ecological limits, these disparities in consumption and impacts are plainly unjust and unsustainable, and they begin to indicate the degree of reductions needed to achieve sustainability and distributive equity. Moreover, what if humanity leaves a fair share of the planet for other species? The "Living Planet Report 2018" (WWF, 2018) concludes that between 1970 and the present, the populations of vertebrate species have declined on average by 60% due to human economic activity, leading to some writers to speak of the "Sixth Great Extinction".

However, the problem of ecological overshoot is far more disturbing when we consider the fundamental commitment (even by rich nations) to ceaseless *growth* in production, consumption, trade, investment, "living standards", wealth and GDP. If the Australian economy continues to grow by 3% p.a. and by 2050 the global population of 9.8 billion people achieved the same levels of GDP per capita, then total world economic output would be around 18 times the present amount. The Global Footprint Network (2018) finds that the *present*

[1] Although the Ecological Footprint metric is an imperfect measure of planetary demand, many critics think it understates the levels of impact, so for present purposes it can be considered a conservative measure of ecological impact and overshoot.

global economy already overshoots sustainable impacts by 70%, and yet the dominant economic paradigm is aiming to grow economic output many times over coming years and decades. This is, as Edward Abbey once said, "the ideology of the cancer cell."

Rejection of the limits to growth position is usually based on the belief that technical advance will deal with the associated problems by way of "decoupling"; that is, by enabling continued increase in production and consumption globally while bringing environmental impacts down to sustainable levels (e.g. Breakthrough Institute, 2015). It is not difficult to show the extreme implausibility of this "technofix" approach. The above figures show the enormity of the reductions that would be required. For the growth economy to operate within the sustainable carrying capacity of the planet and leave a share of the planet for other species, impact rates per unit of GDP would arguably have to be cut by 90% or more by 2050.

What makes the decoupling vision even more implausible is that despite decades of extraordinary technological advance, global energy and resource use continues to rise steadily, not decline as the decoupling theory would imply. A litany of studies on decoupling show that efficiency gains within a globally growing economy have not led to, and will not lead to, reduced overall impacts, let alone reduce them sufficiently to achieve sustainability (Wiedmann et al., 2015b; Alexander, 2015; Ward et al., 2016; Trainer, 2016a).

If economists are serious about designing a sustainable economy that could work well for all people (and species), not just for a minority in the highly developed regions of the world, then the severity of this "limits to growth" perspective must be central in their accounts. It is wrong and indeed dangerous to assume the legitimacy of high-impact modes of living that could not possibly be shared by all people. However, presently ever-rising GDP per capita is the goal to which most people and all nations aspire, puffed up by fantasies of sufficient decoupling. This "green growth" approach is a recipe for ecological and thus humanitarian disaster. Whether we like it or not, humanity will have to deal with the implications of living in an age of pressing ecological limits.

Thus, the essential conclusion to be drawn from the limits to growth position is that the magnitude of ecological overshoot is so great that a transition to a sustainable society will require significant "degrowth" in the rich world to levels of per capita resource use that are likely to be in the region of one-tenth of present Australian per capita levels. Across the political spectrum, little attention has been given to the profound implications of this for social, economic, or political theory and practice (Trainer, 2019). Even some prominent figures in the schools of post-growth and steady state economics do not seem to appreciate what this degree of downshifting means in terms of the economics of sustainability, especially when we are talking about timescales needed to avoid dangerous climate change (see Anderson, 2015).

3. Vision and basic principles of the Simpler Way

The following sections outline in general terms the kind of economy that we argue must be adopted in view of the limits to growth predicament. Few existing proposals for a new economy deal satisfactorily or explicitly, if at all, with the essential themes, principles and implications deriving from need to reduce impacts of affluent living standards by 90% or more. Even within the degrowth, eco-socialist and related movements there is a general failure to grasp the enormity of the consequences that follow if the limits to growth analysis and its implications are accepted.

Achieving the reductions necessary will include huge demand-side reductions (consuming less and differently) as well as significant efficiency gains (cleaner production), but primarily the task is to create a Simpler Way society that transcends the growth paradigm and systemically supports ways of living based on frugal but sufficient material living standards. Sometimes the rejection of growth economics or consumerist lifestyles are broadly stated by those who appreciate the limits to growth predicament, but rarely is the alternative form of economy this implies unpacked in any detail. This is problematic, because without a relatively clear vision of where we need to end up, our theories of transition cannot know *to where* we must transition.

3.1. *Globalised production and distribution vs. the Simpler Way*

In order to begin unpacking our vision of an alternative society, consider a simple example of egg production, from which broader lessons can be derived. A study by Trainer, Malik, and Lenzen (2018) presents an analysis comparing the supply of eggs via the normal industrial / supermarket path with a local cooperative supply path. The dollar and energy costs of the former were found to be in the order of 50 to 200 times larger than those of the latter. Such achievements are due to the smallness of scale, proximity, self-sufficiency, cooperation and integration of communities, which makes it possible to totally eliminate many costly inputs, to recycle "wastes", to benefit from "co-products", and to administer via spontaneous and informal social interaction.

For instance, the typical supermarket egg has vast and complex global supply chains involving distant shipping fleets, agribusiness, trucking and thus roads and petroleum, warehousing, chemicals for disease control, infrastructures, supermarkets, storage, packaging, marketing, legal services, the finance, advertising and insurance industries, waste removal and dumping, computers, a commuting workforce, and expensively trained personnel. It also involves damage to ecosystems, especially via carbon emissions and agribusiness effects including the non-return of nutrients to soils.

However, eggs supplied via worker cooperatives within integrated settlements based on local production can avoid almost all these costs, while enabling immediate use of all "wastes". Permaculture design ensures that elements in a system are highly self-maintaining, perform many functions at the same time, and that all "wastes" become valued resources (Holmgren, 2018). For instance, recycling of kitchen scraps along with free ranging can meet total poultry nutrient needs. Poultry and other animal manures, including human, can be directly fed into nearby methane digesters or compost heaps, thereby eliminating the need for inputs to village food production from the fertiliser industry. Essentially no transport is required, given local production is geared to meeting local needs, using mostly local resources. Poultry perform multiple functions automatically, including providing meat, fertilising orchards, cleaning up garden beds and fruit fly lavae, producing the next generation of chicks, and maintaining fire breaks. Monitoring and maintenance of systems can be totally informal, via spontaneous discussion and action within the community. In addition, cooperative care of poultry and other animals adds to amenity and leisure resources and facilitates community bonding.

This example is indicative of how localised modes of production, supply, and consumption can radically reduce energy and resources demands, while providing various benefits (e.g. healthier food, better treatment of animals, enjoyable production, etc.). The same approach can be applied to most other domains of social and economic life, including other food items, dwelling construction, clothing supply, many services, as well as providing leisure, entertainment and education. The "Remaking Settlements" study (Trainer, 2016b) explores the potential when applied to an outer suburb of Sydney, finding that radical restructuring might enable almost all food needs to be met within the suburb while enabling many other material and social benefits. These include eliminating unemployment, avoiding the need for a sewer system, dramatically reducing the need for transport for work and income, and creating thousands of person-hours of community improvement per week via voluntary working bees and committees.

The foregoing discussion has been of theoretical possibilities, but the claims are supported by evidence from actual communities functioning along the lines being advocated. Lockyer's (2017) study of the Dancing Rabbit Eco-village in Missouri found the following per capita use rates compared with national US averages. Car use, 8%; distance driven, 10%; liquid fuel use, 6%; solid waste generated, 18%; proportion of solid waste recycled on site, 34%; electricity use 18%, with three times as much electricity sent to the grid as is used; water use, 23%, with two-thirds of this collected from village roofs. Furthermore, the community's quality of life indicators align with the evidence from other studies of eco-villages; e.g., 81% of respondents rate happiness with life situation at 7/10 or better, and almost all say life had improved since moving to the community. (For similar findings see Grinde et al., 2017; Alexander and Ussher, 2012.) Another way of putting these findings is that it takes remarkably little in the way of monetary or resource expenditure to ensure a highly satisfying lifestyle, if communities accept materially simple ways and are organised to this end. A materially simple life can be a spiritually richer life. (For further supporting argument and evidence, see Trainer 2016b; 2013.) We argue that a just and sustainable society cannot be achieved unless these highly communal, localised, self-sufficient and frugal ways become the norm.

4. Implications for specific aspects of the Simpler Way economy

The following sub-sections briefly sketch various ways that sectors of an integrated Simpler Way economy could function to yield high levels of sustainability, economic justice and quality of life. It focuses primarily on aspects of the neighbourhood and town levels. Implications for the national and international levels are briefly considered later.

4.1. The local geography of the economy

Dealing adequately with the dual challenges of climate change and fossil energy depletion implies significantly reducing energy demand compared to current levels in the rich nations and meeting that reduced demand with renewables (Alexander and Floyd, 2018). Given the close connection between energy and economy, an energy descent future implies moving away from the energy-intensive modes of globalised production and distribution and toward highly local production, using mostly local resources.

Given the commitment to materially simple lifestyles and to simplified local supply systems, far less would need to be produced than at present (in rich nations) and most of the goods and services one would need in everyday life would come from within at most a few kilometres of where one lived. Most people could get to work on foot or on a bicycle, although a few would need to travel further, mostly in buses or trains powered by electricity or biofuels (the greatly reduced distances and regularity of travel would make these modes financially and ecologically affordable). Community vehicles might be available for occasional use when needed (e.g. moving bricks or timber to a construction site). Every household owning a private automobile, even if electric, has no place in a sustainable economy (Alexander and Gleeson, 2019). Because very little transport would be needed, many roads and car parks could be dug up or converted to different uses, significantly increasing space in cities for local gardens, orchards, forests or woodlots, animal pens, or fish ponds. Very little air travel would be needed.

Economic self-sufficiency (or rather collective sufficiency) should be seen in terms of concentric circles. In the centre will be the most important economic and social unit of all, the highly productive and highly self-sufficient household (see Holmgren, 2018; Trainer, 2016b). Even in the present economy more than half the work done takes place there, although most of it is unpaid. The household would be embedded within the neighbourhood economy, involving free goods from commons, sharing/gifting surpluses, mutual assistance, cooperatives, swapping and bartering, working bees and committees, and including many small enterprises, co-operatives and household hobby production.

The next circles out would include the suburb, and then the town or city (Alexander and Gleeson, 2019). Ideally the area surrounding the settlement would contain its dairy, timber plantation, grain and grazing lands. Several towns might be found within a region, maybe ten km across, including a few factories producing items such as fridges, bicycles, and radios for local use and for export to the wider economy. As noted further below, few resources or commodities would need to come from the state and national economic sectors, and very little from overseas, perhaps only items such as high tech medical or computer equipment or resources needed for essential energy infrastructure (e.g. solar panels). There would need to be no reduction in research and development of appropriate technology in a Simpler Way society, since so many resources would be saved by avoiding wasteful consumption and production. For example, there would be little need for marketing expenditure in a Simpler Way society, so the trillion dollars per annum currently spent on advertising (WARC, 2016) could be redirected to better uses, such as funding the transition to renewable energy. No effort would be made to persuade people to purchase and consume.

4.2. Sectors of the new economy

One of the (overlapping) sectors of the new economy would still use money. It could include a large domain in which market forces were allowed to operate (although in the long-term future we might decide we do not need the market.) One sector would be fully planned and under participatory social control (e.g. for provision of water, energy, housing, and health care). One would involve non-profit community cooperatives. One large sector would not involve money. It would include household production, barter arrangements, mutual aid, working bees, cooperatives, the gifting of surpluses, and free goods from the commons.

The commons would be extensive, including the public fruit and nut orchards, clay pits, fish ponds, herb patches, woodlots and various supply systems, for instance via poultry, honey and fish farming cooperatives. Many buildings such as community centres, craft rooms, store sheds, pond and windmills and especially neighbourhood workshops would be constructed and managed as community property. As many of these as needed would be crammed into space created by the elimination of most roads and car parks, and also located just outside settled areas.

Probably the largest sector would be made up of small privately-owned firms and farms. These would best be thought of as the tools which people used to make their social contribution and draw a stable, sufficient income via a secure livelihood. They would not be capitalist enterprises. They would not involve investment of capital by people who need to do no work, to make money, to invest again. Their value would be in giving many families and small groups the freedom to enjoy owning their venture and exercising their skills, and running it in the way they wished. Privately owned firms would be required to keep within socially determined guidelines monitored by the town assembly.

The zero-growth economy we are describing would automatically eliminate almost all the financial sector, primarily because, as few realise, interest could no longer be charged on loans. If a loan is to be repaid plus interest the borrower must create more "value" than was borrowed, meaning there will be economic growth. Thus rich people could no longer receive large incomes from "investments" and without having to do any work for them. Banks would therefore do little more than hold savings and lend some of them out for an administrative fee. A corollary is that banks would no longer create money via the fractional reserve system. In a zero-growth economy no new money needs to be created and put into circulation.

It should therefore come as no surprise that the economy being described is one in which there is no aggregate growth, and all important functions are decided by rational social decisions. Cooperation not competition is the supreme principle, there is no interest paid, and little is left to market forces. In short, a degrowth transition to a Simpler Way society is not compatible with a capitalist economy.

4.3. The craft economy

Many goods would be produced in households and very small firms by artisan producers. The main reason for this is that artisan or craft production is enjoyable; producing in factories is not. Given that the volume of production in a frugal degrowth economy would be far lower than it is now, people who delight in making pottery, furniture, clothing, toys etc. might provide most of the goods the community needs.

Production of most of these goods would only need to replace breakages and wear, meaning production would not need to meet the shifting needs of fashion. It would make sense for some items to be mass-produced in factories, such as work shoes and boots. Automated corporations can produce much more "cheaply", but the required scale necessitates long distance transport of inputs and finished goods, and provides little or no satisfaction for workers. Living simply and obtaining many things without the need for money makes it more likely that we would be happy to pay more for well-made and durable goods.

4.4. Social control of the economy

There cannot be a satisfactory economy unless there is the collective power to make sure that the things produced are those that should be produced, so that all people have their needs met, social cohesion and morale are kept in good condition, the environment is cared for, and development is socially desirable. A free enterprise or capitalist economy will not do these things. Markets ignore need, justice, morality, social benefit and ecological values. They can do no other because they allocate scarce things according to "effective demand", i.e., those who offer to pay more money for an item get it. Millions of people die every year not due to scarcity of resources but because the provision of food and water is determined by what will maximise the profits of those who supply them. Markets therefore are incapable of doing anything other than allocating scarce resources to the rich and developing the most profitable industries as distinct from those that are most needed.

Polanyi was among those who have stressed this fundamental mistake in the economy we have had since *The Great Transformation* (1944). In ancient and in medieval times all one's behaviour was expected to be governed by a general moral code, which for instance prohibited taking advantage of another person's misfortune. But Polanyi and others have pointed out that with the advent of capitalism the rules governing economic behavior became separated from the general code, enabling entrepreneurs to ruin their competitors or pounce on a fire sale, and to concern themselves only with whether or not monetary benefits to themselves could be secured. Damage caused to others or to society or to ecosystems need no longer be considered. Polanyi pointed out that unless "market forces" were kept under social control, society and the environment would eventually be destroyed. That explains most of what is wrong with the world today.

Therefore, especially in a context of very scarce resources, desirable outcomes will not be possible unless the main economic functions are under social control, that is, determined by careful discussion and deliberate, rational decision making and collectively defined regulation. However, it will be emphasised below that this must be carried out by participatory democratic means, not by centralised bureaucracies.

The town or suburb, through informal deliberations and committees, and ultimately its local assemblies, must be in control of shaping the local economy. The citizens will be clearly aware that their fate depends on whether or not their settlement works well and has established the arrangements that enable it to provide basic necessities. If some needs are not being met then the town will realise that it must take action, for instance set up a cooperative to plug the gap. If private enterprise and the market can meet the need then the town would be relieved not to have to deal with the problem. Thus, although the town council or committee would have a lot to attend to, it would be wise to take a minimalist approach, dealing with and fixing only those things that need attention.

The over-riding concerns would be to maximise town self-sufficiency and thus resilience, and to ensure that many essential needs are provided for as locally as possible, using local resources. Thus the town must work out how to make sure that everybody has a livelihood, a way of making a valued contribution through enjoyable productive activity. There is no excuse for unemployment; that is only found in deeply flawed societies. There is much to do and many cooperatives and working bees needing people. Indeed, the main form of tax payment would be giving time to the working bees building and maintaining local infrastructures, harvesting from the commons, and providing many services. Only a little tax in the form of money would be needed to pay for maintenance of town systems, and for national government revenue.

Significant monetary inequality would cease to be an issue. It would be of no consequence whether or not one had a high income or great monetary wealth, and there would be little or no point in seeking these. One's "wealth" – the richness of one's life – would depend primarily on one's community resources and all would have equal access to these.

Only if the town has control over its own fate can decisions take into account all the relevant considerations and prevent outcomes being determined solely by what will maximise monetary returns to those with money to spend or invest. Often it will be obvious that considerations such as equity, environment, aesthetics and the public good should outweigh those that a market system would focus on.

4.5. How will the control be exercised?

Few if any would want the social control and regulation of the economy to be exercised by large, authoritarian, centralised state bureaucracies, but in the economy under discussion that is not only avoidable, it is impossible. In a world of very limited resources, a sustainable society will have to be made up mostly of many small and highly self-sufficient and self-governing community economies. These will have to be run as participatory democracies – they cannot be run effectively any other way. They will not make viable decisions unless people familiar with the situation make those decisions. There will be few paid bureaucrats or councils, because in a world of scarce resources not much paid government will be affordable. Therefore, most policy formation and management of "public works" will have to be carried out by local citizens, by committees and town assemblies, and by voluntary working bees. The right decisions for the town will not be made or implemented unless arrived at and supported by conscientious citizens. Fortunately the small and simplified economies which do not grow and are largely cooperative will make it easier for the social control to be exercised via participatory democratic processes; that is, by spontaneous informal discussion leading to decisions by town assemblies.

Many and possibly most of the economic activities would not need any formal control. Informal discussion would sort out whether too many tomatoes were planted last year. In addition the many committees would study and advise on issues such as the best apple varieties to plant when the parking lot is dug up. Adjustments, repairs and research would primarily be carried out spontaneously by citizens who knew their fate depended on good maintenance of systems and who enjoyed keeping their town in good shape. These are the kinds of "administrative" practices that make many eco-villages function well without paid politicians or bureaucrats.

If it became clear that, for instance, there were more bakeries than was sensible, the community would have to work out the best solution for all concerned. This might include helping some people to transfer to other necessary activities. No firm would be allowed to go bankrupt; the town needs to look after all its members and make good use of all its resources. It would have its own bank and "business incubator" i.e. a panel of experienced advisers, and its working bees would help firms to set up or modify premises. All would understand that their town could not thrive unless all take collective responsibility for making sure that it does, and unless it looks after everyone. Thus, there would be a profoundly different ethos compared with today; it would be seen as "our economy" – ours to keep in good shape and to provide for the community.

Obviously there would be a need to think carefully about desirable limits to social control, and to minimise interference with freedoms. Ideally the concern would be to take remedial action only where social needs were not being met. This would mostly be a matter of plugging gaps rather than blocking or coercing. Two factors would make the control task easier: firstly, the economy would not be complex; and secondly, it would be driven not by profit or market forces but by citizens who prioritised the common good.

4.6. Work

The corollary of radical economic degrowth is that far less work and production would take place, especially in the formal economy, and many of the things that were needed would be produced in far less resource-expensive ways. For example, food supply would require little work to produce trucks to bring food to towns. As has been explained, there would be no involuntary unemployment. The town would make sure that all who wanted work had a share of the work that needed doing.

Because the amount of work to be done for money would be dramatically reduced, most people might only need to do work one or two days per week, although much of their time would be going into productive use within the household, commons and working bees and committees. Some people, such as professionals, could choose to work full time for money, perhaps paying more montetary tax than those who paid by working bee contributions.

Consumer-capitalist society has destroyed work. For most it is more or less a burden that has to be endured. It is hardly fulfilling or a source of anything valuable aside from the pay packet. In the new economy, alienation as Marx described it would be largely eliminated. Indeed, it could be argued that "work" as such could easily be totally eliminated in a Simpler Way society. In eco-villages, members generally produce via their enjoyable small farm or craft activities and contributions to working bees, which are usually experienced as forms of leisure activity. Potters see their products benefiting and being appreciated by others. Working bees can turn work that would be boring or otherwise unpleasant in capitalist society into enjoyable and socially bonding festivals. Thus the distinction between work and play might completely disappear. People would generally be more active in their working lives, contributing to fitness and health and therefore minimising need for expensive healthcare. It is notable that Cuban's have better health than citizens of the US, despite spending around one tenth on healthcare (O'Hanlon and Harvey, 2017).

4.7. The economy beyond the town or suburb

The foregoing discussion has focused on the local economy, arguing that in a resource-scare world this must be the new basic economic unit if per capita impacts are to be reduced to sustainable levels. National and international economies will have a necessary role but they will be far smaller than at present and of much less significance. Their main function will be to enable the local economies to thrive, by providing those relatively few things they need but cannot conveniently produce in their local bioregion.

Despite the productive power of the local economy there will always be a relatively small number of important items the towns cannot produce for themselves, and some of these will have to come from large industries far away. How then might economic systems beyond the town level be organised? The basic concept must be the location within or close to towns of some small capacity to produce items for export to other towns within the national economy. For instance, it is not feasible for each town to produce its own refrigerators, so a few towns would specialise in the production of these or of components, making them available for purchase by other towns, thereby earning the export income needed to purchase similar items from the national economy.

The main economic role of the (remnant, small) "state" will be to organise the distribution and functioning of these industries, to ensure that all towns have the capacity to earn sufficient income from their exports into the national economy to pay for their purchases from it. It is unlikely that it would make sense for these exporting firms to be other than town cooperatives. The state and national governments would also have to attend to provision of infrastructures and systems enabling these supply chains, such as roads and railway networks, and finance, insurance and legal systems.

Some firms at this level would be large, such as those producing steel, cement and railway equipment. In a zero-growth economy with a very small GDP it is not clear why it would make sense for these firms to be privately owned. This would mean that issues to do with investment, innovation, efficiency, restructuring etc. would have to be dealt with via public ownership and management; i.e., some form of socialism.

4.8. What role for the state in the transition?

If enough communities manage to establish commitment to taking control of increasingly self-sufficient and self-managed local economies then before long their dependence on the national economy for those few but crucial inputs will force them to demand and then initiate radical change in that economy. They will insist that governments focus their attention on restructuring the national economy so that its primary role is to make sure the towns and settlements can get the basic inputs they must have but can't produce for themselves. If financial crises resurface in coming years, as is likely, national governments will be faced with the task of urgently enabling devastated workers and towns to grope towards local self-sufficiency, given that the industries they used to work in would have collapsed and could not be restored. Thus national governments will be forced to gear available national productive capacity to building the required alternative local structures and systems, including especially the training and demonstration facilities.

Central governments would quickly realise that they cannot implement the required changes. They could not construct or run viable local economies. They would have few resources and more importantly the right decisions can only be made and implemented by people who understand their specific conditions and have the necessary ideas, attitudes, social cohesion and morale.

Before long this pressure is likely to shift from submitting requests to the state to making demands on it, and then to taking increasing control of it. There will be increasing insistence that frivolous industries must be phased out so that scarce resources can be devoted to meeting fundamental town and regional needs. Meanwhile towns will be driven by necessity to bypass the centre and take initiatives such as setting up their own farms, energy supplies and factories, thus taking various functions out of the hands of the state. It will be increasingly recognised that the local level is the only one where the right decisions for highly self-sufficient communities can be made. Communities becoming familiar with taking control of their own fate will be growing less inclined to allow the centre to go on calling the shots.

The final step would be the grass roots pushing their way into the centre, that is taking direct control over the remaining state functions through the classic anarchist mechanisms whereby federations and delegations thrash out policy proposals which are brought back down to the town assembly level for endorsement. The increasing interest now in Citizen Initiated Referenda indicates how this is feasible. Some city budgets are now worked out by the CIR process. One of its merits is that decisions are not made, handed down and enforced from the top, so they are more likely to be widely accepted. The ultimate goal is to make sure that endorsement is carried out at the level of the town assemblies, and thus that representative government is replaced by direct participatory government.

Accordingly, a Simpler Way society would not be an eco-socialist economy, if that means a centrally-planned society governed by a strong centralised state. As has been detailed, a Simpler Way society would have to involve thoroughly participatory democratic decision making, spontaneity, subsidiarity, and rejection of hierarchy and professional rulers. Issues beyond the town level would be handled by sending delegates to meetings within federations. And the basic transitional strategy would be "prefiguring" the new world within the shell of the old. This economy can therefore be described as eco-anarchist.

4.9. Is all this too unrealistic?

The common response to this vision is that it is far too utopian; that it fails to recognise that humans are not capable of making such an alternative society work. This response reveals a serious misunderstanding. The argument has not been that this economy is going to be achieved, or that it is likely to be achieved. It is quite unlikely to be achieved. The argument has been that we had better try very hard to achieve it, because whether we like it or not, there is no other way to defuse the major global problems now threatening to destroy us.

Clearly the ways of living outlined in this article would not work in today's society, because they require quite different attitudes and values to those predominant now. In consumer-capitalist society most people are out to maximise their self-interest, see nothing wrong with competitive systems in which they might be one of the winners while many end up with less than enough, and would oppose the enlightened social regulation and restraint needed to ensure satisfactory outcomes for all. It is essential to realise that a satisfactory society cannot be designed for such people. Most make the mistaken assumption that we can have a peaceful, sustainable and just world while we all go on living affluently and competitively and continually striving to increase our "living standards" and GDP without limit. However, these are the core behaviours and goals that are causing global problems.

But are these dispositions too deeply entrenched to alter? An important feature of The Simpler Way is that while it *requires* values like moderation, frugality, sharing and cooperation, it also *rewards* them. A satisfactory life in a thriving community will not be possible unless there is cooperation and willing contribution and concern for the common good, but it will also be a situation in which those values and practices will be enjoyable. If the new ways can be established they will tend to be self-reinforcing. This suggests that, for those people who accept the general vision outlined in this paper, one of the primary goals is to help develop the social consciousness and understandings required for a sustainable economy to emerge. Until that consciousness emerges, the necessary system change to a Simpler Way society will not emerge. Thus what is required is essentially a cultural revolution, which would create the conditions needed to drive the necessary economic and political revolutions.

5. Conclusion

There has been insufficient recognition of the way all the major global problems derive primarily from having exceeded the sustainable limits to growth. For instance the damage being inflicted on ecosystems can only increase unless we move to lifestyles and systems involving far lower per capita consumption than we are familiar with in the rich world. The soon to be 7+ billion people living in poorer countries can only receive a fair share of the planet's resources if those living in the rich countries reduce their consumption dramatically. Most armed conflict is to do with fierce competition to secure scarce resources and markets, meaning that if we insist on remaining affluent we will need to remain heavily armed. And the social cohesion and quality of life even in the richest countries will continue to deteriorate. These problems cannot be defused unless simper lifestyles and systems are willingly embraced.

Several analysts have stressed the fragile house-of-cards nature of the global economy in our age of financial and ecological limits (see, for e.g., Korowicz, 2012; Morgan, 2013; Greer, 2008). Above all is its dependence on debt, now in excess of $250 trillion, three or more times global GDP and far higher than before the GFC. These considerations align with Marx's fundamental insight regarding the self-destructive contradictions built into the foundations of capitalism, even though he did not clearly envisage its resource limits.

In an oil dependent economy, it is highly likely that if the yield from shale oil production falters in the next decade or so a global debt crash of unprecedented proportions will suddenly impact. It might not be the final GFC; some envisage partial recovery initiating a "bumpy road down" or a slow "catabolic collapse" (Greer, 2008). But others foresee the end of civilisation and the die-off of billions. What is to be hoped for is a "Goldilocks depression" that falls short of catastrophic breakdown but is serious enough to jolt large numbers into recognising that the growth and greed system is not going to provide for them.

From the Simpler Way perspective little can be done to influence that trajectory. Strategically it is doubtful whether striving to reform consumer-capitalist society is a good investment of time and resources. It is locked on a suicidal path that its institutions and culture are incapable of recognising let alone altering. What is to be done instead is to work to prepare for the building of sensible systems as the existing system continues its deterioration. This involves what the anarchists call "prefiguring" – that is, building here-and-now the kinds of local lifestyles and systems sketched in the earlier sections of this paper. Many are striving to develop those ways within the Voluntary Simplicity, Eco-village, Permaculture, and Transition Towns movements. A smooth or successful transition is very unlikely, but these movements currently embody the most coherent strategy for working toward a resilient landing.

References

Alexander, S. (2015) *Prosperous Descent: Crisis as Opportunity in an Age of Limits*. Melbourne: Simplicity Institute.

Alexander, S, and Floyd, J. (2018) *Carbon Civilisation and the Energy Descent Future: Life Beyond this Brief Anomaly*. Melbourne: Simplicity Institute.

Alexander, S, and Gleeson, B. (2019) *Degrowth in the Suburbs: A Radical Urban Imaginary*. Singapore: Palgrave Macmillan.

Alexander, S, and Ussher, S. (2012) "The Voluntary Simplicity Movement: A Multi-National Survey Analysis in Theoretical Context." *Journal of Consumer Culture* 12(1), pp. 66-88.

Anderson, K. (2015) "Duality in Climate Science." *Nature Geoscience* 9, pp. 898-90.

Breakthrough Institute, (2015) *An Ecomodernist Manifesto*. Breakthrough Institute. Available at: http://www.ecomodernism.org/ (accessed 10 Feb. 2019).

Ehrlich, P, and Ehrlich, A. (2013) "Can Collapse of Global Civilization be Avoided?" *Proceedings of the Royal Society B*. 280(1754). Available at: https://royalsocietypublishing.org/doi/full/10.1098/rspb.2012.2845 (accessed 10 Feb.2019).

Global Footprint Network (2018) "National Footprint Accounts 2018." Available at: http://data.footprintnetwork.org/#/ (10 Feb. 2019).

Greer, J M. (2008) *The Long Descent: A User's Guide to the End of Industrial Civilisation*. Gabriola Island: New Society Publishers.

Grinde, B et al. (2017) "Quality of Life in Intention Communities." *Social Indicators Research* 137(2), pp. 625-640.

Hickel, J. (2017) *The Great Divide: A Brief Guide to Global Inequality and its Solutions*. London: William Heinemann.

Holmgren, D. (2018) *Retrosuburbia: The Downshifters Guide to a Resilient Future*. Hepburn Springs: Melliodora Publishing.

Korowicz, D. (2012) "Trade Off: Financial System Supply-Chain Cross Contagion – A Study in Global Systemic Collapse." *Feasta* (30th June, 2012). Available at: http://www.feasta.org/wp-content/uploads/2012/10/Trade_Off_Korowicz.pdf (accessed 10 Feb. 2019).

Lane, R. (2000) *The Loss of Happiness in Market Democracies*. New Haven, CT: Yale University Press.

Lockyer, J. (2017) "Community, Commons, and Degrowth at Dancing Rabbit Ecovillage." *Journal of Political Ecology*, 24(1), pp. 519-542.

Morgan, T. (2013) *Perfect Storm: Energy, Finance, and the End of Growth*. Tullet Prebon Report, Strategy Insights (Issue 9). Available at: https://ftalphaville-cdn.ft.com/wp-content/uploads/2013/01/Perfect-Storm-LR.pdf (accessed 10 Feb. 2019).

O'Hanlon, C, and Harvey, M. (2017) "Doing More with Less: Lessons from Cuba's Health Care System." *Georgetown Journal of International Affairs* (2 October 2017). Available at: https://www.georgetownjournalofinternationalaffairs.org/online-edition/2017/10/2/doing-more-with-less-lessons-from-cubas-healthcare-system (accessed 10 Feb. 2019).

Polanyi, K. (1944) *The Great Transformation*. New York: Farrer and Rinehart.

Steffan et al. (2015) "The Trajectory of the Anthropocene: The Great Acceleration." *The Anthropocene Review* 2(1), pp. 81-98.

Trainer, T. (2010) *Transition to a Sustainable and Just Society*. Sydney: Envirobook.

Trainer, T. (2013) "Your Delightful Day: The Benefits of Life in the Simpler Way." *Simplicity Institute Report 13b*, pp.1-8.

Trainer, T. (2016a) "Another Reason Why a Steady-State Economy will not be a Capitalist Economy." *Real-World Economics Review*, 76, pp. 55-64.

Trainer, T. (2016b) "Remaking Settlements: The Potential Cost Reductions Enabled by the Simpler Way." Available at: http://thesimplerway.info/RemakingSettlements.htm (accessed 10 Feb. 2019).

Trainer, T. (2019) "Entering the Era of Limits and Scarcity: The Radical Implications for Social Theory." *Journal of Political Ecology*, 26(1), pp. 1-18.

Trainer, T., Malik, A, and Lenzen, M. (2018) "Comparing the monetary, resource and ecological costs of industrial and Simpler Way local production: Consider egg supply." Working paper available at:
http://thesimplerway.info/Eggs.html (accessed 15 Feb. 2019).

WARC (2016) "Global Marketing Spend to Hit $1tr." Accessed 10 February 2019 from:
https://www.warc.com/NewsAndOpinion/News/Global_marketi ng_spend_to_hit_1tr/37178

Ward, J et al. (2016) "Is Decoupling GDP Growth from Environmental Impact Possible?" *PLOS One* 11(10): e0164733. Available at:
http://dx.doi.org/10.1371/journal.pone.0164733 (accessed 10 Feb 2019).

Wiedmann, T, Schandl, H, and Moran, D. (2015a) "The Footprint of Using Metals: New Metrics of Consumption and Productivity." *Environmental Economics and Policy Studies*, 17(3), pp. 369-388.

Wiedmann, T. et al. (2015b) "The Material Footprint of Nations." *PNAS* 112(20), pp. 6271-6276.

WWF (2018) "Living Planet Report 2018." Available at:
https://wwf.panda.org/knowledge_hub/all_publications/living_ planet_report_2018/ (accessed 10 February 2019).

Made in the USA
Monee, IL
04 November 2026

67381641R10312